THREE-DIMENSIONAL
ELECTRON MICROSCOPY
OF MACROMOLECULAR ASSEMBLIES

THREE-DIMENSIONAL ELECTRON MICROSCOPY OF MACROMOLECULAR ASSEMBLIES

Visualization of Biological Molecules in Their Native State

Joachim Frank

OXFORD

UNIVERSITY PRESS

2006

OXFORD

UNIVERSITY PRESS

Oxford University Press, Inc., publishes works that further
Oxford University's objective of excellence
in research, scholarship, and education.

Oxford   New York
Auckland   Cape Town   Dar es Salaam   Hong Kong   Karachi
Kuala Lumpur   Madrid   Melbourne   Mexico City   Nairobi
New Delhi   Shanghai   Taipei   Toronto

With offices in
Argentina   Austria   Brazil   Chile   Czech Republic   France   Greece
Guatemala   Hungary   Italy   Japan   Poland   Portugal   Singapore
South Korea   Switzerland   Thailand   Turkey   Ukraine   Vietnam

Copyright © 2006 by Oxford University Press, Inc.

Published by Oxford University Press, Inc.
198 Madison Avenue, New York, New York 10016

www.oup.com

Oxford is a registered trademark of Oxford University Press

Library of Congress Cataloging-in-Publication Data

Frank, J. (Joachim), 1940-
Three-dimensional electron microscopy of macromolecular assemblies :
visualization of biological molecules in their Native state / by
Joachim Frank. – 2nd ed.
    p. cm.
Includes bibliographical references and index.

ISBN 978-0-19-518218-7

1. Three-dimensional imaging in biology. 2. Electron microscopy.
I. Title

QH324.9.T45F73 2005
570'.28'25 dc22   2004018762

Printed in the United States of America
on acid-free paper

To Carol, Marial, Hosea, Jody,
Sophie, and Sparkle

Preface

Since the first edition of this book appeared, in 1996, there has been a tremendous surge in applications of cryo-electron microscopy (cryo-EM) in the study of macromolecular assemblies, especially those in single-particle form. Richard Henderson noted at that time that the young field had all that is needed to grow: a Gordon Conference [three-dimensional electron microscopy (3DEM) of macro-molecules, started in 1985], a mailing list ("3DEM," started by Ross Smith, now in the hands of Gina Sosinski), and then an introductory book.

Although the path toward atomic resolution is much more tedious than in electron crystallography of 2D crystals, realization of such resolution for entirely asymmetric molecules is now in close reach. Now that instruments have matured toward high-brightness, high-coherence, digital image readout, and computer control, and now that computational methods have become more standardized, it is tempting to draw a parallel to the surge in the beginning days of X-ray crystallography. Indeed, two important similarities are that density maps of macromolecular assemblies are now being published at an ever accelerating rate, and that their deposition in public data banks is now stipulated by prominent scientific journals.

The present, revised edition of the book is intended to be more comprehensive in several aspects. Even though the main emphasis remains on algorithms and computational methods, a basic introduction into transmission electron micro-scopy and the design of the instrument has now been provided, and specimen preparation is covered in greater depth. Also reflected in the new book is the remarkable convergence of processing methods that has taken place, which has made it possible to sharpen the focus of some sections. Both reconstruction and classification algorithms have advanced. Most significant have been developments in areas that have to do with the interpretation of maps: tools for segmentation, and for the fitting and docking of X-ray structures into cryo-EM density maps. The addition of appendix material describing the most important mathematical tools and theorems of Fourier processing has been found indispensable because of frequent explicit references to these concepts throughout the book.

Even though single-particle techniques have been added to the arsenal for analyzing and processing virtually all ordered structures, it was not possible to include these extended applications. It will, however, become apparent that all techniques for image alignment, classification, reconstruction, and refinement covered here are readily transferred to icosahedral particles as well as pieces of helices and 2D crystals.

Throughout this book, the extent to which mathematical descriptions and derivations are used varies considerably, and some of the sections may provide detail unnecessary for practitioners of routine methods. These sections are marked by an asterisk, for judicious skipping.

In preparing this second edition, I received many suggestions and practical help from a large number of colleagues in the 3DEM field: Francisco Asturias, Timothy Baker, Nicolas Boisset, Jose-Maria Carazo, Luis Comolli, John Craighead, David DeRosier, Jacques Dubochet, Achilleas Frangakis, Daphne Frankiel, Ed Gogol, Peter Hawkes, Martin Kessel, Jianlin Lei, Pawel Penczek, Michael Radermacher, Teresa Ruiz, Trevor Sewell, Fred Sigworth, Ross Smith, Christian Spahn, Terence Wagenknecht, and Sharon Wolf. (It is possible that one name or the other does not appear on the list that appears above—I apologize in advance for such unintentional omissions.)

I particularly thank the members of my group, in particular Bill Baxter, Haixiao Gao, Jianlin Lei, Bob Grassucci, ArDean Leith, Mike Marko, Kakoli Mitra, Bimal Rath, Jayati Sengupta, Tapu Shaikh, and Derek Taylor for preparing some of the pictures and diagrams and assisting me with research for relevant material. I'm indebted to my colleagues at the Wadsworth Center, Charles Hauer, David LeMaster, and Hongmin Li, for helping me in the compilation of the table on sample quantities used in cryo-EM, mass spectroscopy, nuclear magnetic resonance (NMR) and X-ray crystallography. I thank Fred Sigworth for pointing out some errors in the first edition to me, and for specific suggestions on how to make the book more useful. Some of the colleagues I contacted went out of their way to supply me with unpublished material; I have to mention in particular Nicolas Boisset, Trevor Sewell, and Francisco Asturias. I gratefully acknowledge that many colleagues, some of whom are not listed here, have given me permission to reproduce their illustrations, and supplied electronic files or original artwork at short notice. My very special thanks go to Pawel Penczek for a critical reading of the manuscript and a helpful discussion.

In finalizing this book, I was assisted by Neil Swami, who obtained permissions and kept track of the large volume of supporting material. Without his diligent help, the preparations would have easily taken another year. Another student, Jamie LeBarron, has been most helpful with the preparation of examples for correspondence analysis. Michael Watters assisted me patiently and competently in finalizing over 100 illustrations, some of which had to be designed from scratch.

Contents

THREE-DIMENSIONAL ELECTRON MICROSCOPY OF MACROMOLECULAR ASSEMBLIES

1

Introduction

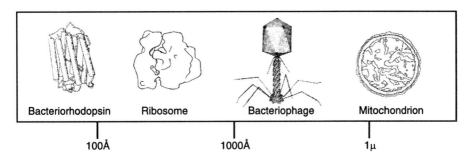

Figure 1.1 The range of biological structures covered by three-dimensional electron microscopy. Drawing of bacteriophage from Kostyuchenko et al. (2003), reproduced with permission of Nature Publishing Group, Inc.

There is, however, a sense in which viruses and chromatin ... are still relatively simple systems. Much more complex systems, ribosomes, the mitotic apparatus, lie before us and future generations will recognise that their study is a formidable task, in some respects only just begun.

— *Aaron Klug, Nobel Lecture 1983*

1. The Electron Microscope and Biology

1.1. General Remarks

The Middle Ages saw the different objects of the world ordered in hierarchies; every living creature, every thing of the animated and unanimated world, was ranked within its own orbit according to its intrinsic power, beauty, and utility

1

to man. Thus, the king of all animals was the lion, the king of all musical instruments the organ, etc. I have no doubt that in those days the electron microscope would have earned itself the name of Queen of all Microscopes, with its awe-inspiring physical size, its unsurpassed resolving power, and its mysterious lenses that work without glass.

Biology today would be unthinkable without the invention of the electron microscope. The detailed architecture of the cell and its numerous organelles, the structure of viruses, and the fine structure of muscle are all beyond the reach of light microscopy; they came into view for the first time, in spectacular pictures, as this new instrument was being perfected. One of the first discoveries made possible by electron microscopy (EM) was the origin of the iridescent colors of butterfly wings by my first academic mentor, Ernst Kinder: they were traced to diffraction of light on ordered arrays of entirely colorless scales (Kinder and Süffert, 1943).

1.2. Three-Dimensional Electron Microscopy

By far the largest number of applications of EM to biology are concerned with interpretations of the image on a qualitative descriptive level. In those cases where quantitative measurements are obtained, they normally relate to distances, sizes, or numbers of particles, etc. Precise measurements of optical densities are rarely relevant in those uses of the instrument. In contrast, there are other studies in which measurements of the images in their entirety are required. In these applications, attempts are being made to form an accurate three-dimensional (3D) representation of the biological object. Here, the term "representation" stands for the making of a 3D map, or image, of the object, which reveals not only its shape but also its interior density variations.

At this point it is worth reflecting on the position of 3D electron microscopy (3DEM) within the overall project of visualizing biological complexity. The biological structure is built up in a hierarchical manner, following in ascending order the levels of macromolecule, macromolecular assembly, cell organelle, cell, tissue, and finally, the whole organism. This hierarchy encompasses an enormous span in physical dimensions. At one end of this span, atomic resolution is required to render a complete description of a macromolecule. At the other end, the scale is macroscopic, on the order of meters for large organisms. Within that vast range, EM bridges the gap, of several orders of magnitude, that is left between the size ranges of objects studied by X-ray crystallography and light microscopy (figure 1.1). Thus, when we take into account the existing techniques of 3D light microscopy and those of macroscopic radiological imaging (i.e., computerized axial tomography, positron emission tomography, and magnetic resonance imaging), the project of visualizing and modeling an organism in its entirety has become, at least in principle, possible. Of course, limitations of data collection, computational speed, and memory will for a foreseeable time prevent the realization of this project for all but the most primitive organisms such as *Escherichia coli*.

Although true three-dimensionally imaging transmission electron microscopes have been conceived, based on an objective lens design that allows unusually large apertures to be used (Hoppe, 1972; Typke et al., 1976), such instruments have never been realized. Three-dimensional information is normally obtained by interpreting micrographs as projections: by combining such projections, taken over a sufficiently large angular range, the object is eventually recovered. In many experiments of practical importance, at not too high resolutions, the interpretation of micrographs as parallel projections of the object is in fact a very good approximation (see Hawkes, 1992).

At the outset, we distinguish the true 3D imaging being considered here from stereoscopic imaging and from serial-section reconstruction. The latter two techniques are unsuitable for imaging macromolecules for several reasons. In stereoscopic imaging, a 3D image is not actually formed until it is synthesized from the two views in the observer's brain. This technique therefore poses obvious difficulties in obtaining an objective description of the structure, and its use is moreover restricted to structures that are isolated, well delineated, and sufficiently large. In serial-section reconstruction, on the other hand, a 3D shape representation is formed by stacking visually selected (or, in some cases, computationally extracted) sets of contours, each derived from a different section. Here, the minimum thickness of sections, usually above 500 Å, precludes application to even the largest macromolecules. (Besides, the material loss and deformation due to mechanical shearing limits the serial-section technique of reconstruction even when applied to much larger subcellular structures.)

Three-dimensional imaging with the electron microscope follows two different methodologies, which are divided essentially according to the size range of the object and — closely related to the physical dimensions—according to its degree of "structural uniqueness" (Frank, 1989, 1992a). On one hand, we have cell components, in the size range of 100–1,000 nm[1], that possess a unique, in a strict sense irreproducible, structure. An example for such an object is the mitochondrion: one can state with confidence that no two mitochondria are the same in the sense of being strictly congruent in three dimensions (figure 1.2a). (Here, the term "congruent" is meant to say that two structures could be brought into precise overlap by "rigid body movement"; i.e., by a movement of the structure, in three dimensions, which leaves the relative distances and angles among all its components intact.) In fact, structures like the mitochondria even lack similarity of a less rigorous kind that would require two structures to be related by a precisely defined geometric transformation, such as an affine transformation.

On the other hand, we have macromolecular assemblies, in the size range of 50–500 nm, which exist in many structurally identical "copies" that are congruent in three dimensions. This implies that such macromolecules will present identical views in the electron microscope when placed on the support in the same orientation (figure 1.2b).

The degree of "structural uniqueness" obviously reflects the way function is realized in a structure. For instance, the function of the mitochondrion as the

---

[1]The nm (nanometer) unit is used here to facilitate comparison of the macromolecular with the subcellular scale. However, throughout the rest of the book, the Å (angstrom) unit (= 0.1 nm) will be used since it is naturally linked to its usage in X-ray crystallography, the field of prime importance for EM as a source of comparative and interpretative data.

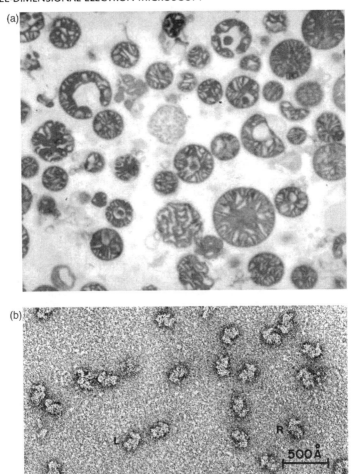

Figure 1.2 Structural uniqueness of a biological system is a function of size. (a) Several rat liver mitochondria (diameter, 0.2–0.8 μm) embedded in Epon and cut into 0.1-μm sections, imaged with a JEOL 1200EX electron microscope operated at 100 kv and electron optical magnification of 25,000 ×. The structure of each mitochondrion is unique, although a general building principle is maintained. Averaging methods cannot be used in the retrieval of the structure (unpublished data, kindly made available by Drs. Luca Scorrano and Stanley Korsmeyer). (b) Field of 40S small subunits of mammalian ribosome (HeLa cells) negatively stained on a carbon film. Particles occur preferentially in two side views, marked "L" (left-facing) and "R" (right-facing). Each image can be understood as a superposition of an underlying unchanged structure and noise. With such a "repeating" structure, averaging methods can be used with success. From Frank et al. (1982) with permission of Elsevier.

power plant of the cell relies critically on the properties of its specific membrane transport proteins and its compartmentation produced by the complex folding of the inner membrane (Tyler, 1992; Mannella et al., 1994), but—and this is the important point—not on the maintenance of a particular exact shape of the entire organelle. In sharp contrast to this, the function of the much smaller ribosome in the synthesis of proteins is tied to the maintenance of a precise conformation that facilitates and constrains the steric interactions of the ribosome's numerous ligands in the process of protein synthesis.

The variation in the degrees to which function dictates, or fails to dictate, the maintenance of a particular structure leads to two fundamentally different approaches to 3D imaging. For objects that have identical structure by functional necessity, powerful averaging techniques can be used to eliminate noise and reduce the electron dose, thereby limiting radiation damage. Alternatively, objects that may vary from one realization to the next without impact on their function can only be visualized as "individual structures" by obtaining one entire projection set per realization. For these latter kinds of objects, it is obviously more difficult to draw generalizations from a single 3D image. This book will exclusively deal with the former problem, while the latter problem has been discussed at some length in an edited volume on electron tomography (Frank, 1992b; see also Frank, 1995).

Methods for the retrieval of structure from nonperiodic objects were first developed in the laboratory of Walter Hoppe at the Max-Planck-Institut in Martinsried (Hoppe, 1974; Hoppe et al., 1974). However, the approach taken by Hoppe's group failed to take advantage of averaging. In fact, it was based on experiments that allowed an exposure of $1000\,e^-/\text{Å}^2$ or more to accumulate during the recording of projections. This approach was rightly criticized because of the high accumulated dose and the absence of a criterion that allowed radiation damage and preparation-related artifacts to be assessed (Baumeister and Hahn, 1975; Frank, 1975a; Crowther, 1976). Averaging over several reconstructions (Knauer et al., 1983; Öttl et al., 1983; Oefverstedt et al., 1994) still does not allow the exposure to be reduced to acceptable levels (below $10\,e^-/\text{Å}^2$). Only the development of computer-controlled electron microscopes (Dierksen et al., 1992, 1993; Koster et al., 1992, 1997) has made it possible to reduce the exposure per projection to extremely low levels, opening the way for a true tomography of macromolecules, where a statistically significant merged reconstruction can in principle be obtained by averaging in three dimensions (Walz et al., 1997b; Nitsch et al., 1998).

2. Single-Particle Versus Crystallographic Analysis

Electron crystallography, either in its pure form or in the single-particle variants discussed in this book, deals with attempts to form precise 3D representations of the objects imaged. The crystallographic approach to structure analysis with the electron microscope follows a distinguished tradition established by Sir Lawrence Bragg, director of the Cavendish Laboratory, which spawned the Molecular Biology Laboratory of the Medical Research Council in Cambridge, England (see, for instance, Watson, 1968). In the 1960s, Klug and Berger (1964) used optical diffraction as a way of evaluating images of periodic structures taken with the electron microscope. Shortly afterward, Klug and DeRosier (1966) developed optical filtration as a way of obtaining noise-free versions of the image of a crystal. From that starting position, the jump to computer applications of Fourier analysis in electron microscopy was obvious and straightforward, as recounted in Klug's Nobel laureate lecture (Klug, 1983).

The use of crystallographic methods to retrieve structural information from electron micrographs (e.g., Hoppe et al., 1969; Erickson and Klug, 1970; Henderson and Unwin, 1975; Unwin and Henderson, 1975) has become known as *electron crystallography* (Glaeser, 1985). To be amenable to this approach, the object must form ordered structures whose dimensions are in the range of the distances traversed by electrons accelerated to 100 kV (in conventional transmission electron microscopes) or 200–400 kV (in intermediate-voltage electron microscopes). More specifically, the thickness range should be such that, for the electron energy used, the chances for multiple scattering are negligible. These structures can be thin crystals, just one unit cell or, at any rate, no more than a few unit cells thick (Unwin and Henderson, 1975; Amos et al., 1982); helical assemblies (DeRosier and Moore, 1970; Stewart, 1988b); or spherical viruses having high symmetry (Crowther et al., 1970).

By being able to deal with ordered structures that fail to form "bulk" crystals amenable to X-ray crystallography, electron crystallography claims an important place in structure research. Among the thin crystals are membrane proteins, most of which have withstood attempts of 3D crystallization (light-harvesting complex, bacteriorhodopsin, and bacterial porin PhoE; see references below). Among the helical structures are the T4 phage tail (DeRosier and Klug, 1968), bacterial flagella (Morgan et al., 1995), and the fibers formed by the aggregation of the acetylcholine receptor (Toyoshima and Unwin, 1988b; Mlyazawa et al., 2003). There are now numerous examples for spherical viruses whose structure has been explored by EM and 3D reconstruction (Crowther and Amos, 1971; Stewart et al., 1993; Cheng et al., 1994; Böttcher et al., 1997; see also the volume on structural biology of viruses edited by Chiu et al., 1997). For molecules that exist as, or can be brought into, the form of highly ordered crystalline sheets, close-to-atomic resolution can be achieved (bacteriorhodopsin: Henderson et al., 1990; Subramaniam and Henderson, 1999; bacterial porin: Jap, 1991; light-harvesting complex: Kühlbrandt et al., 1994; tubulin: Nogales et al., 1998; see also the earlier survey in Cyrklaft and Kühlbrandt, 1994). The entire methodology of electron crystallography is presented in a forthcoming volume (R.M. Glaeser et al., in preparation).

From the outset, it is clear that in nature, highly ordered objects, or objects obeying perfect symmetry, are the exception, not the rule. Indeed, maintenance of precise order or perfectly symmetric arrangement does not have a functional role for most biological macromolecules. Obvious exceptions are bacterial membranes where structural proteins form a tight, ordered network or systems that involve cooperativity of an entire array of ordered molecules in contact with one another. Examples are the helical assemblies underlying the architecture of bacterial flagella (Morgan et al., 1995; Yonekura et al., 2000) and muscle (Rayment et al., 1993; Schröder et al., 1993; Chen et al., 2002b). Storage in the cell is another circumstance where highly ordered aggregates may be formed. An example is the two-dimensional (2D) ordered arrangement of 80S ribosomes in lizard oocytes (Milligan and Unwin, 1986).

If molecules fail to form ordered arrays in the living system, they can under certain conditions be induced to do so in vitro. However, even now that sophisticated methods for inducing crystallinity are available (see Kornberg and Darst, 1991, and the comprehensive review of techniques by Jap et al., 1992), the likelihood of success is still unpredictable for a new protein. What factors into the success is not just the ability to obtain crystals, but to obtain crystals sufficiently well ordered to be amenable to the crystallographic approach of 3D reconstruction. As a result, the fraction of macromolecular assemblies whose structures have been solved to atomic resolution by methods of electron crystallography is still quite small.

Thus, in conclusion, electron crystallography of crystals and ordered assemblies has a number of limitations, the most obvious of which is the limited availability of crystals. Another is the difficulty in dealing with crystals that are imperfect. Finally, one must also realize that the electron crystallographic approach has another shortcoming: the crystalline packing constrains the molecule so that it may assume only a small range of its possible physiologically relevant conformations.

3. Crystallography without Crystals

The fundamental principle of crystallography lies in the use of redundancy to achieve a virtually noise-free average. Redundancy of structural information is available in a crystal because the crystal is built from a basic structural unit, the *unit cell*, by translational repetition (helical and icosahedral structures are generated by combinations of translations and rotations). The unit cell may be composed of one or several copies of the molecule, which are in a strict geometrical arrangement, following the regime of symmetry. Electron crystallography is distinguished from X-ray crystallography by the fact that it uses *images* as primary data, rather than diffraction patterns. Translated into Fourier lingo, the availability of images means that the "phase problem" known in X-ray crystallography does not exist in EM: the electron microscope is, in Hoppe's (1983) words, a "phase-measuring diffractometer." Another important difference between electron and X-ray crystallography is the fact that electrons interact with matter more strongly, hence even a single-layered (2D) crystal no larger than a few micrometers in diameter provides sufficient contrast to produce, upon averaging, interpretable images, while X-ray diffraction requires 3D crystals of macroscopic dimensions for producing statistically acceptable measurements.

The catchphrase "crystallography without crystals,"[2] lending the title to this section, alludes to the fact that there is little difference, in principle, between the image of a crystal (seen as a composite of a set of translationally repeating images of the structure that forms the unit cell) and the image of a field containing such structures in isolated form, as so-called single particles.[3] In both situations,

---

[2]Title of a lecture given by the author at the Second W. M. Keck Symposium on Computational Biology in Houston, 1992.

it is possible to precisely align and superimpose the individual images for the purpose of forming an average, although the practical way of achieving this is much more complicated in the case of single particles than in the case of the crystal. (Indeed, two entire chapters, chapters 3 and 4, of this book are devoted to the problem of how to superimpose, classify, and average such single-particle images.) Another complication arises from the fact that macromolecules in single-particle form are more susceptible to variations in their microenvironment, in terms of stain thickness (in case stain is being used), hydrophilic properties of the support grid, etc. Because of the limited accuracy of alignment and the presence of structural variability, attainment of atomic resolution poses formidable hurdles.

However, with a concentrated effort in data collection, algorithm development, and employment of parallel computing, *routine* attainment of information on secondary structure, at a resolution between 5 and 10 Å, is a definite possibility. Individual α-helices were first visualized by Henderson and Unwin (1975) for highly ordered 2D crystals of bacteriorhodopsin. The same was achieved 12 years later for viruses (Böttcher et al., 1997; Conway et al., 1997), and more recently for molecules entirely without symmetry such as the ribosome (Spahn et al., 2005). In fact, going by physical principles, Henderson (1995) does not foresee fundamental problems preventing achievement of atomic (∼3 Å) resolution for nonsymmetric particles, provided that they are beyond a certain size that is required for accurate alignment.

In summary, the two types of electron crystallography have one goal in common: to suppress the noise and to extract the 3D structure from 2D images. Both are distinguished from X-ray crystallography by the fact that true images are being recorded, or—in the language of crystallography—phases are being measured along with amplitudes. Both types of electron crystallography also employ the same principle: by making use of the fact that many *identical copies* of a structure are available. The difference between the two methods of electron crystallography is threefold: (i) "true" crystallography, of crystals, does not require alignment; (ii) the molecules forming the unit cell of a crystal exhibit considerably less variation in structure; and (iii) only crystallography of crystals is able to benefit from the availability of electron diffraction data that supply highly accurate amplitudes, often to a resolution that exceeds the resolution of the image. For these reasons, single-particle methods, even if they have the potential of achieving atomic resolution (Henderson, 1995), are often confined to the exploration of "molecular morphology" (a term used by Glaeser, 1985). However, the value of this contribution cannot be overestimated: besides yielding information on quaternary structure, the low-resolution (e.g., 10–20 Å) map of a large macromolecular assembly also provides powerful constraints useful for phasing X-ray data.

---

[3]The term "single particles" stands for "isolated, unordered particles with—in principle—identical structure." Later on, well after this term with this particular meaning was coined, it was used in a different context: of measurements (such as by atomic force or fluorescence resonance energy transfer studies) carried out on individual macromolecules.

Table 1.1 Sample Quantities and Concentrations for Common Imaging and Analytical Techniques.

	Cryo-EM	X-Ray	NMR	MS
Sample volume	$5\,\mu l^a$	$10\,\mu l$	$400\,\mu l$	$0.1–10\,\mu l$
No. of samples/exp	5–10	1 crystal	1 ($< 20\,kDa$)[b] several (20–40 kDa)	1
Concentration	50 nM	$50\,\mu M^c$	0.5–1.0 nM	
Total amount	0.25 pmol	$500\,pmol^d$	0.2–0.4 μmol	10 fmol[e] 10–100 fmol[f] 100 fmol–10 pmol[g]
Total time for data collection	1–4 days	2 h (synchrotron) 48 h (in-house)	1–2 weeks ($< 20\,kDa$) 6–8 weeks (20–40 kDa)	seconds (spectra) hours (LC–MS/MS)

[a] Droplet on grid, prior to blotting.
[b] For stable proteins of this size, a single U-13C, 15N sample is likely sufficient.
[c] Based on 10 mg/ml of a 100 kDa protein.
[d] Final amount required for the crystal. However, a multiple of this amount is required to try out different crystallization conditions.
[e] Quantity of *digested* protein required to obtain information on ~20% of total sequence. Amount of initial protein is 10 times that amount.
[f] Proteomics/protein identification.
[g] Posttranslational modification analysis.
Abbreviations: Cryo-EM = cryo-electron microscopy; LC = liquid chromatography; MS = mass spectroscopy; NMR = nuclear magnetic resonance; X-ray = X-ray crystallography.

One important consideration in assessing the pros and cons of the different techniques of molecular visualization, so far not mentioned, is the quantity of sample required. The concentration of molecules in the cell varies widely with the type of molecule, making it extremely difficult in some cases to get hold of the minimum sample volume by cell extracts and purification. A comparison of sample requirements for typical experiments (table 1.1) makes it clear that cryo-electron microscopy (cryo-EM) is the technique with the smallest quantity required. (Mass spectroscopy, which is not a visualization technique, is included as a yardstick for highest sensitivity.)

4. Toward a Unified Approach to Structural Analysis of Macromolecules

Notwithstanding the existence of cross-fertilizing contributions of a few pioneering groups with expertise in both fields, EM and X-ray crystallography have largely developed separately. The much younger field of electron microscopic image analysis has benefited from adopting the tools and working methods of statistical optics (e.g., O'Neill, 1969), the theory of linear systems (e.g., Goodman, 1968), and multivariate data analysis and classification (Lebart et al., 1977). In recent years, a convergence has taken place, brought about mainly by the development of cryo-techniques in EM. It has now become possible to relate the 3D density distribution of a molecule reconstructed by EM quantitatively to its low-resolution X-ray map. Quantitative comparisons of this kind (e.g., Jeng et al., 1989; Rayment et al., 1993; Stewart et al., 1993; Boisset et al., 1994b;

Roseman et al., 2001; Gao et al., 2003; Ludtke et al., 2004) are becoming more common as instruments with high coherence and stability become more widely available. Even with negative staining, it is often startling to see the agreement between 2D or 3D maps computed from X-ray data and EM in minute features at the boundary of the molecule; see, for instance, Lamy (1987), de Haas et al. (1993), and Stoops and coworkers (Stoops et al., 1991a; Kolodziej et al., 1997). However, with cryo-EM data, the agreement even extends to the interior of the molecule. The ability to compare, match, and merge results from both types of experimental sources offers entirely new possibilities that are beginning to be realized in collaborations between practitioners of cryo-EM and X-ray crystallography, as in much of the work cited above.

Cryo-electron crystallography of both forms of specimens (crystals and single particles), when combined with results from X-ray crystallography, presents a way of resolving very large structures to atomic resolution. Often components of a macromolecular assembly can be induced to form crystals suitable for X-ray crystallography, but the same may not be true for the entire assembly because of its size or inherent flexibility. On the other hand, electron crystallography of such a structure using cryo-methods may make it possible to obtain its 3D image at low resolution. By matching the high-resolution X-ray model of the component to the 3D cryo-EM image of the entire macromolecule, high positional accuracies beyond the EM resolution can be obtained, and atomic resolution within the large structure can be reached, or at least approximated (e.g., Wang et al., 1992; Stewart et al., 1993; Stewart and Burnett, 1993; Cheng et al., 1994; Boisset et al., 1994b, 1995; Volkmann and Hanein, 1999; Wriggers et al., 2000; Gao et al., 2003; Ludtke et al., 2004; Rossmann et al., 2005).

The most recent development of the spray-mix method in cryo-EM (Berriman and Unwin, 1994; Walker et al., 1994; White et al., 1998), the use of caged compounds (e.g., Ménétret et al., 1991; Rapp, 1998; Shaikh et al., 2004), and the potential of newer methods of single-particle reconstruction from sorted heterogeneous populations of molecules in different states (e.g., Heymann et al., 2003, 2004; Gao et al., 2004) strengthen the contribution of EM to structural biology as these new techniques are capable of providing dynamic information on conformational changes or functionally relevant binding reactions.

5. Single-Particle Reconstruction, Macromolecular Machines, and Structural Proteomics

After the successful charting of the entire genomes of several species, including that of *Homo sapiens*, the postgenomic era was narrowly defined initially by the opportunities offered through comparisons of genes and their expression patterns (spatial and temporal) in the cell. What received initially little attention was the analysis of proteins made according to genetic instructions, and, in particular, the analysis of their multiple interactions in the cell. *Proteomics* is the term that has been coined for the systematic study of proteins by themselves, but a word for a field of study whose aim is to look at their multiple interactions is so far missing.

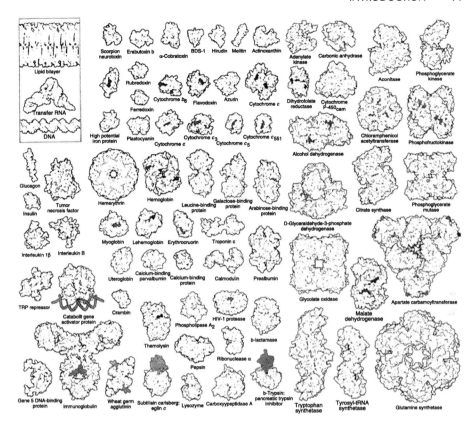

Figure 1.3 Sizes and shapes of protein complexes in the cell. From Goodsell and Olson (1993), reproduced with permission of Elsevier.

Meanwhile, evidence that molecules interact in the cell in a compact, spatially localized, highly choreographed way has accumulated in many areas of cell biology. The remarkable diversity of shapes and sizes of proteins whose structure is known (figure 1.3) is a reflection of the specificity with which they interact, often in a lock-and-key fashion. Almost every process in the cell, from DNA repair, transcription, splicing, translation, folding, degradation, to the formation of vesicles, has associated with it an entity, a machine, whose name typically ends with "some." Examples are spliceosome, ribosome, and aptoptosome, which are engaged, respectively, in splicing mRNA, synthesizing protein, and initiating programmed cell death. The word "machine" offers itself since the components follow mechanical movements such a rotation, gliding, and lever motions.

One of the defining characteristics of such a system is *processivity*. By this it is understood that a component or ligand, after joining the assembly, stays with it for a certain time, carrying out its function from the beginning to the end. Typically, in such a machine, many proteins and RNA components, often utilized in multiple ways, are acting in concert, as though following a precise script. Transcription, for instance, involves more than a dozen factors and cofactors. During the successive phases of translation (initiation, elongation, termination, and recycling), the ribosome interacts with mRNA, tRNA, and a host of factors with catalytic or regulatory roles. Bruce Alberts, in an oft-cited article heralding a special issue of the journal *Cell*, speaks of the cell as a "collection of macromolecular machines" (Alberts, 1998). This idea represents a radical departure from the view, prevalent only a decade before his article appeared, that within the various compartments separated by membranes, molecules are interacting more or less by chance.

Recently, the systematic analysis of the contents of the yeast cell by a method of molecular tagging and mass spectroscopy (Gavin et al., 2002; Ho et al., 2002) has uncovered the remarkable number of complexes formed by the association of two or more proteins. This number was found to be in the range of 200 to 300. Many proteins are "modules" that are multiply used in the formation of different complexes. For most of the complexes found in this inventory, the function is as yet unknown. However, there can be no doubt that they participate in key processes in the living cell, and that an understanding of the structure and dynamics of the interactions of the constituent proteins is required as a necessary step toward a full understanding of life processes at the cellular level.

The large size of the complexes formed in multiple associations and their likely dynamical variability in the functional context takes most of these complexes out of the reach of X-ray crystallography, and leaves cryo-EM as one of the few techniques of structural investigation (Abbott, 2002). Single-particle techniques, in particular, are preferred since they leave molecular interactions unconstrained. (To be sure, results obtained by cryo-EM will continue to be most useful when X-ray structures of components are available.) The program of systematically studying these complexes in their stepwise associations and ligand interactions for several model organisms ("Structural Proteomics": see, for instance, Sali et al., 2003) will be a large undertaking that is likely to occupy biology for a good part of this century. It will fuel further developments in instrumentation and automation, as well as further advances in algorithms and the computational techniques described in this book.

Another challenge, beyond the project of looking at structures in multiple associations within in vitro systems, is the task of visualizing and identifying such complexes within the context of the cell. Again, there is reason to believe that EM will make seminal contributions in this endeavor (Baumeister and Steven, 2000; Böhm et al., 2000; McEwen and Frank, 2001; McEwen and Marko, 2001; McIntosh, 2001; Frank et al., 2002; Schliwa, 2002; Wagenknecht et al., 2002; Grünewald et al., 2003; Sali et al., 2003). Even though electron tomographic imaging of cellular structures is outside the agenda of this book, occasional references will be made to this methodology, since some of the problems to be solved, such as 3D reconstruction, motif search, segmentation, and denoising, are common to tomography and single-particle reconstruction.

6. The Electron Microscope and the Computer

The development of the computer has had the most profound impact on the way science is done. This revolution has particularly affected all areas of research that involve the interpretation of images: it has marked the change from a qualitative to a quantitative description of objects represented, be it a star, an amoeba, or a ribosome. For a long time, image processing in EM has relied on a tedious procedure by which the image is first recorded on a photographic film or plate and then is scanned sequentially by a specialized apparatus—the microdensitometer. For each image element, the transmission is recorded and converted to optical density, which is then converted to a digital value that can be processed by the computer. In 1968, when the first images were digitized in the laboratory of my mentor, Walter Hoppe, at the Max-Planck-Institut für Eiweiss-und Lederforschung (later part of the Max-Planck-Institut für Biochemie in Martinsried), they were stored on paper tape. The scanning took all night. Twenty-five years later, the same job can be done within one second with a video scanner. Moreover, the new types of electron microscopes allow images to be read directly into the computer (Brink et al., 1992; Downing et al., 1992; Brink and Chiu, 1994; Carragher et al., 2000; Potter et al., 2002), entirely bypassing what Zeitler (1992) ironically called "the analog recorder"—the photographic plate. With an instantaneous readout of the image into the computer, feedback control of the instrument became available in the 1990s, along with the more outlandish possibility of running a microscope standing in California from another continent (e.g., Fan et al., 1993).

As a consequence of slow-scan CCD (charge-coupled device) cameras and microcircuitry becoming faster and more powerful while falling in price, totally integrated instruments with the capability of automated data collection were developed a decade ago (Koster et al., 1989, 1990, 1992; Dierksen et al., 1992, 1993) and are now offered commercially. The high degree of integration between the electron microscope, its electronic control circuitry, its hardware for online readout, and the computer has given rise to an instrument of an entirely new type, a revolution best compared with the introduction of automated diffractometers in X-ray crystallography decades ago.

2

Electron Microscopy of Macromolecular Assemblies

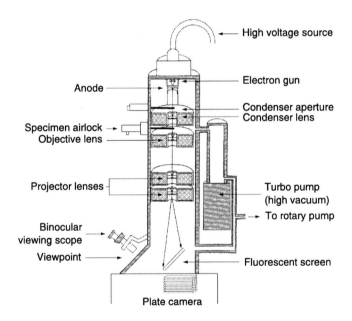

High voltage source

Electron gun

Anode

Condenser aperture
Condenser lens

Specimen airlock
Objective lens

Projector lenses

Turbo pump
(high vacuum)

To rotary pump

Binocular
viewing scope

Viewpoint

Fluorescent screen

Plate camera

Figure 2.1 Schematic diagram of a transmission electron microscope. Adapted from a drawing provided by Roger C. Wagner, University of Delaware (Web site: www.udel.edu/ Biology/Wags/b617/b617.htm).

1. Principle of the Transmission Electron Microscope

The transmission electron microscope uses high-energy (100 keV or higher) electrons to form an image of very thin objects.[1] Electrons are strongly scattered

---

[1]This section is not meant as a substitute for a systematic introduction, but serves mainly to provide a general orientation, and to introduce basic terms that will be used in the text.

15

by matter. For instance, for 120 kV electrons, the mean free path for elastic scattering in vitreous ice is in the region of 2800 Å, and for amorphous carbon 1700 Å (Angert et al., 1996). Values for resins used for plastic embedding and for frozen-hydrated tissue closely match those for vitreous ice. The values for inelastic scattering are 850 and 460 Å, respectively. Their paths can be bent by magnetic lenses so that an image is formed on a fluorescent screen (for direct viewing), photographic film, or scintillator of a charge-coupled device (CCD camera) (for recording). The much shorter wavelength of electrons, as compared to light, carries the promise of much higher resolution. However, because of lens aberrations, the resolution of an electron microscope falls far short of what one would expect from the de Broglie wavelength (0.037 Å for electrons accelerated to 100 kV). Instead, even today's best electron microscopes have resolutions in the range 1–2 Å.

Since electrons are also scattered by the molecules of the air, the electron microscope column needs to be permanently kept at high vacuum. The need to maintain the vacuum makes it necessary to introduce specimens and the photographic plates through special airlock mechanisms. The problem posed by the basic incompatibility between the vacuum requirements and the naturally hydrated state of biological specimens has only recently been solved by cryo-electron microscopy (cryo-EM) techniques: the specimen is kept frozen, hydrated throughout the experiment, at a temperature (liquid nitrogen or helium) at which the vapor pressure of water is negligible. Before the introduction of practical cryo-techniques, in the beginning of the 1980s (Dubochet et al., 1982), molecular specimens were usually prepared in a solution of heavy metal salt such as uranyl acetate or uranyl molybdate, and subsequently air-dried, with consequent loss of high-resolution features and interior detail (see sections 2.2 and 2.3). Despite these shortcomings, this technique of *negative staining* is still used in the initial phases of many projects of molecular imaging.

For the general principle and theory of scattering and image formation, and the many applications of the transmission electron microscope, the reader is referred to the authoritative works by Reimer (1997) and Spence (1988). In the following, we will make reference to a schematic diagram of a typical transmission electron microscope (figure 2.1) to explain the essential elements.

Cathode. Electrons are emitted by the cathode, which is kept at a high negative voltage, while the rest of the microscope is at ground potential. There are two types of cathode that have gained practical importance: the thermionic cathode and the field-emission gun (FEG). The thermionic cathode works on the principle that on being heated to sufficiently high temperature, metals allow electrons to exit. Typically, a hairpin-shaped wire made of tungsten is used. Use of a material with low work function such as lanthanum hexaboride (LaB_6) as tip enhances the escape of electrons and leads to a brightness higher than that of tungsten. In the FEG, the exit of electrons is enhanced (by lowering the potential barrier) and directionally focused by the presence a strong electric field at the cathode tip. To this end, a voltage in the region of 2 kV is applied to a so-called *extractor electrode* situated in close vicinity to the tip. As a consequence, extraordinary brightness is attained. *Brightness* is directly related to *spatial coherence*: to achieve high coherence, which is required for high resolution (see section 3), one has to restrict the emission from the cathode to a very small angular range (or, to use the term to be introduced later on, *effective source size*; see section 3.3.1). Unless the brightness of the cathode is high, this restriction leads to unacceptably low intensities and long exposure times. In the 1990s, commercial instruments equipped with a FEG came into widespread use.

Wehnelt cylinder. The cathode filament is surrounded by a cup-shaped electrode, with a central approx. ~1-mm bore, that allows the electrons to exit. Kept on a potential that is slightly more (~500 V) negative than that of the cathode, the Wehnelt cylinder is specially shaped such that it ushers the electrons emitted from the cathode toward the optic axis, and allows a space charge formed by an electron cloud to build up. It is the tip of this cloud, closest to the Wehnelt cylinder opening, that acts as the actual electron source.

Condensor lenses. These lenses are part of the illumination system, defining the way the beam impacts on the specimen that is immersed in the objective lens. Of practical importance is the spot size, that is, the physical size of the beam on the specimen, and the angle of beam convergence, which determines the coherence of the illumination. There are usually two condensor lenses, designated CI and CII, with CI situated immediately below the Wehnelt cylinder. This first condensor lens forms a demagnified image of the source. The size of this source image ("spot size") can be controlled by the lens current. The second condensor is a weak lens that is used to illuminate the specimen with a variable physical spot size and angle of convergence. At one extreme, it forms an image of the source directly on the specimen (small spot size; high angle of convergence). At the other extreme, it forms a strongly blurred image of the source (large spot size; low angle of convergence). Additional control over the angle of convergence is achieved by variation of the CII apertures.

Objective lens. This lens, the first image-forming lens of the electron microscope, determines the quality of the imaging process. The small focal length (on the order of millimeters) means that the specimen has to be partially immersed in the bore of the lens, making the designs of mechanisms for specimen insertion, and specimen support and tilting quite challenging. Parallel beams are focused in the back focal plane, and, specifically, beams scattered at the same angle are focused in the same point. Thus, a diffraction pattern is formed in that plane. The *objective lens aperture*, located in the back focal plane, determines what portion of the scattered rays participate in the image formation. Contrast formed by screening out widely scattered electrons by the objective lens aperture is termed *amplitude contrast*. More important in high-resolution electron microscopy (EM) is the *phase contrast*, which is contrast formed by interference between unscattered and elastically scattered electrons that have undergone a phase shift due to wave aberrations and defocusing.

Objective lens aberrations, foremost (i) the third-order spherical aberration, (ii) chromatic aberration, and (iii) axial astigmatism, ultimately limit the resolution of the microscope. *Spherical aberration* has the effect that for beams inclined to the optic axis, the focal length is less than for those parallel to the axis. As a consequence, the image of a point becomes a blurred disk. This aberration can be partially compensated by operating the lens in underfocus, which has just the opposite effect (i.e., the focal length being greater for beams inclined to the axis), though with a different angular dependency (see section 3 where this important compensation mechanism utilized in bright-field microscopy is discussed). *Axial astigmatism* can be understood as an "unroundness" of the lens magnetic field, which has the effect that the way specimen features are imaged depends on their orientation in the specimen plane. Axial astigmatism can be readily compensated with special corrector elements that are situated directly underneath the objective lens.

Finally, *chromatic aberration* determines the way the lens focuses electrons with different energies. A lens with small chromatic aberration will focus electrons slightly differing in energy into the same plane. The importance of this aberration is twofold: electrons may have different energies because the high voltage fluctuates, or they may have different energies as a result of the scattering (elastic versus inelastic scattering, the latter accompanied by an energy loss). Modern electron microscopes are well stabilized, making the first factor unimportant. However, a small chromatic aberration constant also ensures that electrons undergoing moderate amounts of energy loss are not greatly out of focus relative to zero-loss electrons.

Intermediate lenses. One or several lenses are situated in an intermediate position between objective and projector lens. They can be changed in strength to produce overall magnifications in a wide range (between 500 and 1 million), though in molecular structure research a narrow range of 35,000 to 60,000 is preferred, as detailed below. Particular settings are available to switch the electron microscope from the imaging to the diffraction mode, a mode important mainly for studying two-dimensional (2D) crystals, and not useful in conjunction with single particles. (Note that in one nomenclature, both the intermediate lenses and the projector lens described below are summarily referred to as "projector lenses.")

Projector lens. This lens projects the final image onto the screen. Since the final image is quite large, the part of the EM column between the projector lens and screen is sensitive to stray fields that can distort the image, and special magnetic shielding is used.

Fluorescent screen. For direct viewing, a fluorescent screen is put into the image plane. The fluorescent screen is a metal plane coated with a thin (range 10–20 μm, depending on voltage) layer of a fluorescent material that converts impinging electrons into bursts of light.

Photographic recording. Upon impact on a photographic emulsion, electrons, just as light, make silver halide grains developable, and an image is formed by grains of metallic silver. Depending on emulsion thickness and electron energy, for a fraction of electrons each produces a cascade of secondary electrons, and the trail of grains in the emulsion is plume-shaped, normal to the plane of the emulsion, with a significant lateral spread.

CCD recording. Cameras using CCD cameras are made of a large array of photosensitive silicon diodes that is optically coupled with a *scintillator*.

The optical coupling is achieved by fiber optics or, less frequently, by a glass lens of large numerical aperture. The scintillator, like the fluorescent screen used in direct observation, converts the energy set free on impact of an electron into a local light signal. The signals from the individual array elements are read out and integrated over a brief period to form a digital image that is directly stored in the computer. Apart from offering the convenience of direct recording, CCD cameras have the advantage over photographic films in that they cover a much wider dynamic range. The disadvantage is that they are still quite expensive, and that they cover a much smaller field of view.

2. Specimen Preparation Methods

2.1. Introduction

Ideally, we wish to form a three-dimensional (3D) image of a macromolecule in its entirety, at the highest possible resolution. Specimen preparation methods must be designed to perform the seemingly impossible task of stabilizing the initially hydrated molecule so that it can be placed and observed in the vacuum. In addition, the contrast produced by the molecule itself is not normally sufficient for direct observation in the electron microscope, and various contrasting methods have been developed. Negative staining (section 2.2) with heavy metal salts such as uranyl acetate produces high contrast and protects, at least to some extent, the molecule from collapsing. However, the high contrast comes at a hefty price: instead of the molecule, with its interior density variations, only a cast of the exterior surface of the molecule is imaged, and only the molecule's shape can be reconstructed. To some extent, stain may penetrate into crevices, but this does not alter the fact that only the solvent-accessible boundary of the molecule is visualized.

As sophisticated averaging methods were developed, it became possible to "make sense" of the faint images produced by the molecule itself, but biologically relevant information could be obtained only after methods were found to "sustain" the molecule in a medium that closely approximates the aqueous environment: these methods are embedment in glucose (section 2.3), tannic acid (section 2.4), vitreous ice ("frozen-hydrated"; section 2.5) and cryo-negative staining (section 2.6). It is by now clear, from comparisons with structures solved to atomic resolution, that embedment in vitreous ice leads to the best preservation of interior features, all the way to atomic resolution. For this reason, the main emphasis in this book, in terms of coverage of methods and applications, will be on the processing of data obtained by cryo-EM.

For completeness, this section will conclude with a brief assessment of gold-labeling techniques, which have become important in some applications of 3DEM of macromolecules (section 2.7), and with a discussion of support grids (section 2.8).

2.2. Negative Staining

2.2.1. Principle

The negative staining method, which was introduced by Brenner and Horne (1959), has been widely used to obtain images of macromolecules with high contrast. Typically, an aqueous suspension is mixed with 1 to 2% uranyl acetate and applied to a carbon-coated specimen grid. The excess liquid is blotted away and the suspension is allowed to dry. Although, to some extent, the stain goes into crevices and aqueous channels, the structural information in the image is basically limited to the shape of the molecule as it appears in projection. As tilt studies and comparisons with X-ray diffraction data on wet suspensions reveal, the molecule shape is distorted due to air-drying (see Crowther, 1976; Kellenberger and Kistler, 1979; Kellenberger et al., 1982). Nevertheless, negative staining has been used with great success in numerous computer reconstructions of viruses and other large regular macromolecular assemblies. The reason why such reconstructions are legitimate is that they utilize symmetries which allow many views to be generated from the view least affected by the distortion. Therefore, as Crowther (1976) argues correctly, "reconstructions of single isolated particles with no symmetry from a limited series of tilts (Hoppe et al., 1974) are therefore somewhat problematical." As this book shows, the problems in the approach of Hoppe and coworkers to the reconstruction of negatively stained molecules can be overcome by the use of a radically different method of data collection and interpretation.

Even in the age of cryo-EM, negative staining is still used in high-resolution EM of macromolecules as an important first step in identifying character-istic views and assessing if a molecule is suitable for this type of analysis (see also the re-evaluation done by Bremer et al., 1992). Efforts to obtain a 3D reconstruction of the frozen-hydrated molecule almost always involve a negatively stained specimen as the first step, sometimes with a delay of several years [e.g., 50S ribosomal subunit: Radermacher et al. (1987a, 1992b); *Androctonus australis* hemocyanin: Boisset et al. (1990b, 1992a); nuclear pore complex: Hinshaw et al. (1992); Akey and Radermacher (1993); skeletal muscle calcium release channel: Wagenknecht et al. (1989a); Radermacher et al. (1994b); RNA polymerase II: Craighead et al., 2002]. A useful compilation covering different techniques of negative staining, including cryo-negative staining, is found in Ohi et al. (2004).

2.2.2. Artifacts Introduced by Negative Staining

2.2.2.1. One-sidedness and variability   Biological particles beyond a certain size are often incompletely stained; parts of the particle farthest away from the carbon film "stick out" of the stain layer ("one-sided staining"; figure 2.2). Consequently, there are parts of the structure that do not contribute to the projection image; one speaks of *partial projections*. A telltale way by which to recognize partial projections is the observation that "flip" and "flop" views of a particle (i.e., alternate views obtained by flipping the particle on the grid) fail to obey the expected mirror relationship. A striking example of this effect is offered by the 40S subunit of the eukaryotic ribosome with its characteristic R and L side

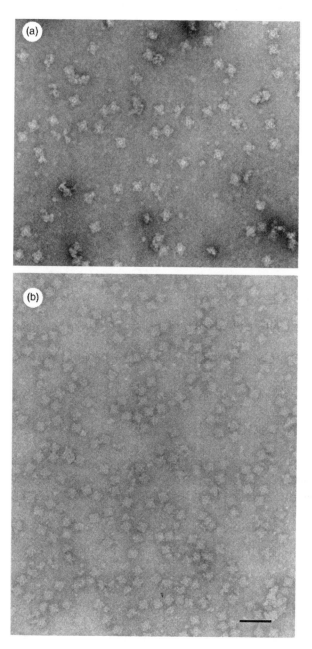

Figure 2.2 Appearance of negatively stained particles in single versus double carbon layer preparations, as exemplified by micrographs of the calcium release channel of skeletal fast twitch muscle. (a) Single-layer preparation, characterized by crisp appearance of the molecular border and white appearance of parts of the molecule that "stick out." From Saito et al. (1988), reproduced with permission of The Rockefeller University Press. (b) Double-layer preparation, characterized by a broader region of stain surrounding the molecule and more uniform staining of interior. From Radermacher et al. (1992a), reproduced with permission of the Biophysical Society.

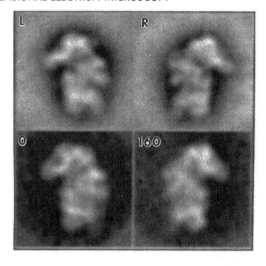

Figure 2.3 One-sidedness of staining in a single-carbon layer preparation produces strong deviations from the mirror relationship between flip/flop related projections. The average of 40S ribosomal subunit images showing the L view (top left) is distinctly different from the average of images in the R view (top right). This effect can be simulated by computing incomplete projections, that is, projections through a partially capped volume (e.g., upper half removed), from a reconstruction that was obtained from particles prepared by the double-layer technique (bottom panels). From Verschoor et al. (1989); reproduced with permission of Elsevier.

views (figure 2.3; Frank et al., 1981a, 1982): the two views are quite different in appearance—note, for instance, the massive "beak" of the R view which is fused with the "head," compared to the narrow, well-defined "beak" presented in the L view. (Much later, cryo-EM reconstructions as well as X-ray structures confirmed this difference in shape between the intersubunit and the cytosolic side of the small ribosomal subunit; see, for instance, Spahn et al., 2001c.)

Since the thickness of the stain may vary from one particle to the next, "one-sidedness" of staining is frequently accompanied by a high variability in the appearance of the particle. Very detailed—albeit qualitative—observations on these effects go back to the beginning of image analysis (Moody, 1967). After the introduction of multivariate data analysis into EM (see chapter 4), the stain variations could be systematically studied as these techniques of data analysis allow particle images to be ordered according to stain level or other overall effects. From such studies (van Heel and Frank, 1981; Bijlholt et al., 1982; Frank et al., 1982; Verschoor et al., 1985; Boisset et al., 1990a; Boekema, 1991) we know that the appearance of a particle varies with stain level in a manner reminiscent of the appearance of a rock partically immersed in water of varying depth (figures 2.4 and 2.5). In both cases, a contour marks the level where the isolated mass raises above the liquid. Depending on the depth of immersion, the contour will contract or expand. (Note, however, that despite the striking similarity between the two situations, there exists an important difference: for the negatively stained particle, seen in projection, only the immersed part is visible in the image, whereas for the

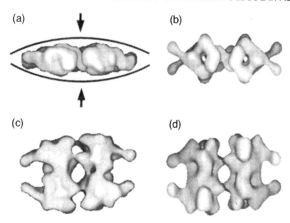

Figure 2.4 The effect of negative staining (uranyl acetate) on the particle shape, illustrated with reconstructions of a hemocyanin molecule from *Androctonus australis* labeled with four Fabs. (a, c) Side and top views of the molecule reconstructed from single-particle images of a negatively stained preparation. (b, d) Side and top views of the molecule from single-particle images obtained by cryo-EM. Comparison of the side views (a, b) shows the extent of the compression that occurs when the molecule is negatively stained and air-dried. While in the cryo-EM reconstruction (b), the orientations of the Fab molecules reflect the orientations of the protein epitope in the four copies of the corner protein; they are all bent in the stained preparation into a similar orientation, so that they lie near-parallel to the specimen grid. However, comparison of the top views (c, d) shows that the density maps are quite similar when viewed in the direction of the beam, which is the direction of the forces (indicated by arrows) inducing flattening. (a, c) from Boisset et al. (1993b), reproduced with permission of Elsevier. (b, d) From Boisset et al. (1995), reproduced with permission of Elsevier.

rock, seen as a surface from above, only the part sticking out of the water is visible.)

Computer calculations can be used to verify the partial-staining model. Detailed matches with observed molecule projections have been achieved by modeling the molecule as a solid stain-excluding "box" surrounded by stain up to a certain z-level. With the 40S ribosomal subunit, Verschoor et al. (1989) were able to simulate the appearance (and pattern of variability) of experimental single-layer projections by partial projections of a 3D reconstruction that was obtained from a double-layer preparation.

Because partial projections do not generally allow the object to be fully reconstructed, the single-carbon layer technique is unsuitable for any quantitative studies of particles with a diameter above a certain limit. In the case of uranyl acetate, this limit is in the range of 150 Å or so, the maximum height fully covered with stain as inferred from the comparison between actual and simulated projections of the 40S ribosomal subunit (Verschoor et al., 1989). For methyl-amine tungstate used in conjunction with Butvar films, on the other hand, much larger stain depths appear to be achieved: fatty acid synthetase, with a long dimension of ∼300 Å, appeared to be fully covered (Kolodziej et al., 1997).

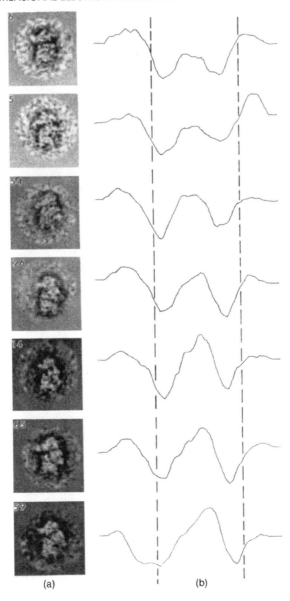

(a)   (b)

Figure 2.5 (a) Images of the 40S ribosomal subunit from HeLa cells, negatively stained in a single-carbon layer preparation. The images were sorted, in one of the first applications of correspondence analysis, according to increasing levels of stain. (b) Optical density traces through the center of the particle. From Frank et al. (1982), reproduced with permission of Elsevier.

The double-carbon layer method of staining [Valentine et al. (1968); Tischendorf et al. (1974); Stöffler and Stöffler-Meilicke (1983); see also Frank et al. (1986), where the relative merits of this technique as compared to those of the single-layer technique are discussed] solves the problem of partial stain immersion by providing staining of the particle from both sides without information loss. This method has been used for immunoelectron microscopy (Tischendorf et al., 1974; Lamy, 1987; Boisset et al., 1988) precisely for that reason: antibodies attached to the surface facing away from the primary carbon layer will then be visualized with the same contrast as those close to the carbon film.

Under proper conditions, the double-carbon layer method gives the most consistent results and provides the most detailed information on the surface features of the particle, whereas the single-layer method yields better defined particle outlines but highly variable stain thickness, at least for conventional stains. Outlines of the particles are better defined in the micrograph because a stain meniscus forms around the particle border (see Moody, 1967), producing a sharp increase in scattering absorption there. In contrast to this behavior, the stain is confined to a wedge at the particle border in sandwich preparations, producing a broader, much more uniform band of stain. Intuitively, the electron microscopist is bound to prefer the appearance of particles that are sharply delineated, but the computer analysis shows that images of sandwiched particles are in fact richer in interior features. Thus, the suggestion that sandwiching would lead to an "unfortunate loss of resolution" (Harris and Horne, 1991) is based mainly on an assessment of the visual appearance of particle borders, not on quantitative analysis.

For use with the random-conical reconstruction technique (see chapter 5, section 5), the sandwiching methods (Valentine et al., 1968; Tischendorf et al., 1974; Boublik et al., 1977) have been found to give most reliable results as they yield large proportions of the grid being double-layered. There is, however, evidence that the sandwiching technique increases the flattening of the particle (i.e., additional to the usual flattening found in single-layer preparations that is due to the drying of the stain; see below.)

2.2.2.2. Flattening   Specimens are flattened by negative staining and air-drying (Kellenberger and Kistler, 1979). As a side effect of flattening, and due to the variability in the degree of it, large size variations may also be found in projection (figure 2.6; see Boisset et al., 1993b).

The degree of flattening can be assessed by comparing the z-extent (i.e., in the direction normal to the support grid) of particles reconstructed from different views showing the molecule in orientations related by 90° rotation. Using such a comparison, Boisset et al. (1990b) found the long dimension of the *Androctonus australis* hemocyanin to be reduced to 60% of its size when measured in the x–y plane (figure 2.4). The z-dimension of the calcium release channel is reduced to as little as 30–40% when the reconstruction from the negatively stained preparation (Wagenknecht et al., 1989a) is compared with the reconstruction from cryo-images (Radermacher et al., 1994a,b). Apparently, in that case, the unusual degree of the collapse is due to the fragility of the dome-shaped structure on the transmembranous side of the molecule.

Evidently, the degree of flattening effect depends strongly on the type of specimen. It has been conjectured, for instance (Knauer et al., 1983), that molecular assemblies composed of RNA and protein, such as ribosomes, are more resistant to mechanical forces than those made entirely of protein. Indications

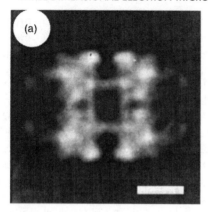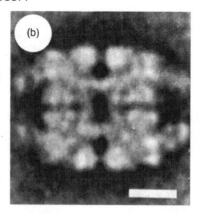

Figure 2.6 Size variation of *A. australis* hemocyanin–Fab complex prepared with the double-carbon layer, negative staining method. (a) Average of small, well-defined molecules that are encountered at places where the surrounding stain is deep; (b) average of large, apparently squashed molecules seen at places where the stain is shallow. The molecules were classified by correspondence analysis. Scale bar, 100 Å. From Boisset et al. (1993b), reproduced with permission of Elsevier.

for particularly strong resistance were the apparent maintenance of the shape of the 30S subunit in the aforementioned reconstructions of Knauer and coworkers and the result of shadowing experiments by Kellenberger et al. (1982), which indicated that the carbon film yields to, and partially "wraps around," the ribosomal particle.

The behavior of biological structures subjected to mechanical forces might be easiest to understand by considering the specific features of their architecture, which includes the presence or absence of aqueous channels and cavities. Indeed, empty shells of the turnip yellow mosaic virus were shown to be totally collapsed in the sandwiched preparation while maintaining their spherical shape in the single-layer preparation (Kellenberger et al., 1982). Along similar lines, the flattening of the 70S ribosome mentioned above was observed when the particle was oriented such that the interface cavity between the two subunits could be closed by compression. Also, the calcium release channel is a particularly fragile structure that includes many cavities and channels, as evidenced later on by cryo-EM reconstructions (Radermacher et al., 1994a,b; Samsó and Wagenknecht, 1998; Sharma et al., 1998).

As a general caveat, any comparisons of particle dimensions in the z-direction from reconstructions have to be made with some caution. As will become clear later on, reconstructions from a single random-conical projection set are to some extent elongated in the z-direction, as a result of the missing angular data, so that the amount of flattening deduced from a measurement of the z-dimension actually leads to an underestimation. By applying restoration (see chapter 5, section 10) to the published 50S ribosomal subunit reconstructed from stained specimens, Radermacher et al. (1992b) found that its true z-dimension was in fact considerably smaller than the apparent dimension. A comparison of the structures shown in this abstract suggests a factor of approximately 0.7.

Negative staining enjoys currently a revival of sorts in the form of *cryo-negative staining*. A separate brief section (section 2.6) will be devoted to this technique, which combines the contrast enhancement due to heavy-metal salt with the shape preservation afforded by ice embedment. Many other variants of negative staining have been recently reviewed and critically assessed by Harris and Scheffler (2002). Several unconventional stains not containing metal atoms were systematically explored by Massover and coworkers (Massover and Marsh, 1997, 2000; Massover et al., 2001). These authors explored light atom derivatives of structure-preserving sugars as a medium, among other compounds, to preserve the structure of catalase used as test specimen (Massover and Marsh, 2000).

Among the variety of stains tested, methylamine tungstate (MAT) appears to occupy a special place (Stoops et al., 1991b; Kolodziej et al., 1997). The combination of low charge and neutral pH apparently minimizes the deleterious interaction of this stain with the molecule. Following Stoops et al. (1991b), the specimen is applied to a Butvar film by the spray method, forming a glassy layer of stain entirely surrounding the molecule. (Some additional references attesting to the performance of this staining method with regard to preservation of the molecule architecture are found in chapter 5, section 12.2; see also figure 5.26 for the distribution of orientations.)

2.3. Glucose Embedment

The technique of preparing unstained specimens using glucose was introduced by Unwin and Henderson (1975). In this technique, the solution containing the specimen in suspension is applied to a carbon-coated grid and washed with a 1% (w/v) solution of glucose. The rationale for the development of this technique was, in the words of the authors, "to replace the aqueous medium by another liquid which has similar chemical and physical properties, but is non-volatile in addition." Other hydrophilic molecules initially tried were sucrose, ribose, and inositol. X-ray evidence indicated that this substitution leaves the structure unperturbed up to a resolution of 1/3 to $1/4 \text{ Å}^{-1}$. Although highly successful in the study of bacteriorhodopsin, whose structure could ultimately be solved to a resolution of $1/3 \text{ Å}^{-1}$ (Henderson et al., 1990), glucose embedding has not found widespread application. One reason for this failure to take hold is that the scattering densities of glucose and protein are closely matched, resulting in extremely low contrast, as long as the resolution falls short of the 1/7 to $1/10 \text{ Å}^{-1}$ range where secondary structure starts to become discernable. (It is this disadvantage that virtually rules out consideration of glucose embedment in imaging molecules in single-particle form.) Aurothioglucose, a potential remedy that combines the sustaining properties of glucose with high contrast of gold (Kühlbrandt and Unwin, 1982), has been discounted because of its high sensitivity to radiation damage (see Ohi et al., 2004). The other reason has been the success of frozen-hydrated electron microscopy (*cryo-electron microscopy*) starting at the beginning of the 1980s, which has the advantage, compared to glucose embedment, that the scattering densities of water and protein are sufficiently different to produce contrast even at low resolutions with a suitable choice of defocus (see section 3.3).

2.4. Use of Tannic Acid

Tannic acid, or tannin, has been used with success by some researchers to stabilize and preserve thin ordered protein layers (Akey and Edelstein, 1983). It was found to be superior in the collection of high-resolution data for the light-harvesting complex II (Kühlbrandt and Wang, 1991; Wang and Kühlbrandt, 1991; Kühlbrandt et al., 1994). For tannin preservation, the carbon film is floated off water, transferred with the grid onto a 0.5% (w/v) tannic acid solution, and adjusted to pH 6.0 with KOH (Wang and Kühlbrandt, 1991). These authors in fact found little differences between the preservation of the high-resolution structure when prepared with vitreous ice, glucose, or tannin, with the important difference being the high success rate of crystalline preservation with tannin versus the extremely low success rate with the other embedding media. The authors discuss at length the role of tannin in essentially blocking the extraction of detergent from the membrane crystal, which otherwise occurs in free equilibrium with the detergent-free medium. Arguably the most successful application of this method of embedment has been in the solution of the structure of the αβ-tubulin dimer, to atomic resolution, by electron crystallography of zinc-induced tubulin sheets, where a mixture of tannin and glucose was used (Nogales et al., 1998).

2.5. Ice-Embedded Specimens

The development of cryo-EM of samples embedded in vitreous ice (Taylor and Glaeser, 1976; Dubochet et al., 1982; Lepault et al., 1983; McDowall et al., 1983; Adrian et al., 1984; Wagenknecht et al., 1988b; see also the reviews by Chiu, 1993 and Stewart, 1990) presented a quantum leap in biological EM as it made it possible to obtain images of fully hydrated macromolecules. The specimen grid, on which an aqueous solution containing the specimen is applied, is rapidly plunged into liquid ethane, whereupon the thin water film vitrifies (figure 2.7). The rapid cooling rate (see below) prevents the water from turning into crystalline ice. It is the volume change in this phase transition that normally causes damage to the biological specimens.

The extremely high cooling rate obtained (estimated at 100,000°/s) is the result of two factors: the small mass (and thus heat capacity) of the grid and the high speed of the plunging. Measurements of DNA persistence lengths by cryo-EM have shown that the time during which a 0.1 μm thick water film (the typical thickness obtained by blotting) vitrifies is in the microsecond range (Al-Amoudi et al., 2004).

The detailed sequence of steps is as follows (Stewart, 1990):

(i) Apply a few microliters of aqueous specimen (buffer composition is not important) to a grid made hydrophilic by glow discharge. The grid could be bare, covered with holey carbon, or with a continuous carbon film.

(ii) Blot for a few seconds, to remove excess buffer, allowing only a thin (∼1000 Å or less) layer to remain.

(iii) Immediately after step (2), submerge grid in a cryogen—usually liquid ethane supercooled by a surrounding bath of liquid nitrogen.

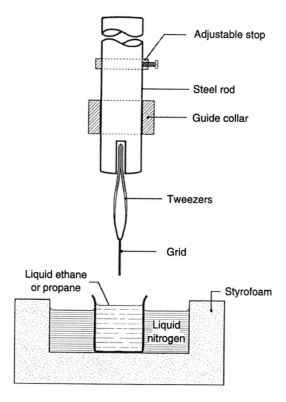

Figure 2.7 Schematic diagram of a freeze-plunger. A steel rod is mounted such that, once released and propelled by gravity, it can move through the guide collar until the adjustable stop hits the collar. At its end, a pair of tweezers are fastened that support the EM grid. For the preparation, a small droplet of the buffer containing the molecules is applied to the grid. Excess liquid is blotted so that a very thin film of the specimen remains on the grid. After blotting, the steel rod is released, and thereby the grid is plunged into the cryogen, normally liquid ethane. The grid is then transferred into liquid nitrogen and mounted onto the cooled specimen holder. From Stewart (1990).

(iv) Transfer grid from cryogen to liquid nitrogen bath containing cryo-holder. Grid must then be kept in nitrogen at all time.

(v) Transfer cryo-holder to transmission electron microscope (TEM) (keeping grid shielded from atmosphere at all time) and record micrographs using low-dose techniques. Maintain temperature well below $-133°C$ (common range is -160 to $-175°C$).

The number of manual steps involved in the whole procedure makes it an "art"; that is, the success rate strongly depends on the skill of the experimenter. For instance, the duration and evenness of blotting controls the thickness and evenness of the ice layer. Apparatuses such as the Vitrobot (Frederick et al., 2000; Braet et al., 2003; now marketed commercially) have been specially designed to achieve high-quality cryo-specimens reproducibly under controlled conditions. High humidity is particularly important since it prevents fast evaporation during the period between blotting and immersion of the grid in the cryogen, thus eliminating one of the chief sources of variability in the water film thickness.

As with glucose embedment, the advantage of frozen-hydrated specimen preparation is that specimen collapse is avoided and that the measured image contrast is related to the structure of the biological object itself, rather than to an extraneous contrasting agent, such as heavy atom salts in negative staining. Thus, by combining cryo-EM with 3D reconstruction, a quantitative, physically meaningful map of the macromolecule can be obtained, enabling direct comparisons with results from X-ray crystallography.[2] Another advantage of cooling is the greater resistance of organic material to radiation damage at low temperatures (see Chiu et al., 1986), although initial estimates have proved overly optimistic. The reason for the reduction in damage is that free radicals produced by ionization during electron irradiation are trapped under these conditions, preventing, or at least reducing, the damage to the structure. About 30 years ago, Taylor and Glaeser (1974) showed that the crystalline order is preserved in thin platelets of catalase cooled down to liquid-nitrogen temperature. Subsequent investigations of a number of protein crystals found general improvements in radiation resistance by a factor between two and six (Chiu et al., 1986; see also the summary given by Dubochet et al., 1988).

An additional reduction of radiation damage is thought to occur when the temperature is further reduced from liquid-nitrogen temperature ($\sim 170\,\mathrm{K}$) to a range close to the temperature of liquid helium. Reports on this effect are, however, inconsistent and contradictory. Measurements by Fujiyoshi (1998) indicated that the temperature of the specimen area rises upon irradiation, potentially to a range where the initially substantial dose protection factor becomes quite small. There are also indications that the size of the effect varies, depending on the aggregation form of the specimen (crystalline or single-particle; Y. Fujioshi, personal communication). One report (Grassucci et al., 2002) suggests that some of the gain in radiation protection may be offset by a reduction in contrast, which would be consistent with earlier observations that the density of beam-exposed ice increases by $\sim 20\%$ when the temperature drops below $32\,\mathrm{K}$ (Heide and Zeitler, 1985; Jenniskens and Blake, 1994). However, more definite experiments showing the actual preservation of a molecule under liquid helium versus liquid nitrogen temperature have yet to be conducted.

Cryo-electron microscopy is not without problems. Proteins have a scattering density that is only slightly higher than the scattering density of vitreous ice, hence the contrast, against an ice matrix, of unstained molecular assemblies formed by proteins is quite low. The intrinsically low contrast of cryo-EM micrographs leads to several problems: (i) it causes difficulties in spotting macromolecules visually or by some automated procedures, especially at low defocus; (ii) it poses problems in aligning particles; and (iii) it leads to the requirement of large numbers of particles

---

[2]Strictly speaking, the image obtained in the TEM is derived from the projected Coulomb potential distribution of the object, whereas the diffraction intensities obtained by X-ray diffraction techniques are derived from the electron density distribution of the object.

for the statistical fortification of results. Another, more hidden problem is the high proportion of inelastic scattering in ice when compared to other support media (such as carbon film). This leads to a high "inelastic background" in the image, or a high portion of Fourier components, in the low spatial frequency range, that do not obey the linear contrast transfer theory (see section 3.3).

Despite these shortcomings, however, cryo-EM has proved to be highly successful and accurate in depicting biological structure, as can be seen from a growing number of comparisons with structures solved by X-ray crystallography (e.g., GroE: Roseman et al., 2001; Ludtke et al., 2004; 70S *E. coli* ribosome: Gabashvili et al., 2000; Spahn et al., 2005).

As a note on nomenclature, both the terms "cryo-electron microscopy" and "electron cryo-microscopy" are being used in the field. It would be desirable at this stage to have one term only. Which one is preferable? In one, the prefix "cryo-" modifies the method "electron microscopy" (not "electron", which might have raised a possible objection); in the other, the prefix "electron" is a modifier of "cryo-microscopy." While "electron microscopy" is a defined term that still allows all kinds of specifications, "cryo-microscopy" by itself does not exist as a meaningful generic term, and this is the reason why "cryo-electron microscopy" (and its universally accepted abbreviation "cryo-EM") is the preferable, more logical name for the technique described here.

2.6. Hybrid Techniques: Cryo-Negative Staining

With ice embedment, the only source of image contrast is the biological material itself. Practical experience has shown that for alignment (and all operations that depend on the total structural contents of the particle) to work, molecules have to be above a certain size, which is in the region of 100 Å. For globular molecules, this corresponds to ~400 kD molecular weight (see formula in Appendix of Henderson, 1995). This is well above the minimum molecular weight (100 kD) derived on theoretical grounds by Henderson (1995) (see section 5.5 in chapter 3). Stark et al. (2001) obtained a cryo-EM reconstruction of the spliceosomal U1 snRP, in the molecular weight range of 200 kD, probably on the border line of what is feasible.

Thus, the technique of 3D cryo-EM of unstained, ice-embedded molecules faces a practical hurdle as the size of the molecule decreases. It is possible to extend this range to molecules that are significantly smaller, by adding a certain amount of stain to the aqueous specimen prior to plunge-freezing. The added contrast of the molecule boundary against the more electron-dense background, rather than the increase of contrast in the molecule's interior, is instrumental in achieving alignment even for small particles. It is therefore to be expected that the size or molecular weight limit is lower for strongly elongated molecules, and higher for the more globular kind, which provides little purchase for rotational alignment algorithms (see chapter 3, section 3 for details on these algorithms).

Cryo-negative staining was introduced by Adrian et al. (1998). The procedure they described is as follows: 4 µl of sample solution is applied to gold sputter-coated holey carbon film. After about 30 s, the grid is placed onto a 100-µl droplet of 16% ammonium molybdate on a piece of paraffin, and allowed to float for 60 s.

Then the grid is removed for blotting and freeze-plunging according to the usual protocol. The pH range of the stain (7–8) avoids the denaturing effects associated with the low pH of more common staining agents such as uranyl acetate.

Later de Carlo et al. (2002) reported that application of this specimen preparation technique, apart from giving the benefit of higher contrast [a 10-fold increase in signal-to-noise ratio (SNR)], also resulted in a reduction of beam sensitivity in single molecules of GroEL. Following an exposure of 10 to $30\,e^-/\mathring{A}^2$, the particles visible in the micrograph were seen to disintegrate and lose contrast in ice, while they remained virtually unchanged in cryo-negative stain. The results obtained by these authors also indicate that the resolution as measured by the Fourier shell correlation of single-particle reconstructions (a measure of reproducibility in Fourier space, to be introduced in later chapters (section 5.2.4, chapter 3; section 8.2, chapter 5) improves substantially over that with ice (figure 2.8). There is, however, a question if this improvement might merely be a reflection of the increased definition of the molecule's boundary, rather than an improvement in the definition of the molecule's interior. At any rate, the stain surrounding the molecule appears to be stabilized under the impact of the beam, producing a crisper appearance and a well-reproducible boundary as indicated by the Fourier shell correlation (de Carlo et al., 2002).

Using the method of Adrian et al. (1984), as well as a variant of it employing Valentine's carbon sandwiching method (Valentine et al., 1998), Golas et al. (2003) were able to reconstruct the multiprotein splicing factor SF3b, which is in the molecular weight region of 450 kD. The same technique (under the name

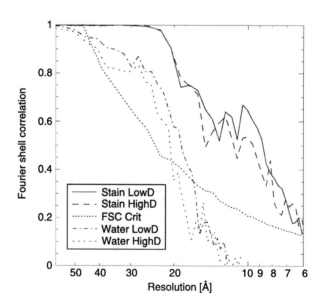

Figure 2.8 Fourier shell correlation plots for different doses and specimen preparation conditions. LowD, low dose, $10\,e/\mathring{A}$; highD, high-dose, $30\,e/\mathring{A}$; water, unstained particles; stain, cryo-negatively stained particles; FSC Crit, 3-σ FSC criterion. From de Carlo et al. (2002), reproduced with permission of Elsevier.

of "cryo-stain sandwich technique") was used by Boehringer et al. (2004) and Jurica et al. (2004) for the 3D visualization of spliceosomal complexes B and C, respectively.

2.7. Labeling with Gold Clusters

The use of selective stains to mark specific sites or residues of a molecule has been explored early on for some time (see review by Koller et al., 1971). The idea of using compounds incorporating single heavy atoms received particular interest. Even sequencing of nucleic acid sequences was once thought possible in this way (Beer and Moudrianakis, 1962). The difficulty with such single-atom probes was the relatively low contrast compared to the contrast arising from a column of light atoms in a macromolecule and from the support. Subsequently, a number of different heavy-atom clusters were investigated for their utility in providing specific contrast (see the detailed account in Hainfeld, 1992). *Undecagold*, a compound that incorporates 11 gold atoms, is clearly visible in the scanning transmission electron microscope (STEM) but not in the conventional electron microscope. Subsequently, Hainfeld and Furuya (1992; see also Hainfeld, 1992) introduced a much more electron-dense probe consisting of a 55-gold atom cluster (*Nanogold*; Nanoprobe Inc., Stony Brook, NY), which forms a 14-Å particle that is bound to a single maleimide site. This compound can be specifically linked to naturally occurring or genetically inserted cysteine residues exposed to the solvent.

From theoretical considerations and from the first experiments made with this compound, the scattering density of the gold cluster is high enough to outweigh the contribution, to the EM image, by a projected thick (200–300 Å) protein mass. Because of the presence of the strong amplitude component, which is transferred by $\cos \gamma$, where γ is the wave aberration function (see section 3.4), the Nanogold cluster stands out as a sharp density peak even in (and, as it turns out, especially in) low-defocus cryo-EM images where the boundaries of macromolecules are virtually invisible (Wagenknecht et al., 1994). The usual highly underfocused images show the cluster somewhat blurred but still as prominent "blobs" superimposed on the molecule. Applications of this method are steadily increasing (e.g., Milligan et al., 1990; Boisset et al., 1992a; Braig et al., 1993; Wagenknecht et al., 1994; Wenzel and Baumeister, 1995; Montesano-Roditis et al., 2001). Boisset and coworkers (1992, 1994a) used the Nanogold cluster to determine the site of thiol ester bonds in human α_2-macroglobulin in three dimensions. The cluster stands out as a "core" of high density in the center of the macromolecular complex. Wagenknecht et al. (1994) were able to determine the calmodulin-binding sites on the calcium release channel/ryanodine receptor. Using site-directed Nanogold labeling, Braig et al. (1993) succeeded in mapping the substrate protein to the cavity of GroEL.

2.8. Support Grids

Carbon films with holes 1–2 μm in diameter, placed on 300 mesh copper or molybdenum grids, provide support and stability to the specimen. The sample

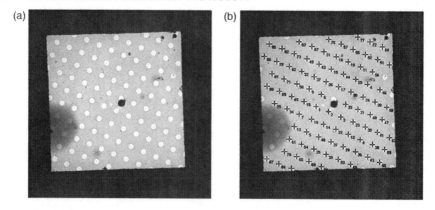

Figure 2.9 EM grid square (60 µm × 60 µm) with Quantifoil film bearing a frozen-hydrated specimen of ribosomes, at 712 × magnification. (a) CCD image. (b) Same as (a), with marks of the data acquisition program indicating the positions of high-quality data. (From Lei and Frank, unpublished results.)

is either directly deposited onto the holey carbon or onto an additional thin but continuous layer of carbon placed onto the holey carbon.

For the "manual" preparation of the holey carbon films (Baumeister and Seredynski, 1976; Baumeister and Hahn, 1978; Toyoshima, 1989), formvar is dissolved in chloroform, and then both acetone and glycerin are added. Sonication of the resulting mixture for 5 min will produce glycerin droplets of the correct size range within 30 min. The formvar is applied to a clean glass slide and dried. Glycerin droplets leave holes in the dried film. This holey formvar film is floated off the slide and transferred onto EM grids. Grids are then coated with carbon in an evaporator. The formvar is removed by placing the grids overnight in contact with filter paper soaked with chloroform or amyl acetate.

The introduction of regular, perforated carbon support foils (Ermantraut et al., 1998), now commerically produced under the trademark *Quantifoil* (Jena, Germany), has greatly aided in making data collection more uniform, reproducible, and convenient. These foils are spanned across EM grids and have circular holes arranged on a regular lattice (figure 2.9), facilitating automated data collection (Potter et al., 1999; Carragher et al., 2000; Zhang et al., 2001; Lei and Frank, unpublished data).

3. Principle of Image Formation in the Transmission Electron Microscope

3.1. Introduction

Image formation in the electron microscope is a complex process; indeed, it is the subject for separate books (e.g., Spence, 1988; Reimer, 1997). In the later chapters of this volume, there will be frequent references to the *contrast transfer function* (CTF) and its dependence on the defocus. It is important to understand the principle of the underlying theory for two reasons: first, the image is not necessarily a faithful representation of the object's projection (see Hawkes, 1992), and hence the same can be said for the relationship between the 3D reconstruction computed from such images and the 3D object that it is supposed to represent. It is, therefore, important to know the imaging conditions that lead to maximum resemblance, as well as the types of computational correction (see section 3.9 in this chapter and section 9 in chapter 5) that are needed to recover the original information. Second, the contrast transfer theory is only an approximation to a comprehensive theory of image formation (see Rose, 1984; Reimer, 1997), and attains its simplicity by ignoring a number of effects whose relative magnitudes vary from one specimen to the other. An awareness of these "moving boundaries" of the theory is required to avoid incorrect interpretations, especially when the resolution of the reconstruction approaches the atomic realm.

However, some of these sections may provide detail unnecessary for practitioners of routine methods, and are marked accordingly (*) for judicious skipping.

3.2. The Weak-Phase Object Approximation*

The basis of image formation in the electron microscope is the interaction of the electrons with the object. We distinguish between *elastic* and *inelastic* scattering. The former involves no transfer of energy; it has fairly wide angular distribution, and gives rise to high-resolution information. The latter involves transfer of energy, its angular distribution is narrow, and it produces an undesired background term in the image. Because this term has low resolution (implied in the narrow scattering distribution), it is normally tolerated, although it interferes with the quantitative interpretation of the image (see also section 4.3 on energy filtering).

In the wave-optical picture, the "elastic" scattering interaction of the electron with the object is depicted as a phase shift $\Phi(\mathbf{r})$ of the incoming wave traveling in the z-direction by

$$\Phi(\mathbf{r}) = \int C(\mathbf{r}, z)\, dz \tag{2.1}$$

where $\mathbf{r}$ is a 2D vector, which we will write as a column vector, $\mathbf{r} = \begin{bmatrix} x \\ y \end{bmatrix}$ or $[x, y]^{\mathrm{T}}$, and $C(\mathbf{r}, z)$ is the 3D Coulomb potential distribution within the object. Thus, the incoming plane wave $\Psi_0 = \exp(ikz)$ is modified according to (figure 2.10)

$$\Psi(\mathbf{r}) = \Psi_0 \exp[i\Phi(\mathbf{r})] \tag{2.2}$$

The weak-phase approximation assumes that $\Phi(\mathbf{r}) \ll 1$, enabling the expansion

$$\Psi(\mathbf{r}) = \Psi_0 \left[1 + i\Phi(\mathbf{r}) - \frac{1}{2}\Phi(\mathbf{r})^2 + \cdots \right] \tag{2.3}$$

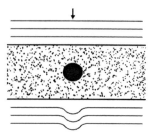

Figure 2.10 A pure phase object and its effect on an incoming plane wave. The phase of the incoming wave is locally shifted by an amount that is proportional to the integral of the potential distribution in the direction of the wave's propagation. Here, the object is shown as a globule embedded in another medium, such as ice. The incoming wave traveling in the z-direction is uniformly shifted everywhere except in x, y places, where it has intersected the globular object as well. In those regions, the shift is proportional to the thickness traversed. This has the effect that the exit wave front is deformed. The phase shift shown here is very large, on the order of the wavelength (the distance between successive crests in this drawing). However, note that for a weak phase object, the shift is quite small compared to the wavelength. (Drawing by Michael Watters.)

which is normally truncated after the second term. Note that this form implies a decomposition of the wave behind the object into an "unmodified" or "unscattered wave" (term 1), and a "scattered wave" [terms $i\Phi(\mathbf{r})$ and following]. If we disregard the higher-order terms, then we can say that the scattered wave is out of phase by 90° with respect to the unscattered wave.

The Frauenhofer approximation of the diffraction theory (Goodman, 1968) is obtained by assuming that the observation is made in the far distance from the object and close to the optical axis—assumptions that are always fulfilled in the imaging mode of the transmission electron microscope. In that approximation, the wave function in the back focal plane of the objective lens is—in the absence of aberrations—the Fourier transform of equation (2.2) or of the approximated expression in equation (2.3). However, the lens aberrations and the defocusing have the effect of shifting the phase of the scattered wave by an amount expressed by the term

$$\gamma(\mathbf{k}) = 2\pi\chi(\mathbf{k}) \tag{2.4}$$

which is dependent on the coordinates in the back focal plane, which are in turn proportional to the scattering angle and the spatial frequency, $\mathbf{k} = (kx, ky)$. The term $\chi(\mathbf{k})$ is called the *wave aberration function* (figure 2.11). In a polar coordinate system with

$$k = |\mathbf{k}|, \quad \phi = a\tan(k_x, k_y)$$

$$\chi(k, \phi) = -\frac{1}{2}\lambda\left[\Delta z + \frac{z_a}{2}\sin 2(\phi - \phi_0)\right]k^2 + \frac{1}{4}\lambda^3 C_s k^4 \tag{2.5}$$

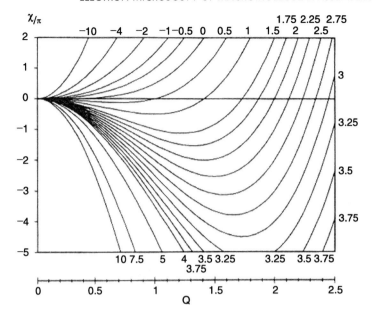

Figure 2.11 Wave aberration function of the transmission electron microscope. The curves give the function $\sin \chi(\Delta \hat{z}; \hat{k})/\pi = \hat{k}^4/2 - \Delta \hat{z} \hat{k}^2$, which is a function of the generalized spatial frequency $\hat{k} = k(C_s\lambda^3)^{1/4}$, for different choices of the generalized defocus $\Delta \hat{z} = \Delta z/(C_s\lambda)^{1/2}$. From Hawkes and Kasper (1994), reproduced with permission of Elsevier.

where λ is the electron wavelength; Δz, the defocus of the objective lens; z_a, the focal difference due to axial astigmatism; ϕ_0, the reference angle defining the azimuthal direction of the axial astigmatism; and C_s, the third-order spherical aberration constant.

An ideal lens will transform an incoming plane wave into a spherical wave front converging into a single point on the back focal plane (figure 2.12a). Lens aberrations have the effect of deforming the spherical wave front. In particular, the spherical aberration term acts in such a way that the outer zones of the wave front are curved stronger than the inner zones, leading to a *decreased* focal length in those outer zones (figure 2.12b).

In summary, the wave function in the back focal plane of the objective lens, $\Psi_{bf}(\mathbf{k})$, can be written as the Fourier transform of the wave function $\Psi(\mathbf{k})$ immediately behind the object, multiplied by a term that represents the effect of the phase shift due to the lens aberrations:

$$\Psi_{bf}(\mathbf{k}) = \mathfrak{F}\{\Psi(\mathbf{k})\} \exp[i\gamma(\mathbf{k})] \tag{2.6}$$

Here and in the following, the symbol $\mathfrak{F}\{\cdot\}$ will be used to denote the Fourier transformation of the argument function, and $\mathfrak{F}^{-1}\{\cdot\}$ its inverse. Also, as a notational convention, we will use lower case letters to refer to functions in real

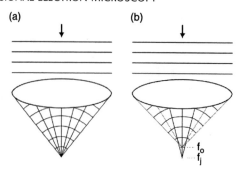

Figure 2.12 The effect of the spherical aberration. (a) In the ideal lens ($C_s = 0$), an incoming plane wave is converted into a spherical wave converging in the focal plane (here in the focal point f_i on the optic axis). (b) In a real lens with finite C_s, the focal length is smaller for the peripheral regions of the lens than the central regions, hence the parts of the plane wave intersecting the periphery are focused in a point f_0 closer to the lens. (Drawing by Michael Watters.)

space and the corresponding capital letters to refer to their Fourier transforms. Thus, $\mathfrak{F}\{h(\mathbf{r})\} = H(\mathbf{k})$, etc.

Next, the wave function in the image plane is obtained from the wave in the back focal plane, after modification by an aperture function $A(\mathbf{k})$, through an inverse Fourier transformation:

$$\psi_i(\mathbf{r}) = \mathfrak{F}^{-1}\{\mathfrak{F}\{\psi'(\mathbf{k})\}A(\mathbf{k})\exp[i\gamma(\mathbf{k})]\} \tag{2.7}$$

$$A(\mathbf{k}) = \begin{cases} 1 & \text{for } |k| = \theta/\lambda \leq \theta_1/\lambda \\ 0 & \text{elsewhere} \end{cases} \tag{2.8}$$

where θ_1 is the angle corresponding to the radius of the objective aperture. Finally, the observed intensity distribution in the image plane (ignoring irrelevant scaling factors) is

$$I(\mathbf{r}) = |\psi_i(\mathbf{r})|^2 \tag{2.9}$$

If the expansion of equation (2.3) is broken off after the second term ("weak phase object approximation"), we see that the image intensity is dominated by a term that results from the interference of the "unmodified wave" with the "scattered wave." In this case, the imaging mode is referred to as *bright-field electron microscopy*. If the unmodified wave is blocked off in the back focal plane, we speak of *dark-field electron microscopy*.

Bright-field EM is the imaging mode most frequently used by far. Because of the dominance of the term that is linear in the scattered wave amplitude, bright-field EM has the unique property that it leads to an image whose contrast is—to a first approximation—*linearly* related to the projected object potential. The description of the relationship between observed image contrast and projected object potential, and the way this relationship is influenced by electron optical parameters, is the subject of the *contrast transfer theory* (see Hansen, 1971; Lenz, 1971; Spence, 1988; Hawkes, 1992; Wade, 1992; Hawkes and Kasper, 1994). A brief outline of this theory is presented in the following section.

3.3. The Contrast Transfer Theory *

3.3.1. The Phase Contrast Transfer Function

If we (i) ignore terms involving higher than first orders in the projected potential $\Phi(\mathbf{r})$ and (ii) assume that $\Phi(\mathbf{r})$ is real, equation (2.9) yields a linear relationship between $O(\mathbf{k}) = \mathfrak{F}\{\Phi(\mathbf{r})\}$ and the Fourier transform of the image contrast, $\mathfrak{F}\{I(\mathbf{r})\}$:

$$F(\mathbf{k}) = O(\mathbf{k})A(\mathbf{k}) \sin \gamma(\mathbf{k}) \tag{2.10}$$

[A factor of 2 following the derivation of equation (2.10) is omitted here (cf. Frank, 1996), to allow for a convenient definition of the CTF.] Mathematically, the appearance of a simple scalar product in Fourier space, in this case with the factor $H(\mathbf{k}) = A(k) \sin \gamma(\mathbf{k})$, means that in real space the image is related to the projected potential by a convolution operation:

$$I(\mathbf{r}) = \int \Phi(\mathbf{r}')h(\mathbf{r} - \mathbf{r}') \, d\mathbf{r}' \tag{2.11}.$$

$$= \Phi(\mathbf{r}) \circ h(\mathbf{r}) \tag{2.12}$$

where $h(\mathbf{r})$ is called the point spread function. [The notation using the symbol "$\circ$" (e.g., in Goodman, 1968) is sometimes practical when expressions involving multiple convolutions need to be evaluated.]

The function $\sin \gamma(\mathbf{k})$ (figures 2.13a and 2.13e) is called the *phase contrast transfer function (CTF)*. It is characterized, as $k = |\mathbf{k}|$ increases, by a succession of alternately positive and negative zones, which are rotationally symmetric provided that the axial astigmatism is fully compensated [i.e., $z_a = 0$ in equation (2.5)]. The zones have elliptic or more complicated shapes for $z_a \neq 0$.

3.3.2. Partial Coherence Effects: The Envelope Function

In equation (2.10), which is derived assuming completely coherent illumination with monochromatic electrons (i.e., electrons all having the same energy), the resolution is limited by the aperture function. In the absence of an aperture, information is transferred out to high spatial frequencies, even though the increasingly rapid oscillations of the CTF make it difficult to exploit that information. In practice, however, the illumination has both finite divergence (or, in other words, the source size is finite) and a finite energy spread. The resulting *partial coherence* dampens the CTF as we go toward higher spatial frequencies, and ultimately limits the resolution. The theoretical treatment of

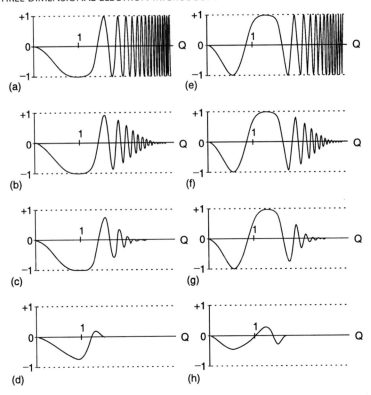

Figure 2.13 The influence of a finite illumination angle on the contrast transfer function (CTF) for two defocus values, $\Delta\hat{z} = 1$ (a–d) and $\Delta\hat{z} = \sqrt{3}$ (e–h). (a, e) Undampened CTF; (b, f) $\hat{q}_0 - 0.05$; (c, g) $\hat{q}_0 = 0.1$; (d, h) $\hat{q}_0 = 0.5$. (For the definition of generalized defocus and source parameters $\Delta\hat{z}$ and $\hat{q}_0$, see section 3. From Hawkes and Kasper (1994), reproduced with permission of Elsevier.

these phenomena is somewhat complicated, and the resulting integrals (e.g., Hanszen and Trepte, 1971a,b; Rose, 1984) make it difficult to gauge the effects of changing defocus, illumination divergence, or energy spread. Approximate envelope representations have been developed (energy spread: Hanszen, 1971; partial spatial coherence: Frank, 1973a; Wade and Frank, 1977), which have the advantage that they reveal the influence of these parameters in a mathematically easily tractable form:

$$H_{pc}(k) = E(k)A(k)\sin\gamma(k) \tag{2.13}$$

where $E(k)$ is the "compound envelope function":

$$E(k) = E_i(k)E_e(k) \tag{2.14}$$

consisting of the term $E_i(k)$, the envelope function due to partially coherent illumination, and the term $E_e(k)$, the envelope function due to energy spread.

(For simplicity, only the radial dependence is considered here. It is straight-forward to write down the full expression containing the polar coordinate dependency in the case $z_a \neq 0$.) The effect of the partially coherent illumination alone is shown in figures 2.13b–d and 2.13f–h: increasing the source size (descri-bed by the parameter q_0, a quantity of dimension 1/length specifying the size of the source as it appears in the back focal plane, or with the generalized parameter $\hat{q}_0$ that will be introduced in the following section) is seen to dampen the high spatial frequency range increasingly. The range of validity for this product representation has been explored by Wade and Frank (1977). The first term is

$$E_i(k) = \exp\left[-\pi^2 q_0^2 (C_s \lambda^3 k^3 - \Delta z \lambda k)^2\right] \tag{2.15}$$

for a Gaussian source distribution, and

$$E_i(k) = 2 \frac{J_1[2\pi q_0 (C_s \lambda^3 k^3 - \Delta z \lambda k)]}{[2\pi q_0 (C_s \lambda^3 k^3 - \Delta z \lambda k)]} \tag{2.16}$$

for a "top hat" distribution (Frank, 1973a), with J_1 denoting the first-order Bessel function. The recurring argument in both expressions $(C_s \lambda^3 k^3 - \Delta z \lambda k)$ is the gradient of the wave aberration function (Frank, 1973a). It is evident from equa-tions (2.15) and (2.16) that $E_i(k) = 1$ wherever this gradient vanishes.

The envelope due to the energy spread is (Hanszen, 1971; Wade and Frank, 1977)

$$E_e(k) = \exp\left[-(\pi \delta z \lambda k^2 / 2)^2\right] \tag{2.17}$$

where δz is the defocus spread due to lens current fluctuations and chromatic aberration in the presence of energy spread. The distinguishing feature of $E_i(k)$ is that it is defocus dependent, whereas $E_e(k)$ is independent of defocus. The combined effect of the two envelopes can be conceptualized as the action of two superimposed virtual apertures, one of which changes with changing defocus and one of which remains constant. For example, if $E_i(k)$ cuts the transmission of information off at k_1 and $E_e(k)$ cuts it off at $k_2 < k_1$, then $E_i(k)$ has no effect whatsoever. From this "product veto rule," an important clue can be derived by looking at the power spectra of a defocus series: if the band limit produced by $E(k) = E_i(k)E_e(k)$ is independent of defocus in a given defocus range, then the energy spread is the limiting factor in that range. In the case where the band limit is observed to be defocus dependent, one part of the defocus range may be dependent and thus ruled by $E_i(k)$, and the other part may be independent, and limited by $E_e(k)$. In instruments with a field-emission gun, which are increasingly used, both illumination spread and energy spread are quite small, and so both envelopes $E_i(k)$ and $E_e(k)$ remain close to unity up to high spatial frequencies, offering the opportunity for reaching atomic resolution (O'Keefe, 1992; Zhou and Chiu, 1993).

In figure 2.14, the effect of the partial coherence envelope function on the band limit is illustrated on a modern instrument equipped with a field emission

(a)   (b)   (c)

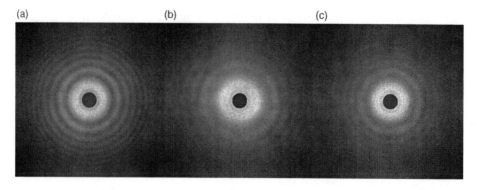

Figure 2.14 Effect of partial coherence on the resolution, as reflected by the extent of Thon rings in the diffraction pattern. The experimental conditions under which the images are taken differ by the "spot size" setting of the Condensor I lens (see section 1), which controls the demagnification of the physical source produced by this lens. A *small* image of the source (or *small* numerical setting of CI) results in high coherence, i.e., *little* damping of the information in Fourier space by the envelope function, or *large* extent of Thon rings, meaning *high* resolution. (a) CI setting = 6; (b) CI setting = 7; (c) CI setting = 8. (Images were of a cryo-specimen of ribosomes, taken on an F30 Polara transmission electron microscope operated at liquid-nitrogen temperature and a voltage of 200 kV. Power spectra were calculated using the method described in section 3.6 of this chapter. Data were provided by Bob Grassucci.)

source. Even though the instrument is built for high coherence, incorrect settings of the condenser system will result in a damping by the envelope function, and thus diminished resolution.

Finally, it should be mentioned that, unless compensated, axial astigmatism, which was left out in equations (2.15–2.17) for notational convenience, will create an azimuthal dependence of the effective band limit through the action of the defocus-dependent illumination envelope, which has then to be written as a function of a vector argument, namely as $E_i(\mathbf{k})$.

3.3.3. The Contrast Transfer Characteristics

When the value of $\sin \gamma(k)$ is plotted as a function of both spatial frequency k and defocus Δz, we obtain a pattern called the *contrast transfer characteristics* (Thon, 1971) of the electron microscope. If we again ignore the effect of axial astigmatism, the characteristics are determined entirely by the values of the remaining parameters in equation (2.5), namely λ (electron wavelength) and C_s (third-order spherical aberration coefficient). Following the convention introduced by Hanszen and Trepte (1971a,b), we can introduce dimensionless variables (see Frank, 1973a):

$$\Delta \hat{z} = \frac{\Delta z}{[C_s \lambda]^{1/2}} \quad \text{("generalized defocus")} \tag{2.18}$$

and

$$\hat{k} = [C_s\lambda^3]^{1/4}k \quad \text{("generalized spatial frequency")} \tag{2.19}$$

For completeness, the generalized source size will also be introduced here:

$$\hat{q}_0 = [C_s\lambda^3]^{1/4}q_0 \tag{2.19a}$$

With these generalized variables, one obtains the *standard characteristics*, which are independent of voltage and the value of the third-order spherical aberration constant, and hence are the same for all transmission electron microscopes:

$$CTF(\hat{k}; \Delta\hat{z}) = \sin\left[-\pi\Delta\hat{z}\hat{k}^2 + \frac{\pi}{2}\hat{k}^4\right] \tag{2.20}$$

This important diagram is shown in figure 2.14. The use of generalized coordinates means that this diagram is universal for all transmission electron microscopes. It makes it possible to determine the optimum defocus setting that is required to bring out features of a certain size range or to gauge the effect of axial astigmatism.

Features of a size range between d_1 and d_2 require the transmission of a spatial frequency band between $1/d_1$ and $1/d_2$. One obtains the defocus value, or value range, for which optimum transmission of this band occurs, by constructing the intersection between the desired frequency band and the contrast transfer zones (figure 2.15). The effect of axial astigmatism can be gauged by moving back and forth along the defocus axis by the amount $z_a/2$ from a given defocus position, bearing in mind that this movement is controlled by the azimuthal angle ϕ according to the behavior of the function $\sin(2\phi)$: going through the full 360° azimuthal range leads to two complete oscillations of the effective defocus value around the nominal value. Evidently, small values of astigmatism lead to an elliptic appearance of the contrast transfer zones, whereas large values may cause the defocus to oscillate beyond the boundaries of one or several zones, producing quasi-hyperbolic patterns.

3.3.4. The Effects of the Contrast Transfer Function

3.3.4.1. Band-pass filtration   The main degrading effects of the underfocus CTF on the image, as compared to those of ideal contrast transfer [i.e., $CTF(k) \equiv$ const. $\equiv 1$], can be described as a combined low-pass (i.e., resolution-limiting) and high-pass filtration, in addition to several reversals of contrast. The effective low-pass filtration results from the fact that, in the underfocus range (by convention, $\Delta z > 0$), the CTF typically has a "plateau" of relative constancy followed by rapid oscillations. In this situation, the high-frequency boundary of the plateau acts as a virtual band limit. The use of information transferred beyond that limit, in the zones of alternating contrast, requires some type of restoration, that is, *CTF correction*.

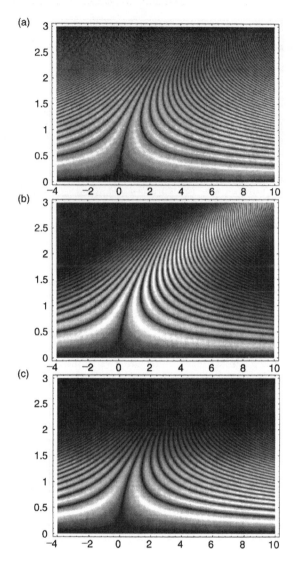

Figure 2.15 Representation of the CTF characteristics in the presence of partial coherence, $E(\Delta\hat{z}; \hat{k})\sin\gamma(\Delta\hat{z}; \hat{k})$, showing the effect of the envelope functions. The vertical axis is the spatial frequency, in generalized units $\hat{k}$, and the horizontal axis is the generalized defocus $\Delta\hat{z}$ [see equations (2.18) and (2.19) for definition of both quantities]. The gray value (white = max.) is proportional to the value of the function. A section along a vertical line $\Delta\hat{z} = \text{const.}$ gives the CTF at that defocus, and allows the effective resolution to be gauged as the maximum spatial frequency (or upper transfer limit), beyond which no information transfer occurs. (a) CTF with $E(\Delta\hat{z}; \hat{k}) = 1$ (fully coherent case); (b) partially coherent illumination with the generalized source size $\hat{q}_0 = 0.5$ but zero defocus spread. In this case, the practical resolution is clearly defocus dependent. At high values of underfocus, not only the resolution is limited (upper boundary), but a central band is also eliminated by the effect of the envelope function. (c) Defocus spread $\delta\hat{z} = 0.125$ in generalized units, in the case of a point source. The practical resolution has no defocus dependence in this case. From Wade (1992), reproduced with permission of Elsevier.

Information transferred outside the first zone is of little use in the image, unless the polarities of the subsequent, more peripheral zones are "flipped" computationally, so that a continuous positive or negative transfer behavior is achieved within the whole resolution domain. More elaborate schemes employ restoration such as Wiener filtering (Kübler et al., 1978; Welton, 1979; Lepault and Pitt, 1984; Jeng et al., 1989; Frank and Penczek, 1995; Penczek et al., 1997; Grigorieff, 1998), in which not only the polarity (i.e., the phase in units of 180°) but also the amplitude of the CTF is compensated throughout the resolution range based on two or more micrographs with different defocus values (see section 3.9 in this chapter and section 9 in chapter 5). The fidelity of the CTF correction, or its power to restore the original object, is limited by (i) the presence of resolution-limiting envelope terms, (ii) the accuracy to which the transfer function formalism describes image formation, and (iii) the presence of noise.

The well-known band-pass filtering effect of the CTF is a result of the CTF having a small value over an extended range of low spatial frequencies. The effect of this property on the image of a biological particle is that the particle as a whole does not stand out from the background, but its edges are sharply defined, and short-range interior density variations are exaggerated.

In figure 2.16, a CTF has been applied to a motif for a demonstration of the contrast reversal and edge enhancement effects. One is immediately struck by the "ghost-like" appearance of the familiar object and the diminished segmentation, by density, between different parts of the scene and the background. This effect is caused by the virtual absence of low-resolution information in the Fourier transform. Similarly, low-contrast objects such as single molecules embedded in ice are very hard to make out in the image, unless a much higher defocus (e.g., in the range 3–6 μm or 30,000–60,000 Å) is used that forces the CTF to rise swiftly at low spatial frequencies.

3.3.4.2. Point-spread function   Another way of describing the effects of the CTF (or any filter function) is by the appearance of the associated point spread function, which is the Fourier transform of the CTF and describes the way a single point of the object would be imaged by the electron microscope. Generally, the more closely this function resembles a delta function, the more faithful is the image to the object. (Simple examples of filter functions and associated point-spread functions are provided in appendix 2.) In the practically used range of defocus, the typical point-spread function has a central maximum that is sometimes barely higher than the surrounding maxima, and it might extend over a sizable area of the image. The long-range oscillations of the point-spread function are responsible for the "ringing effect," that is, the appearance of Fresnel fringes along the borders of the object.

3.3.4.3. Signal-to-noise ratio   To consider the effect of the CTF on the SNR, we make use of *Parseval's theorem* (chapter 3, section 4.3) to express the signal variance:

$$\text{SNR} = \frac{1}{\text{var}(N)} \int_B |A(k)|^2 |\text{CTF}(k)|^2 \, dk = \int_B |A(k)|^2 |\sin \gamma(k) E(k)|^2 \, dk$$

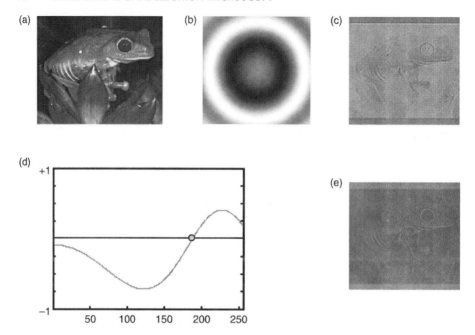

Figure 2.16 Demonstration of the effect of an electron microscopic transfer function on a real-life motif. (a) The motif. (b) CTF with the profile shown in (d). The first, innermost transfer interval conveys negative contrast, the following transfer interval positive contrast. (c) Motif after application of the CTF. Compared to the original, the distorted image is characterized by three features: (i) inversion of contrast; (ii) diminished contrast of large regions; (iii) edge enhancement (each border is now sharply outlined). With higher defocus, an additional effect becomes apparent: the appearance of fringes (Fresnel fringes) with alternating contrast along borders. As a rule, the resolution of the degraded image is determined by the position of the first zero of the CTF. (e) Degraded image (c) after contrast inversion. (From N. Boisset, unpublished lecture material.)

where $A(k)$ is the radial amplitude falloff of the object's transform, $E(k)$ is the envelope function, and var(N) is the noise variance. The squaring in the integrand means that the SNR is insensitive to the degree of fidelity by which the signal is rendered. For instance, any flipping of phases in a spatial frequency band of the CTF has no effect on the signal variance and, by extension, on the SNR. It is easy to see that the SNR oscillates as a function of defocus (Frank and Al-Ali, 1975), since extra lobes of the CTF migrate into the integration area as Δz decreases (i.e., as the underfocus becomes larger). Most pronounced is the effect of the SNR variation when low underfocus values are considered. The slow start of sin $\gamma(k)$ for low underfocus values will mean that a substantial portion of the Fourier transform in the low spatial frequency region makes no contribution to the SNR. Low-defocus micrographs are, therefore, notorious for poor visibility of unstained particles embedded in a layer of ice.

For a "white" signal spectrum without amplitude falloff [i.e., $A(k) \equiv 1$ throughout], the SNRs of two images with defocus settings Δz_1 and Δz_2 are related as (see Frank and Al-Ali, 1975)

$$\text{SNR}_1/\text{SNR}_2 = \frac{\int_B |\text{CTF}_1(k)|^2 \, dk}{\int_B |\text{CTF}_2(k)|^2 \, dk} = \frac{\int_B |\sin \gamma_1(k) E_1(k)|^2 \, dk}{\int_B |\sin \gamma_2(k) E_2(k)|^2 \, dk} \qquad (2.20a)$$

Note in passing that the integral over the squared CTF that appears in equation (2.20a) can be recognized as one of Linfoot's criteria (Linfoot, 1964) characterizing the performance of an optical instrument (Frank, 1975b): it is termed structural contents. Another criterion by Linfoot, the flat integral over the CTF, is termed correlation quality as it reflects the overall correlation between an object with "white" spectrum and its CTF-degraded image.

3.4. Amplitude Contrast

Amplitude contrast of an object arises from a virtual—and locally changing—loss of electrons participating in the "elastic" image formation, either by electrons that are scattered outside of the aperture or by those that are removed by inelastic scattering (see Rose, 1984). These amplitude components are, therefore, entirely unaffected by energy filtering. Rose writes in his account of information transfer: "It seems astonishing at first that a (energy-) filtered bright-field image, obtained by removing all inelastically scattered electrons from the beam, represents an elastic image superposed on a inelastic 'shadow image.' " The shadow image that Rose is referring to is produced by amplitude contrast. Since the detailed processes are dependent on the atomic species, the ratio of amplitude to phase contrast is itself a locally varying function. Formally, the amplitude component of an object can be expressed by an imaginary component of the potential in equation (2.3). The Fourier transform of the amplitude component is transferred by $\cos \gamma(\mathbf{k})$, which, unlike the usual term $\sin \gamma(\mathbf{k})$, starts off with a maximal value at low spatial frequencies. The complete expression for the image intensity thus becomes

$$I(\mathbf{k}) = O_r(\mathbf{k}) \sin \gamma(\mathbf{k}) - O_i(\mathbf{k}) \cos \gamma(\mathbf{k}) \qquad (2.21)$$

where $O_r(\mathbf{k})$ and $O_i(\mathbf{k})$ are the Fourier transforms of the real (or weak phase) and imaginary (or weak amplitude) portions of the object, respectively (Erickson and Klug, 1970; Frank, 1972c; Wade, 1992). Equation (2.21) is the basis for heavy/light atom discrimination using a defocus series (Frank, 1972c, 1973b; Kirkland et al., 1980; Typke et al., 1992; Frank and Penczek, 1995), following an original idea by Schiske (1968). The reason is that the ratio between amplitude and phase contrast is greater for heavy atoms than for light atoms.

Only for a homogeneous specimen (i.e., a specimen that consists of a single species of atom; see Frank and Penczek, 1995) is it possible to rewrite equation (2.21) in the following way:

$$I(\mathbf{k}) = O_r(\mathbf{k})[\sin \gamma(\mathbf{k}) - Q(\mathbf{k}) \cos \gamma(\mathbf{k})] \qquad (2.22)$$

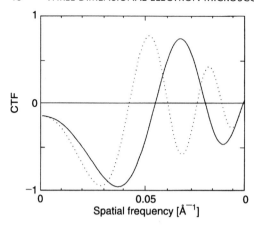

Figure 2.17 Electron-optical contrast transfer function ($C_s = 2$ mm) for a mixed phase/amplitude object ($Q_0 = 0.15$), for two defocus values: $\Delta z = -0.9$ μm (solid line) and $\Delta z = -1.5$ μm (dotted line). From Frank and Penczek (1995), reproduced with permission of Urban and Fischer Verlag.

Here, $Q(\mathbf{k}) = O_i(\mathbf{k})/O_r(\mathbf{k})$ is a function characteristic for each atomic species, but it is often assumed that $Q(\mathbf{k})$ is (i) the same for all atoms in the specimen and (ii) constant within the small spatial frequency range of practical interest in most cryo-EM applications (see Toyoshima and Unwin, 1988a,b; Stewart et al., 1993). (A more accurate formulation is found in section 3.5.) With these approximations, it is again possible to speak of a single contrast transfer function:

$$H'(\mathbf{k}) = \sin \gamma(\mathbf{k}) - Q_0 \cos \gamma(\mathbf{k}) \tag{2.23}$$

Compared with the function obtained for a pure phase object, the function described by equation (2.23) has the zeros shifted toward higher radii (figure 2.17). However, the most important change lies in the fact that, at low spatial frequencies, the transfer function starts off with a nonzero term, $H'(0) = -Q_0$. Thus, the cosine-transferred term mitigates the pronounced band-pass filtering effect produced by $\sin \gamma$, brought about by the deletion of Fourier components at low spatial frequencies.

The value of Q_0 is usually determined by recording a defocus series of the specimen and measuring the positions of the zeros of $H'(\mathbf{k})$ in the diffraction patterns of the micrographs. These diffraction patterns can be obtained either by optical diffraction or by computation (see section 3.6). Other measurements of Q_0 were done by following the amplitudes and phases of reflections in the computed Fourier transform of a crystal image as a function of defocus (Erickson and Klug, 1970) or by observing the lines of zero contrast transfer in optical diffraction patterns of strongly astigmatic images (Typke and Radermacher, 1982). Toyoshima and Unwin (1988a) and Toyoshima et al. (1993) obtained Q_0 measurements by comparing micrographs that were taken with equal amounts of underfocus and overfocus.

Averaged values for Q_0 of negatively stained (uranyl acetate) specimens on a carbon film range from 0.19 (Zhu and Frank, 1994) to 0.35 (Erickson and Klug, 1970, 1971). The wide range of these measurements reflects not only the presence of considerable experimental errors, but also variations in the relative amount and thickness of the stain compared to that of the carbon film. For specimens in ice, the values range from 0.07 (ribosome: Zhu and Frank, 1994) to 0.09 (acetylcholine receptor: Toyoshima and Unwin, 1988a) and 0.14 (tobacco mosaic virus: Smith and Langmore, 1992).

3.5. Formulation of Bright-Field Image Formation Using Complex Atomic Scattering Amplitudes

A treatment of image formation that takes account of the atomic composition of the specimen is found in Frank (1972). We start by considering the wave in the back focal plane, as in equation (2.6), but now trace its contributions from each atom in the object. The scattering contribution from each atom is the atomic scattering amplitude, $f_j(\mathbf{k})$, multiplied with a phase term that relates to the atom's position in the specimen plane, $\mathbf{r}_j$:

$$\Psi_{bf}(\mathbf{k}) = \sum_j f_j(\mathbf{k}) \exp[-2\pi i \mathbf{k} \mathbf{r}_j] \tag{2.6a}$$

Amplitude effects are accounted for by the fact that the atomic scattering amplitudes are complex:

$$f_j(\mathbf{k}) = f'_j(\mathbf{k}) + i f''_j(\mathbf{k}) \tag{2.23a}$$

with the real part relating to the phase, and the imaginary part relating to the amplitude (absorption) properties. The complex electron atomic scattering amplitudes were tabulated by Haase, based on Hartree–Fock (Haase, 1970a) and Thomas–Fermi–Dirac potentials (Haase, 1970b), for voltages used in transmission electron microscopy. As stated in equation (2.21), the Fourier transform of the image intensity in the linear approximation is $I(\mathbf{k}) = O_r(\mathbf{k}) \sin(\mathbf{k}) - O_i(\mathbf{k}) \cos \gamma(\mathbf{k})$ but now the terms $O_r(\mathbf{k})$ and $O_i(\mathbf{k})$ relating, respectively, to the phase and amplitude portions of the object can be explicitly written as

$$O_r(\mathbf{k}) = \sum_j f'_j(\mathbf{k}) \exp[-2\pi i \mathbf{k} \mathbf{r}_j] \tag{2.23b}$$

$$O_i(\mathbf{k}) = \sum_j f''_j(\mathbf{k}) \exp[-2\pi i \mathbf{k} \mathbf{r}_j] \tag{2.23c}$$

It is clear that it is no longer possible to use the simplification made in equation (2.22), in which $O_i(\mathbf{k})$ was expressed as a quantity proportional to $O_r(\mathbf{k})$, in the form $O_i(\mathbf{k}) = Q(\mathbf{k})O_r(\mathbf{k})$, since each atom species has its own ratio of phase/amplitude scattering. This ratio goes up with the atomic number Z.

This formulation of image formation demonstrates the limitation of approximations usually made. It becomes important in attempts to computationally predict images from coordinates of X-ray structures composed of atoms with strongly different atomic numbers, as in all structures composed of protein and RNA. If these predictions are made on the basis of the earlier equation (2.22) or even without regard to the radial falloff, as in equation (2.23), then the contrast between protein and RNA comes out to be much smaller than in the actual cryo-EM image. For instance, the DNA portion was difficult to recognize in naively modeled images of nucleosomes, while it was readily apparent in cryo-EM images (C. Woodcock, personal communication).

3.6. Optical and Computational Diffraction Analysis—The Power Spectrum

The CTF leaves a "signature" in the diffraction pattern of a carbon film, which optical diffraction analysis (Thon, 1966, 1971; Johansen, 1975) or its computational equivalent is able to reveal. Before a micrograph can be considered worthy of the great time investment that is required in the digital processing (starting with scanning and selection of particles), its diffraction pattern or *power spectrum* should first be analyzed. This can be achieved either by optical or computational means.

The term "power spectrum," which will frequently surface in the remainder of this book, requires an introduction. It is defined in electrical engineering and statistical optics as the inverse Fourier transform of the autocorrelation function of a stochastic process. The expectation value of the absolute-squared Fourier transform provides the means to calculate the power spectrum. Thus, if we had a large number of images that all had the same statistical characteristics, we could estimate the power spectrum of this ensemble by computing their Fourier transforms, absolute-squaring them, and forming the average spectrum. The question is, can we estimate the power spectrum from a single image? The absolute-squared Fourier transform of a single image (also called *periodogram,* see Jenkins and Watts, 1968) yields a very poor estimate. But the answer to our question is still yes, under one condition: often the image can be modeled approximately as a *stationary* stochastic process; that is, the statistical properties do not change when we move a window from one place to the next. In that case, we can estimate the power spectrum by forming an ensemble of subimages, by placing a window in different (partially overlapped) positions, and proceeding as before when we had a large number of images. (An equivalent method, using convolution, will be outlined below in section 3.7.)

In contrast, even though it is often used interchangedly with "power spectrum," the term "diffraction pattern" is reserved for the intensity distribution of a single image, and it therefore has to be equated with the periodogram. A recording of the diffraction pattern on film, during which the area selected for diffraction is moved across the micrograph, would be more closely in line with the concept of a power spectrum.

After these preliminaries, we move on to a description of the optical diffractometer, a simple device that has been of invaluable importance particularly in the beginning of electron crystallography. Even though computational analysis has become very fast, the optical analysis is still preferred for quick prescreening of the micrographs collected. The optical diffractometer consists of a coherent light source, a beam expander, a large lens, and a viewing screen, and allows the diffraction pattern of a selected image area to be viewed or recorded (figure 2.18). In the simplest arrangement, the micrograph is put into the parallel coherent laser

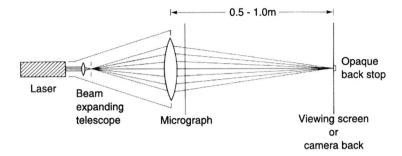

Figure 2.18 Schematic sketch of an optical diffractometer. A beam-expanding telescope is used to form a spot of the coherent laser light on the viewing screen. When a micrograph is placed immediately in front of the lens of the telescope, its diffraction pattern is formed on the viewing screen. From Stewart (1988a), reproduced with permission of Kluwer Academic/Plenum Publishers.

beam. In the back focal plane, the complex amplitude is given by the Fourier transform of the transmittance function, which for low contrast is proportional to the optical density. The latter, in turn, is in good approximation proportional to the image intensity. In other words, the complex amplitude in the back focal plane of the optical diffractometer is given by equation (2.21). Now, the diffraction pattern $D(\mathbf{k})$ observed or recorded on film is the square of the complex amplitude, that is, $D(\mathbf{k}) = |I(\mathbf{k})|^2$. Thus, if we additionally take into account the envelope function $E(\mathbf{k})$ and an additive noise term $N(\mathbf{k})$, ignore terms in which $N(\mathbf{k})$ is mixed with the signal term, and make use of equation (2.23), we obtain

$$D(\mathbf{k}) = |O(\mathbf{k})|^2 E(\mathbf{k})^2 [\sin \gamma(\mathbf{k}) - Q_0 \cos \gamma(\mathbf{k})]^2 + |N(\mathbf{k})|^2 \qquad (2.24)$$

for the part of the image transform that originates from elastic scattering. (Another part of the image transform, ignored in this formula, comes from inelastic scattering.)

We obtain the interesting result, first noted by Thon (1966), that the object spectrum (i.e., the absolute-squared Fourier transform of the object) is modulated by the squared CTF. In order to describe this modulation, and the conditions for its observability, we must make an approximate yet realistic assumption about the object. A thin carbon film can be characterized as an amorphous, unordered structure. For such an object, the spectrum $|O(\mathbf{k})|^2$ is nearly "white." What is meant with "whiteness" of a spectrum is that the total signal variance, which is equal to the squared object spectrum integrated over the resolution domain $\mathbf{B}$ (Parseval's theorem),

$$\mathrm{var}\{o(\mathbf{r})\} = \int_B |O'(\mathbf{k})|^2 \, d\mathbf{k} \qquad (2.25)$$

where $O(\mathbf{k})$ is the Fourier transform of the "floated," or average-subtracted object $[o(\mathbf{r}) - \langle o(\mathbf{r}) \rangle]$, is evenly partitioned within that domain. (If we had an instrument that would image this kind of object with an aberration producing a uniform phase shift of $\pi/2$ over the whole back focal plane, except at the position of the primary beam, then the optical diffraction pattern would be uniformly white.) Hence, for such a structure, multiplication of its spectrum with the CTF in real instruments will leave a characteristic squared CTF trace (the "signature" of which we have spoken before).

Observability of this squared CTF signature actually does not require the uniformity implied in the concept of "whiteness." It is obviously enough to have a sufficiently smooth spectrum, and the radial falloff encountered in practice is no detriment.

We can draw the following (real-space) parallel: in order to see an image on a transparent sheet clearly, one has to place it on a light box that produces uniform, untextured illumination, as in an overhead projector. Similarly, when we image a carbon film in the EM and subsequently analyze the electron micrograph in the optical diffractometer, the carbon film spectrum essentially acts as a uniform virtual light source that makes the CTF (i.e., the transparency, in our analogy, through which the carbon spectrum is "filtered") visible in the Fourier transform of the image intensity (figure 2.19). Instead of the carbon film, a 2D crystal with a large unit cell can also be used: in that case, the CTF is evenly sampled by the fine grid of the reciprocal lattice [see the display of the computed Fourier transform of a PhoE crystal, figure 2.18 in Frank (1996)].

The use of optical diffraction as a means of determining the CTF and its dependence on the defocus and axial astigmatism goes back to Thon (1966). Since then, numerous other electron optical effects have been measured by optical diffraction: drift (Frank, 1969), illumination source size (Frank, 1976; Saxton, 1977; Troyon, 1977), and coma (Zemlin et al., 1978; Zemlin, 1989a). Typke and Köstler (1977) have shown that the entire wave aberration of the objective lens can be mapped out.

Modern optical diffractometers designed for convenient, routine quality testing of electron micrographs recorded on film are equivpped with a CCD camera whose output is monitored on a screen. The entire unit is contained in a light-tight

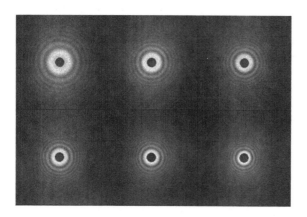

Figure 2.19 Computed power spectra for a number of cryo-EM micrographs in the defocus range 1.2–3.9 µm.

box, and the tedium of working in the dark is avoided (B. Amos, personal communication). On the other hand, electron microscopes are also increasingly fitted with a CCD camera and a fast computer capable of producing instantaneous "diagnostic" Fourier transforms of the image on-line (e.g., Koster et al., 1990). With such a device, the data collection can be made more efficient, since the capture of an image in the computer can be deferred until satisfactory imaging conditions have been established. Provided that the size of the specimen area is the same, optical and computational diffraction patterns are essentially equivalent in quality and diagnostic value.

Another note concerns the display mode. In the digital presentation, it is convenient to display the modulus of the Fourier transform, that is, the square root of the power spectrum, because of the limited dynamic range of monitor screens. Logarithmic displays are also occasionally used, but experience shows that these often lead to an unacceptable compression of the dynamic range, bringing out the background strongly and rendering the zeros of the CTF virtually invisible.

3.7. Determination of the Contrast Transfer Function

The CTF may either be determined "by hand," specifically by measuring the positions of the zeros and fitting them to a chart of the CTF characteristics, or by using automated computer-fitting methods.

In the method of manual fitting, the CTF characteristics of the microscope (see section 3.3) is computed for the different voltages in use (e.g., 100 or 200 kV) and displayed on a hard copy, preferably on a scale that enables direct comparison with the print of the optical diffraction pattern. As long as the lens of the microscope remains the same, the CTF characteristics remain unchanged. (In fact, as was pointed out above, a single set of curves covers all possible voltages and spherical aberrations, provided that generalized parameters and variables are used.) In the simplest form of manual defocus determination, a set of measured radii of CTF zero positions is "slid" against the CTF characteristics until a match is achieved. Since the slope of the different branches of the characteristics is shallow in most parts of the pattern, the accuracy of this kind of manual defocus determination is low, but it can be improved by using not one but simultaneously two or more diffraction patterns of a series with known defocus increments. Such a set of measurements forms a "comb" which can be slid against the CTF characteristics in its entirety.

Manual fitting has been largely superceded by computer-assisted interactive or fully automated methods. In earlier fully automated methods (Frank et al., 1970; Frank 1972c; Henderson et al., 1986), the Fourier modulus $|F(\mathbf{k})|$ (i.e., the square root of what would be called the diffraction pattern) is computed from a field of sufficient size. The theoretical CTF pattern is then matched with the experimental power spectrum using an iterative nonlinear least squares fitting method. The parameters being varied are Δz, z_a, ϕ_0, and a multiplicative scaling factor. Thus, the error sum is (Frank et al., 1970)

$$E(\Delta z, z_a, \phi_0, c) = \sum_j \{G(\mathbf{k}_j; \Delta z, z_a, \phi_0, c) - |F(\mathbf{k}_j)|\}^2 \qquad (2.26)$$

where

$$G(\mathbf{k}_j; \Delta z, z_a, \phi_0, c) = \frac{c}{|\mathbf{k}_j|} |\sin \gamma(\mathbf{k}_j; \Delta z, z_a, \phi_0)| \qquad (2.27)$$

The summation runs over all Fourier elements indexed j, and c is a simple scaling constant.

It has not been practical to include envelope parameters in the 2D fitting procedure. The $1/k$ dependence was also used by other groups (Henderson et al., 1986; Stewart et al., 1993) to match the observed decline in power with spatial frequency. This $1/k$ dependency, obtained for negatively stained specimens, lacks a theoretical basis but appears to accommodate some of the effects discussed by Henderson and Glaeser (1985), such as specimen drift and fluctuating local charging.

The error function [equation (2.26)] has many local minima. The only way to guarantee that the correct *global* minimum is found in the nonlinear least squares procedure is by trying different starting values for Δz. Intuitively, it is clear that smoothing the strongly fluctuating experimental distribution $|F(\mathbf{k})|$ will improve the quality of the fit and the speed of convergence of the iterative algorithm. Smoothing can be accomplished by the following sequence of steps: transform $|F(\mathbf{k})|^2$ into real space, which yields a distribution akin to the Patterson function, mask this function to eliminate all terms outside a small radius, and finally transform the result back into Fourier space (Frank et al., 1970). A smooth image spectrum may also be obtained by dividing the image into small, partially overlapping regions $p_n(\mathbf{r})$ and computing the average of the Fourier moduli $|F_n(\mathbf{k})| = |\mathfrak{F}\{p_n(r)\}|$ (Zhu and Frank, 1994; Fernández et al., 1997; Zhu et al., 1997). This method actually comes close to the definition of the power spectrum, at a given spatial frequency $\mathbf{k}$, as an expectation value of $|F(\mathbf{k})|^2$ (Fernández et al., 1997) (see section 3.6). This way of estimating the power spectrum proves optimal when the areas used for estimation are overlapped by 50% (Zhu et al., 1996; Fernández et al., 1997). A further boost in signal can be achieved by using a thin carbon film floated onto the ice.

More recently, several automated procedures for CTF determination have been described (Huang et al., 2003; Mindell and Grigorieff, 2003; Sander et al., 2003a). In the first approach, the power spectrum, after smoothing and background correction, is being compared with the analytical function for the CTF assuming different values for Δz on a grid, and different settings for the axial astigmatism. In the second approach (Sander et al., 2003a), power spectra of individual particles are classified by *correspondence analysis* followed by hierarchical ascendant classification (regarding both subjects, see chapter 4), and a comparison of the resulting class averages with analytically derived CTFs. Finally, Huang et al. (2003) make use of a heuristic approach, by dividing the power spectrum into two regions with strongly different behavior, and fitting the radial profile in these regions with different polynomial expressions. Both envelope parameters and axial astigmatism are handled with this method, as well. One of the observations made in this study is that the B-factor formalism, often used to summarily describe the decline of the Fourier amplitudes as seen on the power spectrum, is seriously deficient. For a more detailed discussion of this point, see chapter 5, section 9.3.1.

One problem with fully automated methods is that they inevitably contain "fiddle parameters" adjusted to particular kinds of data, as characterized by specimen support, instrument magnification, etc. Manual checks are routinely required. The other class of CTF estimation methods is based entirely on user interaction, and here the concentration is on making the tools convenient and user-friendly. High sensitivity can be obtained by rotationally averaging the squared Fourier transform, and comparing the resulting radial profiles with the theoretical curves. Of course, rotational averaging assumes that the axial astigmatism is well corrected, otherwise the radial profile will be only approximately correct in reflecting the mean defocus.

The first works in this category (Zhou and Chiu, 1993; Zhu and Frank, 1994; Zhu et al., 1997) made the user go through a number of steps, at that time without the benefits of a graphics tool. The one-dimensional profile obtained by rotational averaging is first corrected by background subtraction, then the resulting profile is fitted to a product of the transfer function with envelopes representing the effects of partial coherence, chromatic defocus spread, and other resolution-limiting effects. Background correction is accomplished by fitting the minima of $|F(\mathbf{k})|$ (i.e., regions where the Fourier transform of the image intensity should vanish, in the absence of noise and other contributions not accounted for in the linear bright-field theory) with a slowly varying, well-behaved function of the spatial frequency radius $k = |\mathbf{k}|$. Zhou and Chiu (1993) used a high-order polynomial function, while Zhu et al. (1997) were able to obtain a Gaussian fit of the noise background, independent of the value of defocus, and to determine the effective source size characterizing partial coherence as well.

Later implementations of this approach employ graphics tools. In WEB, the graphics user interface of SPIDER (Frank et al., 1981b, 1996), the rotationally averaged power spectrum is displayed as a 1D profile along with the theoretical CTF curve, which can be tuned to match the former by changing the position of a defocus slider. Figure 2.20a shows the radial profiles of power spectra obtained for cryo-EM micrographs (ribosomes embedded in ice, with a thin carbon film added) with different defocus settings. Figure 2.20b shows an example of a background correction using a polynomial curve obtained by least squares fitting of the minima. The background-corrected curve (bottom in figure 2.20b) can now be interpreted in terms of the square of the CTF modulated by an envelope term that causes the decay at high spatial frequencies. In addition to the tool in SPIDER/WEB, several convenient graphics user tools for interactive CTF estimation have emerged. One, CTFFIND2, is part of the MRC image processing software package (Crowther et al., 1996). Further on the list are tools contained within the packages EM (Hegerl and Altbauer, 1982; Hegerl, 1996), EMAN (Ludtke et al., 1999), IMAGIC (van Heel et al., 1996), and Xmipp (Marabini et al., 1996b).

Axial astigmatism can be taken into account in this fitting process, without the need to abandon the benefit of azimuthal averaging, by dividing the 180° azimuthal range (ignoring the Friedel symmetry-related range) into a number of sectors. For each of these sectors, the defocus is then separately determined.

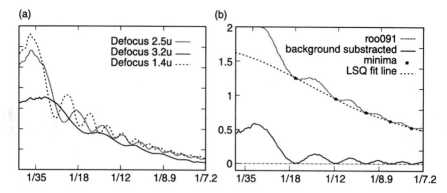

Figure 2.20 Radial profiles of Fourier modulus (i.e., square root of "power spectrum") obtained by azimuthal averaging, and background correction. Vertical axis is in arbitrary units. (a) Profiles for three different defocus settings. (b) Background correction. The minima of the experimental profile (upper part of panel) were fitted by a polynomial using least squares, and the resulting background curve was subtracted, to yield the curve in the lower part. Apart from a region close to the origin, the corrected curve conforms with the expectation of the linear CTF theory, and can be interpreted as the square of the CTF with an envelope term.

However, this method appears to be unreliable when applied to specimens in ice because of the reduced SNR (J. Zhu, personal communication, 1994). Zhou et al. (1996) developed a graphics tool, now integrated into the EMAN package (Ludtke et al., 1999), that allows the user to put circles or ellipses into the zeros of the experimental 2D power spectrum, thus enabling the estimation of axial astigmatism as well. Huang et al. (2003), already mentioned before, report success in the estimation of axial astigmatism with their automated CTF determination procedure, but the success hinges on a careful estimation of the background.

Related to this subject is the measurement of the entire CTF characteristics and, along with it, the parameters characterizing energy spread and partial coherence. The purpose here is not to find the parameters characterizing a single micrograph, but to describe the instrument's electron optical performance over an entire defocus range. Krakow et al. (1974) made a CTF versus Δz plot by using a tilted carbon grid and 1D optical diffraction with a cylindrical lens. Frank et al. (1978b) used this method to verify the predicted defocus dependence of the envelope function for partial spatial coherence (Frank, 1973a). Burge and Scott (1975) developed an elegant method of measurement in which a large axial astigmatism is intentionally introduced. As we have seen in the previous section, the effect of a large amount of axial astigmatism is that, as the azimuthal angle goes through the 360° range, the astigmatic defocus component $z_a \sin[2(\phi - \phi_0)]$ [equation (2.5)] sweeps forward and backward through the CTF characteristics, producing a quasi-hyperbolic appearance of most of the contrast transfer zones. For large enough Δz, the diffraction pattern, either optically derived (Burge and Scott, 1975) or computed (Móbus and Rühle, 1993), will contain a large segment of the CTF characteristics in an angularly "coded" form. Móbus and Rühle

(1993) were able to map the hyperbolic pattern into the customary CTF versus Δz diagram by a computational procedure.

3.8. Instrumental Correction of the Contrast Transfer Function

Many attempts have been made to improve the instrument so that the CTF conforms more closely with the ideal behavior. In the early 1970s, several groups tried to change the wave aberration function through intervention in the back focal plane. Hoppe and coworkers (Hoppe, 1961; Langer and Hoppe, 1966) designed zone plates, to be inserted in the aperture plane, that would block out all waves giving rise to destructive interference for a particular defocus setting. A few of these plates were actually made, thanks to the early development of microfabrication in the laboratory of the late Gottfried Möllenstedt, at the University of Tübingen, Germany. However, contamination build-up, and the need to match the defocus of the instrument precisely with the defocus the zone plate is designed for, made these correction elements impractical.

Unwin (1970) introduced a device that builds up an electric charge at the center of the objective aperture (actually a spider's thread) whose field modifies the wave aberration function at low angles and thereby corrects the transfer function in the low spatial frequency range. Thin phase plates (*Zernike plates*) designed to retard portions of the wave have also been tried (see Reimer, 1997). Typically, these would be thin carbon films with a central bore leaving the central beam in the back focal plane unaffected. These earlier attempts were plagued by the build-up of contamination, so that the intended phase shift could not be maintained for long. The idea was recently revived (Danev and Nagayama, 2001; Majorovits and Schröder 2002a); this time, however, with more promise of success since modern vacuum technology reduces contamination rates several-fold.

Another type of phase plate, the *Boersch plate*, is immune to contamination as an electric field is created in the immediate vicinity of the optic axis by the use of a small ring-shaped electrode supported by, but insulated from, thin metal wires (Danev and Nagayama, 2001; Majorovits and Schröder, 2002a). Also, a very promising development to be mentioned here is an electromagnetic corrector element that can be used to "tune" the wave aberration function, and in the process also to boost the low-spatial frequency response if desired (Rose, 1990).

Corrections that are applied *after* the image has been taken, such as Wiener filtering and the merging of data from a defocus series (see section 2), have traditionally not been counted as "instrumental compensation." However, as the computerization of electron microscopes is proceeding, the boundaries between image formation, data collection, and on-line postprocessing are becoming increasingly blurred. By making use of computer control of instrument functions, it is now possible to compose output images by integrating the "primary" image collected over the range of one or several parameters. One such scheme, pursued and demonstrated by Taniguchi et al. (1992), is based on a weighted superposition of electronically captured defocus series.

3.9. Computational Correction of the Contrast Transfer Function

3.9.1. Phase Flipping

As has become clear, the CTF fails to transfer the object information as represented by the object's Fourier transform with correct phases and amplitudes. All computational corrections are based on knowledge of the CTF as a function of the spatial frequency, in terms of the parameters, Δz, C_s, and z_a, which occur in the analytical description of the CTF [equations (2.4), (2.5), and (2.10)]. Additionally, envelope parameters that describe partial coherence effects must be known. The simplest correction of the CTF is by "phase flipping," that is, by performing the following operation on the image transform:

$$F'(\mathbf{k}) = \begin{cases} -F(\mathbf{k}) & \text{for } H(\mathbf{k}) < 0 \\ F(\mathbf{k}) & \text{for } H(\mathbf{k}) > 0 \end{cases} \tag{2.28}$$

which ensures that the modified image transform $F'(\mathbf{k})$ has the correct phases throughout the resolution domain. However, such a correction leaves the spatial-frequency dependent attenuation of the amplitudes uncorrected: as we know, Fourier components residing in regions near the zeros of $H(\mathbf{k})$ are weighted down by the CTF, and those located in the zeros are not transferred at all.

3.9.2. Wiener Filtering—Introduction

The Wiener filtering approach can be described as a "careful division" by the CTF such that noise amplification is kept within limits:

Given is a micrograph that contains a signal corrupted by additive noise. In Fourier notation,

$$I(\mathbf{k}) = F(\mathbf{k})H(\mathbf{k}) + N(\mathbf{k}) \tag{2.28a}$$

We seek an estimate $\hat{F}(\mathbf{k})$ that minimizes the expected mean-squared deviation from the Fourier transform of the object, $F(\mathbf{k})$:

$$E\{|F(\mathbf{k}) - \hat{F}(\mathbf{k})|^2\} = \min \tag{2.29}$$

where $E\{\cdot\}$ denotes the expectation value computed over an ensemble of images. We now look for a filter function $S(\mathbf{k})$ with the property:

$$\hat{F}(\mathbf{k}) = S(\mathbf{k})I(\mathbf{k}); \tag{2.30}$$

that is, one that yields the desired estimate when applied to the image transform. The solution of this problem, obtained under the assumption that there is no correlation between $F(\mathbf{k})$ and $N(\mathbf{k})$, is the well-known expression:

$$S(\mathbf{k}) = \frac{H^*(\mathbf{k})}{|H(\mathbf{k})|^2 + P_N(\mathbf{k})/P_F(\mathbf{k})} \tag{2.31}$$

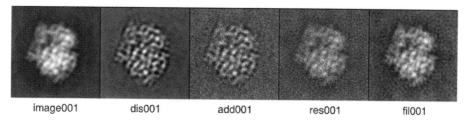

image001   dis001   add001   res001   fil001

Figure 2.21 Demonstration of Wiener filtering. Image001 was produced by projecting a (CTF-corrected) ribosome density map (Gabashvili et al., 2000) into the side-view direction. From this, an "EM image" (dis001) was generated by applying a CTF. Next, noise was added (add001). Application of the Wiener filter assuming high signal-to-noise ratio (SNR = 100) resulted in the image res001. Low-pass filtration of this image to the bandwidth of the original image produces a "cleaned-up" version, fil001. The texture apparent in both particle and background results from a too aggressive choice of the Wiener filter, implicit in the overestimation of the SNR, which has the effect that the noise in the regions around the zeros of the CTF is disproportionately amplified. Still, the low-pass filtered, Wiener-filtered version (fil001) is a close approximation of the original.

where $P_N(\mathbf{k})$ and $P_F(\mathbf{k})$ are the power spectra of noise and object, respectively. It is easy to see that the filter function corrects both the phase [since it flips the phase according to the sign of $H^*(\mathbf{k})$] and the amplitude of the Fourier transform. The additive term in the denominator of equation (2.31) prevents excessive noise amplification in the neighborhood of $H(\mathbf{k}) = 0$.

It is instructive to see the behavior of the Wiener filter for some special cases. First, the term $P_N(\mathbf{k})/P_F(\mathbf{k})$, which could be called the "spectral noise-to-signal ratio," can often be approximated by a single constant, $c = 1/\alpha$, the reciprocal of the signal-to-noise ratio (see chapter 3, section 4.3). Let us first look at the filter for the two extreme cases, $\alpha = 0$ (i.e., the signal is negligible compared with the noise) and $\alpha = \infty$ (no noise present). In the first case, equation (2.31) yields $S(\mathbf{k}) = 0$, a sensible result since the estimate should be zero when no information is conveyed. In the second case, where no noise is assumed, $S(\mathbf{k})$ becomes $1/H(\mathbf{k})$; that is, the restoration (equation 2.30) reverts to a simple division by the transfer function. When the SNR is small, but nonzero, which happens when the Wiener filter is applied to raw data, then the denominator in equation (2.31) is strongly dominated by the term $1/\alpha$, rendering $|H(\mathbf{k})|^2$ negligible. In that case, the filter function becomes $S(\mathbf{k}) \approx \alpha H^*(\mathbf{k})$, that is, restoration now amounts to a "conservative" multiplication of the image transform with the transfer function itself. The most important effect of this multiplication is that it flips the phase of the image transform wherever needed in Fourier space. [In the example above (figure 2.21), high SNR was assumed even though the SNR is actually low. Consequently, the restoration is marred by the appearance of some ripples.]

3.9.3. Wiener Filtering of a Defocus Pair

Thus far, it has been assumed that only a single micrograph is being analyzed. The disadvantage of using a single micrograph is that gaps in the vicinity of the zeros of the transfer function remain unfilled. This problem can be solved, and the gaps filled, by using two or more images at different defocus settings (Frank and Penczek, 1995; Penczek et al., 1997). In the following, we show the case of two images, with Fourier transforms $I_1(\mathbf{k})$ and $I_2(\mathbf{k})$, and two different CTFs $H_1(\mathbf{k})$ and $H_2(\mathbf{k})$. We seek

$$\hat{F}(\mathbf{k}) = S_1(\mathbf{k})I_1(\mathbf{k}) + S_2(\mathbf{k})I_2(\mathbf{k}) \tag{2.31a}$$

and the filter functions ($n = 1, 2$) become

$$S_n(\mathbf{k}) = \frac{H_n^*(\mathbf{k})}{|H_1(\mathbf{k})|^2 + |H_2(\mathbf{k})|^2 + P_N(\mathbf{k})/P_F(\mathbf{k})} \tag{2.32}$$

For a suitable choice of defocus settings, none of the zeros of $H_1(\mathbf{k})$ and $H_2(\mathbf{k})$ coincide, so that information gaps can be entirely avoided. [It should be noted that in the application considered by Frank and Penczek (1995), the images are 3D, resulting from two independent 3D reconstructions, and the spectral noise-to-signal ratio $P_N(\mathbf{k})/P_F(\mathbf{k})$ (assumed to be the same in both images) is actually much less than that for raw images, as a consequence of the averaging that occurs in the 3D reconstruction process.] As in the previous section, where we have dealt with the Wiener filtration of a single image, we can, under certain circumstances, simplify equation (2.32) by assuming $P_N(\mathbf{k})/P_F(\mathbf{k}) = \text{const.} = 1/\text{SNR}$. Application of the Wiener filter to a defocus doublet is shown below for a test image (figure 2.22).

3.9.4. Wiener Filtering of a Defocus Series N > 2

Even more safety in avoiding information gaps is gained by using more than two defocus settings. Generalization of the least-squares approach to N images with different defocus settings leads to the estimate (Penczek et al., 1997):

$$\hat{F}(\mathbf{k}) = \sum_{n=1}^{N} S_n(\mathbf{k})I_n(\mathbf{k}) \tag{2.32a}$$

with the filter functions:

$$S_n(\mathbf{k}) = \frac{\text{SNR}_n H_n^*(\mathbf{k})}{\sum_{n=1}^{N} \text{SNR}_n |H_n(\mathbf{k})|^2 + 1} \tag{2.32b}$$

where SNR_n is the signal-to-noise ratio of each image. [Note that under the assumption that $n = 2$ and $\text{SNR}_1 = \text{SNR}_2$, equation (2.32b) reverts to equation (2.32) for the case of constant SNR.]

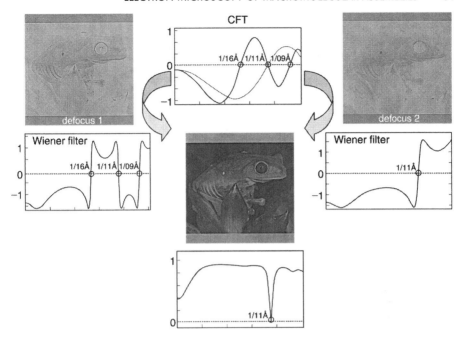

Figure 2.22 Demonstration of Wiener filtering with $N = 2$ degraded images having different amounts of underfocus. Graphs on the left and right show how the Wiener filter amplifies Fourier components close to the zeros, but does not go beyond a certain amplitude, whose size depends on the SNR. The graph below the restored image shows the effective transfer function achieved after combining the information from both images. It is seen that two zeros have been entirely eliminated, while a third one, at $1/11\,\text{Å}^{-1}$, is not, since it is shared by both CTFs. (From N. Boisset, unpublished lecture material.)

3.9.5. Separate Recovery of Amplitude and Phase Components

In the above descriptions of the phase flipping and Wiener filtering approaches to CTF correction, it has been assumed that the specimen has the same scattering properties throughout. If we take into account the fact that there are different atomic species with different scattering properties, we have to go one step back and start with equation (2.21), which describes the different image components relating to the phase [with transform $O_r(\mathbf{k})$] and amplitude portion [with transform $O_i(\mathbf{k})$] of the object:

$$I(\mathbf{k}) = O_r(\mathbf{k}) \sin \gamma(\mathbf{k}) - O_i(\mathbf{k}) \cos \gamma(\mathbf{k}) \qquad (2.32c)$$

In 1968, Schiske posed the question whether these two different object components [one following the Friedel law, the other one an anti-Friedel law; see equation (A.10) in appendix 1] can be separately retrieved, by making use of several measurements of $I(\mathbf{k})$ with different defocus settings; that is, by recording a defocus series of the same object. An advantage of separating the two components is the enhanced contrast between heavy and light atoms ("heavy/light atom discrimination"; see Frank, 1972c, 1973b) that we expect to find in the amplitude component. The solution for $N = 2$ defocus settings is obtained by simply solving the two resulting equations (from 2.32c) for two unknowns, yielding

$$F_r(\mathbf{k}) = \frac{I_1(\mathbf{k}) \cos \gamma_2(\mathbf{k}) - I_2(\mathbf{k}) \cos \gamma_1(\mathbf{k})}{\sin[\gamma_1(\mathbf{k}) - \gamma_2(\mathbf{k})]}$$

$$F_i(\mathbf{k}) = \frac{I_1(\mathbf{k}) \sin \gamma_2(\mathbf{k}) - I_2(\mathbf{k}) \sin \gamma_1(\mathbf{k})}{\sin[\gamma_1(\mathbf{k}) - \gamma_2(\mathbf{k})]}$$

(2.33)

For $N > 2$, there are more measurements than unknowns, and the problem can be solved by least-squares analysis, resulting in a suppression of noise (Schiske, 1968). Following Schiske's general formula, one obtains the following expression for the best estimate of the object's Fourier transform:

$$\hat{F}(\mathbf{k}) = \frac{-i \sum_{m=1}^{N} I_m(\mathbf{k}) \exp[i\gamma_m(\mathbf{k})]\{N - \sum_{n=1}^{N} \exp[2i(\gamma_n(\mathbf{k}) - \gamma_m(\mathbf{k}))]\}}{N^2 - |\sum_{n=1}^{N} \exp[-2i\gamma_n(\mathbf{k})]|^2}$$

(2.34)

which can then be split up into the "phase" and "amplitude" components, by separating the part observing Friedel symmetry (see appendix 1) from the part that does not, as follows:

$$F_r(\mathbf{k}) = [\hat{F}(\mathbf{k}) + \hat{F}^*(-\mathbf{k})]/2$$

(2.35a)

and

$$\hat{F}_i(\mathbf{k}) = [\hat{F}(\mathbf{k}) - \hat{F}^*(-\mathbf{k})]/2$$

(2.35b)

In a first application of the Schiske formula, Frank (1972c) demonstrated enhancement in features of stained DNA on a carbon film. In-depth analyses following a similar approach were later developed by Kirkland and coworkers 1980, 1984), mainly for application in materials science. Interestingly, in their test of the technique, Typke et al. (1992) found that small magnification differences associated with the change in defocus produce intolerably large effects in the restored images and proposed a method for magnification compensation. Figure 2.23 shows the phase and amplitude portion of a specimen field, restored from eight images of a focus series.

In many applications of single-particle reconstruction, CTF correction is carried out as part of the 3D reconstruction procedure. Details will be found in the chapter on 3D reconstruction (chapter 5, section 9). Advantages and disadvantages in carrying out CTF correction at the stage of raw data versus the later stage where volumes have been obtained will be discussed in chapter 5, section 9.2.

3.10. Locally Varying CTF and Image Quality

Under stable conditions we can assume that the entire image field captured by the electron micrograph possesses the same CTF, unless the specimen is tilted and

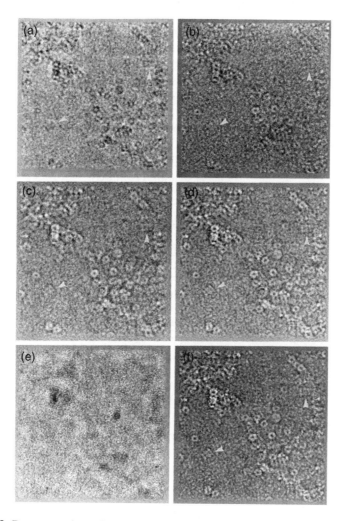

Figure 2.23 Demonstration of Schiske-type restoration applied to a defocus series of eight micrographs (ranging from −5400 to +5400 Å) showing proteasomes on carbon embedded in vitreous ice. (a, b) Two of the original micrographs. (c) Restored phase part of the object, obtained by using the micrographs (a, b). (d) Restored phase part, obtained by using four micrographs. (e, f) Amplitude and phase parts of the specimen, respectively, restored from the entire defocus series. In the phase part, the particles stand out stronger than in (d, e) where fewer images were used. The amplitude part (e) reflects the locally changing pattern in the loss of electrons participating in the elastic image formation, either due to inelastic scattering or due to scattering outside the aperture. Another contribution comes from the fact that a single defocus is attributed to a relatively thick specimen (see Frank, 1973a; Typke et al., 1992). The particles are invisible in this part of the restored object. In the phase part (f), the particles stand out strongly from the ice + carbon background on account of the locally increased phase shift. The arrows point to particles in side-view orientations that are virtually invisible in the original micrographs but now stand out from the background. From Typke et al. (1992), reproduced with permission of Elsevier.

unless the z-position of particles is strongly variable, due to the larger thickness of the ice (cf. van Heel et al., 2000). Moreover, with stable specimens and imaging conditions, there is no reason to expect that parameters characterizing the envelope functions should vary across the image field. Indeed, the single-particle reconstruction strategy outlined in chapter 5, section 7, is based on the assumption that the entire micrograph can be characterized by a single CTF. The fact that this strategy has led to a 7.8-Å reconstruction of the ribosome (Spahn et al., in preparation; see figures 5.24 and 6.6) indicates that these stable conditions can be realized.

However, there are conditions of instability that can lead to substantial variability of the power spectrum. Instability can result from local drift and electric charging, as discussed by Henderson (1992). In addition, local variability can result from changes in the thickness of ice. One way of observing the local variations in the power spectrum is by scanning the aperture defining the illuminated area in the optical diffractometer relative to the micrograph (i.e., sliding the micrograph across the fixed geometry of the optical diffractometer beam), and observing the resulting pattern. The computational analysis offers more options and better quantification: Gao et al. (2002) divided the micrograph into subfields of 512×512, computed the local, azimuthally averaged power spectra, and subjected these to correspondence analysis (see chapter 4, sections 2 and 3). For some micrographs (containing ribosomes in vitreous ice, and a thin carbon film) analyzed, the amplitude falloff varied substantially across the image field. Sander et al. (2003a), using a similar method of analysis, also found local variability of the amplitude falloff, parameterized by a variable B-factor, in analyzing micrographs of spliceosome specimens prepared with cryo-stain. Another study (Typke et al., 2004) came to similar results, but demonstrated that some improvement is possible by metal coating of the support film surrounding the (specimen-bearing) holes of the holey carbon film. Considering the extent of the variability found by all three groups, it might be advisable to use a diagnostic test as part of the routine screening and preprocessing of the data, so as to avoid processing and refining inferior particles.

4. Special Imaging Techniques and Devices

4.1. Low-Dose Electron Microscopy

Concerns about radiation damage led to careful electron diffraction measurements at the beginning of the 1970s (Glaeser, 1971). The results were not encouraging: the electron diffraction patterns of 2D crystals formed by L-valine were found to disappear entirely after exposure to $6\,e^-/\text{Å}^2$ (Glaeser, 1971). Crystals of the less sensitive adenosine ceased to diffract at an exposure of about $60\,e^-/\text{Å}^2$, still much lower than the then "normal" conditions for taking an exposure. To some extent, low temperature affords protection from radiation damage (Taylor and Glaeser, 1974), by trapping reaction products in situ, but the actual gain (five- to eight-fold for catalase and purple membrane crystals at

−120°C) turned out smaller than expected (Glaeser and Taylor, 1978; Hayward and Glaeser, 1979).

At a Workshop in Gais, Switzerland (October, 1973), the prospect for biological EM with a resolution better than $1/30\,\text{Å}^{-1}$ was discussed, and radiation damage was identified as the primary concern (Beer et al., 1975). The use of averaging to circumvent this hurdle had been suggested earlier on by Glaeser and coworkers (1971; see also Kuo and Glaeser, 1975). Even earlier, a technique termed "minimum dose microscopy," invented by Williams and Fisher (1970), proved to preserve, at $50\,\text{e}^-/\text{Å}^2$, the thin tail fibers of the negatively stained T4 bacteriophage that were previously invisible for doses usually exceeding $200\,\text{e}^-/\text{Å}^2$. In their pioneering work, Unwin and Henderson (1975) obtained a $1/7\,\text{Å}^{-1}$ resolution map of glucose-embedded bacteriorhodopsin by the use of a very low dose, $0.5\,\text{e}^-/\text{Å}^2$, along with averaging over a large (10,000 unit cells) crystal field.

In the current usage of the term, "low dose" refers to a dose equal to, or lower than $10\,\text{e}^-/\text{Å}^2$. (Since "dose" actually refers to the radiation absorbed, "low exposure" is the more appropriate term to use.) Experience has shown that exposures larger than that lead to progressive disordering of the material and eventually to mass loss and bubbling. Susceptibility to radiation damage was found to vary widely among different materials and with different specimen preparation conditions. Changes in the structure of negatively stained catalase crystals were investigated by Unwin (1975). The results of his work indicate that rearrangement of the stain (uranyl acetate) occurs for exposures as low as $10\,\text{e}^-/\text{Å}^2$. The radiation damage studies by Kunath et al. (1984) on single uranyl acetate-stained glutamine synthetase (glutamate–ammonia ligase) molecules came to a similar conclusion,[3] comparing single, electronically recorded frames at intervals of $1\,\text{e}^-/\text{Å}^2$.

Frozen-hydrated specimens are strongly susceptible to radiation damage. For such specimens, bubbles begin to appear at exposures in the region of $50\,\text{e}^-/\text{Å}^2$, as first reported by Lepault et al. (1983). A study by Conway et al. (1993) compared the reconstructions of a *herpes simplex* virus capsid (HSV-1) in a frozen-hydrated preparation obtained with total accumulated exposures of 6 and $30\,\text{e}^-/\text{Å}^2$. The authors reported that although the nominal resolution (as obtained with the Fourier ring correlation criterion, see chapter 3, section 5.2.4) may change little, there is a strong overall loss of power in the Fourier spectrum with the five-fold increase of dose. Still, the surface representation of the virus based on a $1/30\,\text{Å}^{-1}$-resolution map shows only subtle changes. The article by Conway et al. (1993), incidentally, gives a good overview over the experimental findings accumulated in the 1980s on this subject. For another review of this topic, the reader is referred to the article by Zeitler (1990).

---

[3]The spot scanning technique was already used by Kunath et al. (1984) in experiments designed to study the radiation sensitivity of macromolecules. It was simply a rational way of organizing the collection of a radiation damage series (which the authors called "movie") from the same specimen field. In hindsight, the extraordinary stability of the specimen (0.05 Å/s) must be attributed to the unintended stabilizing effect of limiting the beam to a 1-μm spot.

Following the work of Unwin and Henderson (1975), several variations of the low-dose recording protocol have been described. In essence, the beam is always first shifted, or deflected, to an area adjacent to the selected specimen area for the purpose of focusing and astigmatism correction, with an intensity that is sufficient for observation. Afterward, the beam is shifted back. The selected area is exposed only once, for the purpose of recording, with the possible exception of an overall survey with an extremely low exposure (on the order of $0.01\ e^-/\text{Å}^2$) at low magnification.

In deciding on how low the recording exposure should be chosen, several factors must be considered: (*i*) *The fog level*: when the exposure on the recording medium becomes too low, the density variations disappear in the background. Control of this critical problem is possible by a judicious choice of the electron-optical magnification (Unwin and Henderson, 1975). Indeed, magnifications in the range $40,000\times-60,000\times$ are now routinely used following the example of Unwin and Henderson (1975). (*ii*) *The ability to align the images of two particles*: the correlation peak due to "self-recognition" or "self-detection" (Frank, 1975a; Saxton and Frank, 1977; see section 3.4 in chapter 3) of the motif buried in both images must stand out from the noisy background, and this stipulation leads to a minimum dose for a given particle size and resolution (Saxton and Frank, 1977). (*iii*) *The statistical requirements*: for a given resolution, the minimum number of particles to be averaged, in two or three dimensions, is tied to the recording dose (Unwin and Henderson, 1975; Henderson, 1995). In planning the experiment, we wish to steer away from a dose that leads to unrealistic numbers of particles. However, the judgment on what is realistic is in fact changing rapidly as computers become faster and more powerful and as methods for automated recording and particle selection are being developed (see chapter 3, section 2.4).

Since the lowering of the exposure comes at the price of lowered signal-to-noise ratio, which translates into lower alignment accuracy, which in turn leads to impaired resolution, the radiation protection provided by low temperatures can be seen as a bonus, allowing the electron dose to be *raised*. Even an extra factor of 2, estimated as the benefit of switching from liquid-nitrogen to liquid-helium temperature (Chiu et al., 1986), would offer a significant benefit (see also section 5.4 in chapter 3).

4.2. Spot Scanning

Recognizing that beam-induced movements of the specimen are responsible for a substantial loss in resolution, Henderson and Glaeser (1985) proposed a novel mode of imaging in the transmission electron microscope whereby only one single small area, in the size range of 1000 Å, is illuminated at a time. This spot is moved over the spectrum field on a regular (square or hexagonal) grid. The rationale of this so-called spot scanning technique is that it allows the beam-induced movement to be kept small since the ratio between energy-absorbing area and supporting perimeter is much reduced. After the successful demonstration of this technique with the radiation-sensitive materials vermiculite and paraffin (Downing and Glaeser, 1986; Bullough and Henderson, 1987; Downing, 1991), numerous structural studies have made use of it (e.g., Kühlbrandt and Downing, 1989; Soejima et al., 1993; Zhou et al., 1994).

Whether or not the spot scanning leads to improvements for single particles as well is still unknown. Clearly, the yield in structural information from crystals is expected to be much more affected by beam-induced movements than from single molecules. On the other hand, to date, single particles without symmetry have not been resolved better than 6 Å resolution, so that the exact nature of the bottlenecks preventing achievement of resolutions in the order of 3 Å is uncertain.

Computer-controlled electron microscopes now contain spot scanning as a regular feature and also allow dynamic focus control (Zemlin, 1989b; Downing, 1992), so that the entire field of a tilted specimen can be kept at one desired defocus setting. The attainment of constant defocus across the image field is of obvious importance for the processing of images of tilted 2D crystals (Downing, 1992), but it also simplifies the processing of single macromolecules following the protocol (chapter 5, sections 3.5 and 5) of the random-conical reconstruction (see Typke et al., 1990).

On the other hand, it may be desirable to retain the focus gradient in more sophisticated experiments, where restoration or heavy/light atom discrimination based on defocus variation are used (see section 3): the micrograph of a tilted specimen essentially produces a focus series of single particles. At a typical magnification of 50,000 and a tilt angle of 60°, the defocus varies by 17,000 Å (or 1.7 μm) across the field captured by the micrograph (assumed to be 50 mm in width), in the direction perpendicular to the tilt axis.

4.3. Energy Filtration

The weak-phase object approximation introduced in section 3.2 enabled us to describe the image formation in terms of a convolution integral. In Fourier space, the corresponding relationship between the Fourier transforms of image contrast and object is very simple, allowing the object function to be recovered by computational methods described in section 3.9. However, this description of image formation is valid only for the bright field image formed by elastically scattered electrons. Inelastically scattered electrons produce another, very blurred image of the object that appears superimposed on the "elastic image". The formation of this image follows more complicated rules (e.g., Reimer, 1997).

As a consequence, an attempt to correct the image for the effect of the contrast transfer function (and thereby retrieve the true object function) runs into problems, especially in the range of low spatial frequencies where the behavior of the inelastic components can be opposite to that expected for the phase contrast of elastic components (namely in the underfocus range, where $\Delta z > 0$). Thus, CTF correction based on the assumption that only elastic components are present will, in the attempt to undo the under-representation of these components, amplify the undesired inelastic component as well, leading to an incorrect, blurred estimate of the object. This problem is especially severe in the case of ice-embedded specimens, for which the cross-section for inelastic scattering exceeds that for elastic scattering.

Another problem caused by inelastic scattering is that it produces a decrease in the SNR ratio of the image to be retrieved (Schröder et al., 1990; Langmore and Smith, 1992). This decrease adversely affects the accuracy of all operations, to be described in the following chapters, that interrelate raw data in which noise is prevalent; for example, alignment (chapter 3, section 3), multivariate data analysis (chapter 4, sections 2 and 3), classification (chapter 4, section 4), and angular refinement (chapter 5, section 7).

The problems outlined above can be overcome by the use of energy-filtering electron microscopes (Schröder et al., 1990; Langmore and Smith, 1992; Smith and Langmore, 1992; Angert et al., 2000). In these specialized microscopes, the electron beam goes through an electron spectrometer at some stage after passing through the specimen. The spectrometer consists of a system of magnets that separate electrons spatially on a plane according to their energies. By placing a slit into this energy-dispersive plane, one can mask out all inelastically scattered electrons, allowing only those electrons to pass that have lost marginal amounts (0–15 eV) of energy ("zero-loss window"). In practice, the energy filter is either placed into the column in front of the projective lens (e.g., the *Omega filter*; Lanio, 1986) or added below the column as final electron optical element (*post-column filter*; Krivanek and Ahn, 1986; Gubbens and Krivanek, 1993). The performance of these different types of filters has been assessed by Langmore and Smith (1992) and Uhlemann and Rose (1994). Examples of structural studies on frozen-hydrated specimens in which energy filtering has been employed with success are found in the work on the structure of myocin fragment S1-decorated actin (~25 Å: Schröder et al., 1993) and the ribosome (25 Å: Frank et al., 1995a,b; Zhu et al., 1997).

Many subsequent studies have demonstrated, however, that both CTF correction and the achievement of higher resolution are possible with unfiltered images, as well. In fact, at increasing resolutions, the influence of the "misbehaved" low spatial frequency terms (i.e., contributions to the image that do not follow CTF theory) is diminished relative to the total structural information transferred. Although there has been steadily increasing interest in energy-filtered instruments, their most promising application appears to be in electron tomography of relatively thick (> 0.1 µm) specimens (see Koster et al., 1997; Grimm et al., 1998). In fact, specimens beyond a certain thickness cannot be studied without the aid of an energy filter.

4.4. Direct Image Readout and Automated Data Collection

Newer electron microscopes are equipped with CCD cameras enabling direct image readout. Fully automated data collection schemes have been designed that have already revolutionized electron tomography, and promise to revolutionize single-particle reconstruction, as well. This is not likely to happen, however, before CCD arrays of sufficient size (8 k × 8 k) and readout speed become widely available and affordable. The algorithmic implementation of low-dose data collection from single-particle specimens with specific features (as defined by shape, texture, average density, etc.) is quite complex, as it must incorporate the many contingencies of a decision logic that is unique to each specific specimen class. These schemes have been optimized for viruses, helical fibers, and globular particles such as ribosomes (Potter et al., 1999; Carragher et al., 2000; Zhang et al., 2001; Lei and Frank, 2005). A systematic search for suitable areas is facilitated by commercially available perforated grids, such as *Quantifoil* (see section 2.8). Recently, the first entirely automated single-particle reconstruction of a molecule (keyhole limpet hemocyanin) was performed as proof of the concept (Mouche et al., 2004).

3

Two-Dimensional Averaging Techniques

Figure 3.1 Averaging applied to the faces of participants of the 1977 EMBO course on Image Processing of Electron Micrographs (at the Biozentrum Basel, March 7–18, 1977, in Basel, Switzerland, organized by U. Aebi and P. R. Smith). (a) Collage of 22 portraits. (b) Image obtained by photographic superimposition of these portraits, produced at the workshop, using the position of the eyes as an alignment aid.

1. Introduction

As we attempt to interpret an electron microscope image in terms of the biological structure that has given rise to it, we are faced with several obstacles: the image lacks clarity and definition, its resolution is limited, and its fidelity to the object is compromised by instrument distortions. Part of the problem, to be addressed by the application of averaging (figure 3.1), relates to the fact that the image contains a large amount of extraneous information not related to the object. In analogy to its definition in one-dimensional (1D) signal processing, the term "noise" is used in image processing to denote all contributions to the image that do not originate with the object.

Two-dimensional (2D) single-particle *averaging*, which preceded single-particle *reconstruction* in electron microscopy (EM), started as an attempt to obtain well-defined views of a molecule, amenable to quantitative measurements. The purpose of this chapter is to describe the theory underlying image averaging, the practical tools used to align images, and frequently used measures of reproducibility and variability. Many of the concepts, such as alignment, resolution, and the definition and interpretation of variance maps, are easily generalized later on when we come to the chapter on three-dimensional (3D) reconstruction (chapter 5).

1.1. The Different Sources and Types of Noise

We distinguish *stochastic* and *fixed-pattern* noise. In the former case, it is impossible to predict the value of the noise contribution to a given pixel. Only its *expectation value* is known—provided that the statistical distribution of the noise is known. In the latter type of noise, by contrast, the value of the noise at a given pixel position is the same every time a measurement is made. Fixed-pattern noise will not be considered in the following because it can be easily eliminated by image subtraction (or division, in the case of multiplicative noise; see below) using an appropriate control image.

We further distinguish between *signal-dependent* and *-independent* noise. Stochastic noise that is signal-dependent has a spatially varying statistical distribution as dictated by the spatially varying signal. An important example for this type of noise, relevant to the subject of this book, is the noise associated with the fluctuations in the number of electrons per unit area, which follows Poisson statistics.

Yet another distinction is important: in the efforts to eliminate noise one must know in what way the noise combines with the signal part of the image, the most important cases being *additive* and *multiplicative* noise. Of the two kinds, additive noise is easier to handle, since it can be readily eliminated by averaging.

It is important to know the imaging step from which the noise originates, or—in another manner of speaking—where the noise source is located. We will first list the various noise sources, following the imaging pathway from the biological object to the digitized image ready for processing. For each source, as we go along, we will indicate the validity of the common additive noise model.[1]

Macromolecules, at least at the beginning of a project of structure research, are still often prepared by the negative staining technique, where a carbon film is used as support. Besides being limited to rendering the shape of the molecule only, this technique entails two sources of noise:

(i) In the process of drying, the heavy metal salts used as stain precipitate in the form of small crystals. The irregular distribution of these crystals, and the variations in thickness of the stain layer as a whole are important sources of noise.

(ii) The carbon layer itself possesses a structure whose image appears super-imposed on the image of the macromolecule. Since the structure of each carbon area is unique, its contribution to the image cannot be eliminated by a simple subtraction technique unless a control image is taken. Within the limits of the weak-phase approximation (see chapter 2, section 3.2), this so-called structural noise is additive, simply because, in the linear approximation of bright-field imaging, the projection of two added structures is equal to the sum of the projections of each component structure. The idea of eliminating the structure of the support by subtraction of two images of the same area, one with and one without the molecule, was briefly discussed in the early 1970s under the name of the *image difference method* (Hoppe et al., 1969; Langer et al., 1970). This idea was motivated by the surprising accuracy with which the images of the same field of carbon could be aligned by cross-correlation (Langer et al., 1970). However, the realization of the image difference method was fraught with difficulties because of the rapid build-up of contamination. Nowadays, the better vacuum of modern instruments and the availability of cryo-stages would give this method a better chance, but the ensuing development of averaging techniques has facilitated the separation of single molecule projections from their background quite efficiently, as this chapter will show, making experimental subtraction techniques obsolete.

The recording process, whether by means of a photographic plate or by a direct image pick-up device, is burdened with another source of noise, which is due to the quantum nature of the electron: the shot noise. This noise portion is caused by the statistical variations in the number of electrons that impinge on the recording target, which follows Poisson statistics. The size of this contribution relative to the signal in a given pixel depends on the local average number of electrons, hence it falls in the category of non-additive signal-dependent noise. Because of the importance of minimizing radiation damage, the shot noise is one of the most serious limitations in the imaging of macromolecules.

The photographic recording of electrons is a complicated process involving scattering in the photographic emulsion, and the ensuing formation of a shower of secondary electrons. After development, every primary electron gives rise to a fuzzy disk-shaped area of silver grains. Photographic noise, or photographic

[1]This section is an expanded version of the section on noise in a review on image processing in electron microscopy (Frank, 1973c).

granularity, is caused by the irregularity of the resulting distribution of these grains in the electron micrograph (see Downing and Grano, 1982; Zeitler, 1992). Assumption of additivity in its effect on the optical density is valid when the contrast is weak.

It is important to realize, in assessing the contribution of photographic noise to the total noise, as well as the deterioration of the image quality due to the photographic recording, that the electron-optical magnification is a free parameter in the experiment: an increase in magnification will cause the object spectrum to contract relative to the noise spectrum—a change in magnification amounts to a spatial frequency transformation. Thus, it is possible (see section 4.3 on signal-to-noise ratio) to choose the magnification such that the effect of the photographic noise is minimal.

The digitization process acts both as a filter—by removing very short-range components of the photographic granularity—and as an additional source of noise, the *digitization noise*, which is due to "density binning," the conversion of a principally continuous optical density signal into one that is discrete-valued. By using a scanner with a sufficiently large dynamic range, this latter type of noise can be virtually eliminated. (For a discussion of the separate subject concerning the conditions for the faithful representation of an image by sampling, see section 2.2.)

An alternative to photographic recording and digitization is the direct electronic readout of the signal with a slow-scan CCD (charge-coupled device) camera. Here, noise is generated in the process of converting the electron into photons on the scintillator, and in the readout amplifier (see Krivanek and Mooney, 1993; de Ruijter, 1995). Some of it is multiplicative fixed-pattern noise, easily eliminated by gain normalization.

1.2. Principle of Averaging: Historical Notes

Averaging of images for the purpose of noise elimination and bringing out common features may be achieved by photographic superimposition, a technique that goes back to Galton (1878). Markham and coworkers (1963, 1964) introduced this method of averaging into EM, applying it to structures with both translational and rotational symmetry (figure 3.2). An example of the use of photographic superposition for symmetry-based enhancement of glutamine synthetase images is found in the work of Valentine et al. (1968).

According to theory, averaging over a correctly aligned set of images is equivalent to the result of Fourier filtration of a regular 2D montage of that set, using appropriately placed delta-functions as mask, to pass only the Fourier components associated with the signal (see Smith and Aebi, 1973). Those components ("reflections") are again arranged in a regular lattice, the "reciprocal lattice." Ottensmeyer and coworkers (1972, 1977) used this principle to obtain averages of single particles: they arranged the individual images (in this case dark-field images of small biological molecules) into a regular gallery and subjected the resulting artificial 2D "crystal" to optical filtration.

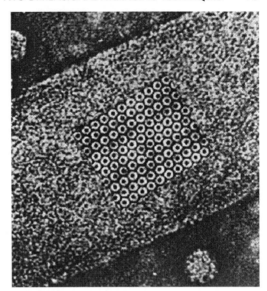

Figure 3.2 Markham's method of photographic superimposition, applied to the noisy image of a 2D crystal. The image was shifted by the repeats in the two directions of the lattice, and the different versions were photographically added together. The insert in the middle shows the result. From Markham et al. (1963), reproduced with permission of Elsevier.

1.3. Equivalence between Averaging and Quasi-Optical Fourier Filtration*

At this point it is instructive to see the relationship between the formation of an average over N noisy realizations $p_i(\mathbf{r})$ of a molecule image $p(\mathbf{r})$, obtained when N equally oriented molecules are projected in the electron microscope:

$$p_i(\mathbf{r}) = p(\mathbf{r}) + n_i(\mathbf{r}); \qquad i = 1\ldots, N \tag{3.1}$$

and the formation of an average by Fourier filtering.[2] Here, $n_i(\mathbf{r})$ denotes the ith realization of an additive noise function. If we arrange the N aligned noisy versions of the projection into a montage of L rows with K images each, so that an image with lexicographic index i is placed at row l and column k according to $i = k + (l - 1) \times K$; $K \times L = N$, then we formally obtain the image of a crystal. This image can be written as a convolution product of the image and an array of delta functions:

$$
\begin{aligned}
m(\mathbf{r}) &= \sum_{l=1}^{L} \sum_{k=1}^{K} p(\mathbf{r}) \circ \delta(\mathbf{r} - k\mathbf{a} - l\mathbf{b}) + n(\mathbf{r}) \\
&= p(\mathbf{r}) \circ \sum_{l=1}^{L} \sum_{k=1}^{K} \delta(\mathbf{r} - k\mathbf{a} - l\mathbf{b}) + n(\mathbf{r})
\end{aligned}
\tag{3.2}
$$

---

[2]The ultimate objective of averaging is the measurement of the structure factor, that is, the 3D Fourier transform of the object. According to the projection theorem (see section 2.1 in chapter 5), each projection contributes a central section to this Fourier transform.

where $n(\mathbf{r})$ is the noise function resulting from placing the N independent noise functions $n_i(\mathbf{r})$ side by side in the montage, and $\mathbf{a},\mathbf{b}$ are orthogonal "lattice vectors" of equal length $|\mathbf{a}| = |\mathbf{b}| = a$. [According to definition (appendix 1), the convolution operation, symbolized by "$\circ$," results in the placement of the motif $p(\mathbf{r})$ into all positions where $\mathbf{r} = k\mathbf{a} + l\mathbf{b}$; $k = 1 \ldots, K$, $l = 1 \ldots, L$].

The Fourier transform of the image defined by equation (3.2) is the sum of the Fourier transform of the signal and that of the noise. The former is, according to the convolution theorem (appendix 1), the product of the Fourier transforms of the projection and the lattice:

$$\mathfrak{F}\{m(\mathbf{r})\} = \mathfrak{F}\{p(\mathbf{r})\} \sum_i \sum_j \delta(\mathbf{k} - i\mathbf{a}^* - j\mathbf{b}^*) + \mathfrak{F}\{n(\mathbf{r})\} \tag{3.2a}$$

where $\mathbf{a}^*$ and $\mathbf{b}^*$ are reciprocal lattice vectors of length $|\mathbf{a}^*| = |\mathbf{b}^*| = 1/a$ that are orthogonal to $\mathbf{a}$ and $\mathbf{b}$, respectively.

On the other hand, the average of the N realizations is

$$\bar{m}(\mathbf{r}) = \frac{1}{N}\left[\sum_{i=1}^{N} p(\mathbf{r}) + \sum_{i=1}^{N} n_i(\mathbf{r})\right] \tag{3.3}$$

and its Fourier transform is

$$\mathfrak{F}\{\bar{m}(\mathbf{r})\} = \mathfrak{F}\{p(\mathbf{r})\} + \mathfrak{F}\left\{\frac{1}{N}\sum_{i=1}^{N} n_i(\mathbf{r})\right\} \tag{3.4}$$

Both Fourier transforms, equations (3.2a) and (3.4), are identical on the points of the reciprocal lattice, defined by multiples of the reciprocal vectors $\mathbf{a}^*$, $\mathbf{b}^*$. (For the purpose of a comparison, the average, equation (3.3), would be placed into an image that has the same size as the montage, and the empty area is "padded" with zeros.)

For a demonstration of the equivalence, figure 3.3a shows a set of noisy 8×8 aligned images of the calcium release channel (Radermacher et al., 1994b) placed into a montage which in turn was padded into a field twice its size. Figure 3.3c shows the average of these 64 images, padded into a field of the same size as in figure 3.3a. The Fourier transform of the artificial "crystal" (figure 3.3b) agrees with the Fourier transform of the average (figure 3.3d) on the points of the reciprocal lattice; the former is simply a sampled version of the latter. By quasi-optical filtration, for example, masking out the reciprocal lattice points, and subsequent inverse Fourier transformation, one obtains an image that contains the average image repeated on the original lattice.

Computer averaging by computational filtration of the reciprocal lattice of a 2D crystal was first accomplished independently by Nathan (1970) and Erickson and Klug (1970) with images of catalase. The mathematically equivalent optical filtration of electron micrographs had been pioneered 4 years earlier by Klug and DeRosier (1966). Glaeser and coworkers (1971) demonstrated, by the use of a computational model of a phantom object, how the repeating motif of a 2D

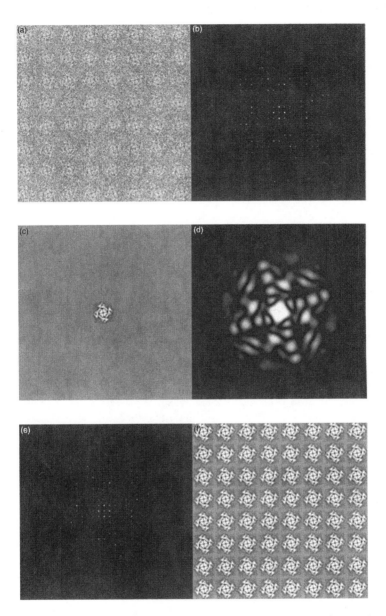

Figure 3.3 Demonstration of the equivalence between Fourier filtering of a pseudo-crystal, generated by arranging the projection of the calcium release channel (Radermacher et al., 1994b) on a regular lattice, and direct averaging of the molecules. (a) Pseudo-crystal containing 8 × 8 noisy 64 × 64 images of the channel; (b) Fourier transform of images in (a) after "padding" into a field twice as large—only the 512 × 512 center of the power spectrum is shown; (c) average of the 8 × 8 images, placed at the center of a 512 × 512 field; (d) Fourier transform of the image in (c) after padding as in (b). It is seen that the Fourier transform in (b) is simply a sample of the continuous Fourier transform of the average image. It represents the average (c) repeated on the lattice on which the images were arranged in (a). Extraction of Fourier components at the precise reciprocal lattice positions (e) and subsequent Fourier synthesis renders the average noise-free (f).

crystal can be recovered from an image taken with extremely low exposure. Although the principle underlying this recovery was well understood, the dramatic visual demonstration helped popularize the concept. Further along, the seminal paper by Unwin and Henderson (1975) provided a milestone in the development of low-dose electron crystallography. From literally invisible raw data, the first image of the seven-helix bundle of bacteriorhodopsin emerged, at 7 Å resolution. It has taken almost a quarter of a century for other groups to come to the same point with viruses (hepatitis B virus: Böttcher et al., 1997; Conway et al., 1997), helically arranged proteins (acetylcholine receptor: Beroukhim and Unwin, 1997), and eventually single molecules lacking symmetry (ribosome: Spahn et al., 2005).

1.4. A Discourse on Terminology: Views Versus Projections

Before coming to the subject of this section, it is necessary to clarify the terms that will be used. In the literature, the term "view" is used in two different ways: the first usage refers to the distinct appearance of a molecule lying in a certain orientation (as in the phrase "the top view of the hemocyanin molecule"), whereas the second usage refers to the actual realization of the image of a molecule presenting that view (e.g., "250 top views of the hemocyanin molecule were averaged"). These two manners of speaking are not compatible with each other. It seems more appropriate to reserve "view" for the generic orientation-related appearance of a molecule, not for each of its realizations. We will, therefore, use the term "view" strictly in the former sense: what we observe in the electron micrograph is an image, or a projection, of a molecule that lies in a certain orientation on the grid, presenting a particular view. Thus, in this manner of speaking, there are exactly as many distinguishable views as there are distinguishable orientations of a molecule on the specimen grid. On a given micrograph, on the other hand, there may be hundreds of molecules presenting the same view.

1.5. The Role of Two-Dimensional Averaging in the Three-Dimensional Analysis of Single Molecules

Historically, 2D averaging of molecules presenting the same view and aligned by cross-correlation (see Frank, 1975; Saxton and Frank, 1977) was an important first step in the development of methods for extraction of quantitative information from single macromolecules (see Frank et al., 1978a, 1981a; van Heel and Frank, 1981; Verschoor et al., 1985). The analysis of a molecule in single-particle form still normally starts with the averaging of molecules occurring in selected, well-defined views (e.g., Davis et al., 2002). This is because averaging presents a fast way of gauging the quality of the data and estimating the potential resolution achievable by 3D reconstruction from the same specimen and under the same imaging conditions. Such an analysis allows the strategy of data collection for 3D reconstruction to be mapped out, since it will answer questions such as: How many distinct views occur with sufficient frequency to allow 3D reconstruction? Is there evidence for uncontrolled variability (from one micrograph to the next, from one grid to the next, etc.)?

The role of averages in the processing leading to a 3D reconstruction depends on the reconstruction strategy used (see chapter 5, section 3). For common-lines based techniques of orientation determination (such as common lines or "angular reconstitution"; chapter 5, section 3.4) to work, they require images with high signal-to-noise ratio; hence, the use of 2D averages is a necessity in this case. If, by contrast, the random-conical technique (chapter 5, section 3.5) is used, averages do not come into play, except as a means to verify the existence of a particle subset presenting a well-defined view, through the formation of class averages.

1.6. Origins of Orientational Preferences

Due to orientational preferences of the molecule on the specimen grid, certain views are often predominant in the micrograph. The degree and physical origins of preferences are quite different for negatively stained and frozen-hydrated specimens, and therefore a separate section is devoted to each. As will be clear later on, both the predominance of certain views and the presence of major gaps in the angular range may cause problems in 3D reconstruction.

Before we come to the discussion of the orientational preferences, a caveat is in order. Data on orientational preference are obtained as the result of an analysis that normally involves a comparison between a 3D reference (i.e., an existing reconstruction of the molecule) and the experimental projection, normally by cross-correlation (chapter 5, section 7.2). For low-dose images, in particular from ice-embedded particles appearing with low contrast, angular assignments and the resulting statistics are subject to a large amount of error since the values of the highest ranking cross-correlation coefficients, each associated with another set of Eulerian angles, are often quite similar.

It is possible, however, that orientational preference or the appearance of angular gaps might be misreported in some instances. Low occurrence of a view as reported by 3D projection matching with a reference was found to be correlated in some specimens with low intrinsic structural content of that view. This could mean that for some specimens, for which the structural content of its projections varies drastically as a function of orientation (as measured by the variance of the projection), the low SNR of the particle images in certain view ranges will produce large errors in angle assignment, leading in turn to the incorrect conclusion that these view ranges are under-represented or absent (D. Frankiel and S. Wolf, personal communication, 2002).

1.6.1. Negatively Stained Specimens

Observations of preferences have been made ever since the negative staining technique was invented. For instance, the 50S ribosomal subunit of *Escherichia coli* is predominantly seen in the crown view and the kidney view (Tischendorf et al., 1974). *Limulus polyphemus* (horseshoe crab) hemocyanin occurs in several distinct orientations, giving rise to a remarkable diversity of views, classified as the pentagonal, ring, cross, and bow-tie view (Lamy et al., 1982) (figure 3.4).

The ratios between the average numbers of molecules observed in the different orientations are constant for a given molecule and preparative method. For instance, a table with the statistics of views observed in the case of the *Androctonus australis* (scorpion) hemocyanin lists 540 particles showing the top view, 726 the side views, and 771 the 45° view (Boisset et al., 1990b). A change in the method of carbon film preparation (e.g., by glow discharging versus untreated) or in the choice of stain used often leads to a change in those ratios. Such observations indicate that the orientational preferences are the result of a complex interplay between the molecule, the sample grid, and the stain. In experiments in which carbon sandwiching is used [see Frank et al. (1988a) and section 2.2 in chapter 2], the interaction between the two layers of carbon introduces an additional force usually favoring orientations that minimize the spatial extension of the particle in the direction normal to the plane of the sample grid.

Both surface charges and features of surface topology are important determinants of orientation. Because large areas of contact are energetically preferred, the stability of the molecule in a particular orientation can often be understood "intuitively," by reference to the behavior of a physical model (an irregular die, as it were) on a horizontal plane under the influence of gravity. Boisset et al. (1990a) developed a computer program that predicts points of stability in an angular coordinate system, provided that a complete description of the shape of the molecule is given. The philosophy behind this calculation is as follows: the molecule is computationally modeled from its subunits. A sphere entirely enclosing the molecule is placed such that its center coincides with the molecule's center of gravity. For 360×360 orientations, covering the entire angular space (except for a trivial azimuthal rotation), a probing plane is placed tangential to the all-enclosing sphere. In each of these orientations, the plane is then moved toward the molecule until it touches it. In the touching position of the probing plane, the (perpendicular) distances between the plane and each of the voxels (volume elements) within the molecule model are summed, yielding a value E that is termed "energy." From this value, an "energy index" I_E is calculated as follows:

$$I_E = \frac{E - E_{\min}}{E_{\max} - E_{\min}} \times 100 \tag{3.5}$$

where $E_{\max}$ and $E_{\min}$ are the maximum and minimum of all E values found.

Figure 3.4 shows some preferred orientations of the *Scutigera coleoptrata* hemocyanin molecule computed by this method. Interestingly, the program predicts all known experimental views satisfactorily, but fails to explain their observed relative frequencies. For instance, two of the views represent more than 95% of all observed molecules, yet the corresponding energy index values are not the lowest on the table computed by the authors. This observation clearly indicates that specific interactions of the molecule with the carbon film are at play. Boisset et al. (1990a) also pointed out that the existence of the so-called 45° view exhibited by chelicerian 4×6-meric hemocyanins, first observed by van Heel et al. (1983), reflects the importance of interactions with the carbon film, since it depicts the molecule in a position that leads to a rather high energy index in terms of the measure given by equation (3.5). In fact, in the case of the *A. australis*

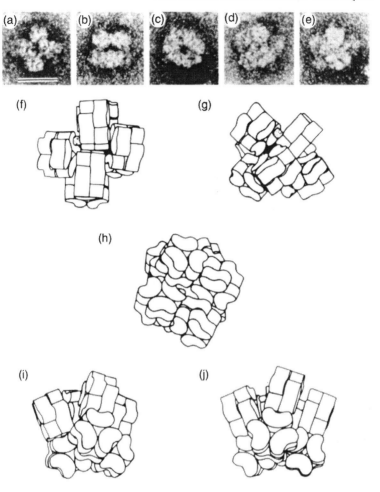

Figure 3.4 Micrographs of *Limulus polyphemus* (horseshoe crab) hemocyanin molecule negatively stained, presenting characteristic views, and explanation of these views in terms of a model placed in different orientations and seen from the top. The views are classified according to the terminology of Lamy et al. (1982) as follows: (a, f) cross view; (b, g) bow tie view; (c, h) ring view; (d, i) asymmetric pentagon view; and (e, j) symmetric pentagon view. From Lamy (1987), reproduced with permission of Elsevier.

hemocyanin, the frequency of the 45° view was found to depend on the hydrophobicity of the carbon film (Boisset et al., 1990a).

1.6.2. Frozen-Hydrated Specimens

In the case of frozen-hydrated preparations, the air–water interface plays a large role in defining the orientation of the molecule (Dubochet et al., 1985, 1988; Wagenknecht et al., 1990; Kolodziej et al., 1998; see figure 5.26a). Judged from the point of view of the simplistic "gravitational" stability model, the

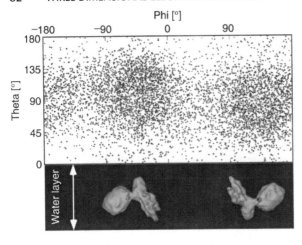

Figure 3.5 Distribution of angles for 6184 particles of bovine NADH: ubiquinone oxidoreductase complex in ice. The two preferred orientations of the molecule in the ice layer (corresponding to the centers of the clusters) are shown in the lower panel. From Griegorieff (1998), reproduced with permission of Elsevier.

A. australis hemocyanin (Boisset et al., 1994b, 1995), the calcium release channel (Radermacher et al., 1994a,b), and other large macromolecular assemblies studied in vitreous ice behave entirely counterintuitively, at least for a fraction of the molecules, in assuming orientations where the molecules "stand on their heads." For example, the *A. australis* hemocyanin is rarely seen in the top view that is characteristic for the molecule in negatively stained preparations, even though the corresponding orientation of the molecule in the water layer leads to maximum contact with the air–water interface. For those molecules, specific interactions with localized surface charges apparently far outweigh other, nonspecific interactions. Such a behavior was documented by Grigorieff (1998) in the case of ubiquinone, an elbow-shaped molecule with a pronounced preference in orientation for two 180° related views (figure 3.5). Oddly, the preferred orientations do not seem to produce maximal contact areas with the air–water interface that would be expected. Another curious phenomenon is the occasional strong asymmetry in the occurrence of "flip" versus "flop" facing molecules. For molecules suspended in a thin water layer, the two boundaries should be exactly equivalent. This phenomenon was first reported by Dubochet et al. (1988), who observed that the "group of nine" fragment of the adenovirus exhibits only a single, left-handed view in the ice preparation.

In experiments with the random-conical reconstruction method (chapter 5, section 5), which relies on the occurrence of particles in some stable views, it is useful to have some control over the orientational preference of the molecule investigated. The inclusion of a thin carbon film in the specimen preparation is one of the options that can be tried. Apparently, the background structure added by the carbon does not appreciably interfere with the alignment (section 3) and classification (chapter 4, section 4), although it inevitably adds some noise. Paradoxically, in the case of the *E. coli* 70S ribosome, inclusion of the carbon film, initially intended to produce stronger orientational preferences for the purpose of the random-conical data collection, had the opposite effect—it was found to make the angular coverage more uniform when compared with pure ice (R. Grassucci, personal communication, 1991).

How accurately are the preferred orientations defined? Given the many factors that affect the molecule's orientation, it is clear that any "defined" orientation in reality possesses some range of uncertainty. There are several reasons for this: the surface of contact between molecule and support is rough and nonplanar on the scale of molecular dimensions (Butt et al., 1991; Glaeser, 1992a,b). Also, as studies in the 1970s indicated (Kellenberger et al., 1982), carbon films below a certain thickness appear to be malleable, forming a curved "seat" for the molecule. In addition, carbon films used in a cryo-environment may warp because of the difference in thermal expansion between carbon and copper (Booy and Pawley, 1992). As a remedy, molybdenum grids with matching expansion coefficients were introduced (Glaeser, 1992b).

Schmutz et al. (1994) invented a very sensitive method of monitoring the flatness of support films, using reflected light microscopy. Investigation of carbon films with this method reveals a high occurrence of wrinkling, even when molybdenum grids are used. However, the maximum angle observed is on the order of 1°, enough to interfere with the attainment of highest resolution from thin 2D crystals, but completely unproblematic for single particles at the present resolutions ($<1/10\,\text{Å}^{-1}$) obtained. Because of these effects, a range of views, rather than a single view, is effectively observed: consequently, the different realizations of a "preferred view" often look as though they come from molecules that are deformed, while they are in fact produced by small changes in the orientation of the molecule.

When the molecule assumes two or more stable positions, related by a rotation around an axis lying parallel to the support plane, the molecule is said to *rock*. For instance, the pattern of variability in the staining of the four building blocks of the *L. polyphemus* hemocyanin half-molecule (and complete molecules from *A. australis*), discovered in the first application of correspondence analysis in EM (van Heel and Frank, 1981; chapter 4, sections 2 and 3), could be explained by the noncoplanar arrangement of the four building blocks.

However, even without rocking effects, the variations in orientation around a preferred view can be pronounced. Only recently, refinement techniques have revealed the extent of these angular deviations from the average view (see chapter 5, section 7.5).

2. Digitization and Selection of Particles

2.1. Hardware for Digitization

2.1.1. The Microdensitometer

The digital microdensitometer samples the optical density of the photographic film on a regular grid and converts the samples into an array of digital numbers. In a flatbed microdensitometer, the film is mounted on a glass plate that moves on a flatbed in the x- and y-directions by the action of stepping motors. With the aid of a short-focal lens, light is focused on the emulsion into a spot whose diameter is matched to the sampling step. Below the emulsion, another lens picks up the transmitted light and conducts it to a photomultiplier, whose analog signals are converted into digital form. Rotating-drum instruments use the same optical arrangement, but here the film is mounted on a cylindrical transparent drum that rotates (producing sampling in the x-direction) while the optical sampling assembly advances in a direction parallel to the cylinder axis (producing sampling in the y-direction).

As an example, a typical choice of sampling increment would be 14 μm, which leads to a pixel size on the object scale of $14/50,000 \times 10,000$ Å, or 2.8 Å, for a magnification of 50,000×. Due to the low contrast of low-dose cryo-EM images, the size of the dynamic range and the availability of a gain adjustment are main considerations in choosing a scanner. In other words, one needs to be sure that the small optical density (OD) range (often ~0.5 OD) is represented by a sufficient number of digital steps. On the other hand, wide-range geometric accuracy, desirable in scanning micrographs of 2D crystals, is obviously relatively unimportant for single-particle reconstruction, where only the accuracy in the measurement of distances on the order of the particle diameter counts.

The measured OD is within the interesting range (up to 1.8 OD for some films) linear with the electron exposure (or number of electrons per unit area) (see Reimer, 1997). This means that the measurements recorded by the microdensitometer can be directly interpreted in terms of electron intensity.

2.1.2. Slow-Scan CCD Camera

Charge-coupled device (CCD) cameras excel with their high dynamic range and exceptional linearity in their response to electrons (Krivanek and Mooney, 1993; de Ruijter, 1995). Unfortunately, the size of the arrays currently commercially offered (up to 4k × 4k) is still not quite in the range that would make CCD recording comparable with photographic recording in terms of total information capacity. A simple calculation shows that the "break-even" point in performance lies even beyond array sizes of 8k × 8k.

First experiences with entirely digital recording, with a 1k × 1k CCD camera, were reported by Stewart et al. (2000). Subsequent pioneering work by Carragher (2000) and Potter et al. (1999) led to the development of fully automated, multiresolution data collection schemes (see also Faruqi and Subramaniam, 2000, and the earlier work by Oostergetel et al., 1998). These schemes rely on a defined arrangement of the holes in recently developed, commercially available perforated grids (Ermantraut et al., 1998). Suitable areas of the grid are "recognized" by the computer at successively increasing levels of magnification, based on statistical measurements such as average and variance, in attempts to emulate the performance of a human operator. Separate optimization of the automated performance is required for different types of specimens; for example, globular single particles, fibers, or patches of 2D crystals (Carragher et al., 2000; Mouche et al., 2003).

2.2. The Sampling Theorem

Image processing rests on the ability to represent the essential information in a continuous image in the form of a finite array without loss. According to the

Whittaker–Shannon sampling theorem (Shannon, 1949), a band-limited continuous function can be represented by a set of discrete measurements ("the samples") taken at regular intervals. It can be shown that the original function can then be completely reconstructed from its samples (see the conceptually simple explanation in Moody, 1990). An image, regarded as a 2D function, requires sampling with a step size of $1/(2B)$ if B is the band limit (i.e., the spatial frequency radius out to which significant information is found) in Fourier space. In the best cases, the resolution[3] of biological specimens in single particle form reaches $1/8 \text{Å}^{-1}$, and so a sampling step equivalent to 4Å on the object scale would already satisfy the stipulation of the sampling theorem. However, such "critical sampling" makes no allowance for the subsequent deterioration of resolution, which is inevitable when the digitized image has to be rotated and shifted, each step involving interpolation and resampling onto a different 2D grid.

Interpolation frequently amounts to the computation of a weighted average over the pixels neighboring the point for which the interpolated value is needed. For example, the four-point bilinear interpolation (e.g., Aebi et al., 1973; Smith and Aebi, 1973) requires the computation of

$$v = a(1 - x)(1 - y) + bx(1 - y) + c(1 - x)y + dxy \tag{3.6}$$

where v is the interpolated value; a, b, c, and d are the values of the pixels surrounding the new point; and (x, y) are the fractional values of its coordinates, with $0 < x < 1$ and $0 < y < 1$. It is clear that the image consisting of interpolated points has a resolution that is approximately only half the original resolution. [Another bilinear interpolation scheme, introduced by Smith (1981) involves weighted averaging of three surrounding points, creating a similar deterioration of resolution.]

In order to keep the resolution loss due to image processing small, it is customary to use a sampling step that is smaller by a factor of 2–3 than required according to the sampling theorem. For example, if a resolution of 8Å were required, then the sampling theorem would require a step size corresponding to

---

[3]Resolution is a quantity in Fourier space and hence has dimension 1/length. This Fourier-based resolution can be linked (see appendix in Radermacher, 1988) to a "point-to-point" distance in real space by the Rayleigh criterion (see Born and Wolf, 1975). Rayleigh considered the diffraction-limited images of two points, which are *Airy disks*, each represented by the intensity distribution given by $[J_1(2\pi rR)/2\pi rR]^2$, where J_1 is the first-order Bessel function, r is the radius, and R is the radius of the diffracting aperture. According to this criterion, two points separated by a distance d_0 are just resolved when the maximum of one Airy disk coincides with the minimum of the second Airy disk. On theoretical analysis, this critical distance d_0 turns out to be $d_0 = 0.6/R$. If we interpret R as the radius of the circular domain within which Fourier terms contribute to the crystallographic Fourier synthesis ("crystallographic resolution"), then we can say that Rayleigh's point-to-point resolution, $1/d_0$, is 1.67 times the crystallographic resolution.

Colloquially, and somewhat confusingly, the real-space quantity 1/resolution is also often referred to as "resolution." Just to eliminate this confusion, we will use the term "resolution distance" when referring to the quantity 1/resolution. Hence, if we compare the distance d_0 and $1/R$, we arrive at the factor 0.6 (Radermacher, 1988), and can state the relationship as follows: the point-to-point resolution distance according to Rayleigh is 0.6 times the inverse of the crystallographic resolution.

$4\,\mathring{A}$, but one would actually scan the micrograph with a step size corresponding to $2\,\mathring{A}$ or even less.

Another issue to be considered is the problem of aliasing. Above we have assumed that the sampling step is smaller or equal to $(1/B)$. When it is chosen larger for some reason (e.g., to accommodate a certain area within an array of given size), then artifacts will arise: the Fourier components outside of the boundaries of $\{+1/s, -1/s\}$, where s is the sampling step, will be "reflected" at the border of the Fourier transforms and show up "aliased" as low spatial frequency components. These "contaminating" terms cannot be removed after the sampling. In principle, a way to solve this problem is by filtering the image (optically) to the desired bandwidth prior to sampling. More practical is the use of a defocused optical probe in the microdensitometer, which performs a convolution operation (appendix 1) "on the fly."

As far as the effective sampling step (on the scale of the object) is concerned, this is determined by the electron optical magnification. The magnification normally used for low-dose images is in the region of 50,000. According to the rationale for this choice provided by Unwin and Henderson (1975), an increase in magnification above 50,000 (while keeping the dose at the specimen constant) would push the exposure close toward, or even below, the fog level of the photographic film. For magnifications significantly lower than 50,000, on the other hand, the modulation transfer function of the photographic film would cause some significant information to be lost [see previous note on the scale transformation (section 1.1) and the study on the properties of electron microscope photographic films by Downing and Grano (1982)].

In the remainder of this book, we make use of the following notation. Each image in a set of N images $\{p_i(\mathbf{r}); i = 1 \ldots, N\}$ is represented by a set of $J = L \times M$ measurements on a regular grid

$$\{r_{\mathrm{lm}}\} = \{l\Delta x, m\Delta y; \ l = 1 \ldots, L; \ m = 1 \ldots, M\} \tag{3.7}$$

where $\Delta x, \Delta y$ are (usually equal) sampling increments in the x- and y-directions, respectively. We will use the lexicographic notation (introduced by Hunt, 1973) when referring to the image elements, that is, a notation referring to a one-dimensionally indexed array. The lexicographic index is

$$j = (m - 1)L + 1 \tag{3.8}$$

and runs from $j = 1$ to $j = LM \ (=J)$. Thus, the discrete representation of an entire image set is given by the array stack $\{p_{ij}; i = 1 \ldots, N; j = 1 \ldots, J\}$. (The term "stack" is used in all instances where a set of related 2D data is consolidated into a single computer file for computational expediency.)

2.3. Interactive Particle Selection

The digitized micrograph is displayed on the screen on the workstation for selection (also referred to as "windowing" or "boxing") of suitable particles. The monitors of present-day workstations cannot easily accommodate the entire array

($\sim$3000 $\times$ 2000 for a standard 100 $\times$ 75 mm film), and therefore one of several measures must be taken: (i) divide the array into subfields that are sufficiently small; (ii) use a utility that allows particle selection with the entire scanned array scrolling through the monitor window; or (iii) reduce the micrograph by sub-sampling to fit the screen. The first option is inconvenient as it leads to logistic and administrative problems. The second option is not very convenient either, as only a fraction of the entire micrograph is seen at a given time, and a method for systematically accessing the whole field must still be used. The third option, using a size reduction, is the most straightforward solution to this problem, although, when the factor becomes too large, the loss in image quality can impair the ability to recognize particular views or eliminate damaged particles.

Size reduction is accomplished by first using "box-convolution," that is, a local averaging operator that replaces the value of each pixel by the average of pixels within a box surrounding it, followed by sampling of the resulting smoothed image. (Instead of the box-convolution, any other low-pass filtering operation can be used, as well.) The ensuing boost in SNR is a consequence of the fact that box convolution performs a low-pass filtration and thereby eliminates the part of the micrograph's Fourier transform, at high spatial frequencies, that has normally a low SNR (see section 4.3). This is the reason for the seemingly paradoxical fact that moderate size reduction (by a factor of 2$\times$ to 3$\times$) of a digitized micrograph improves the apparent contrast of particles.

Contrast and visibility of particles, and our ability to make decisions on the basis of visual judgments, are strongly affected by the electron microscopic defocus. Low values of underfocus, as has become clear from the description of image formation in chapter 2, result in diminished overall contrast of the particle against its background, since a crucial portion of the low-spatial frequency information is suppressed. While box convolution (or low-pass Fourier filtration, with a judicious choice of filter radius) does not restore this part of the Fourier transform, it nevertheless combats noise in the high spatial frequency range, leading to enhanced visibility of the particle for purpose of its visual selection.

Adiga et al. (2004) looked for ways to suppress the background in a field of particles by making use of adaptive filtering techniques. Such preprocessed images can greatly simplify the interactive selection of particles, but could also lead to effective automated means of particle selection (see next section). For completeness, the program QVIEW (Shah and Stewart, 1998) should be mentioned, which was specially designed to facilitate interactive, manual selection of particles.

2.4. Automated Particle Selection

Interactive selection is subjective, tedious, and time-consuming. Inevitably, the push toward higher resolution in 3D reconstruction of single macromolecules (see chapter 5, section 8) necessitates the use of larger and larger numbers of particles. Currently, particle numbers in the range 10,000–30,000 are routinely encountered. In one instance (Spahn et al., 2005), some 130,000 particles were used. Numbers well in excess of 100,000 may have to be handled routinely in the future, as part of the progression of single-particle reconstruction toward atomic resolution. It is clear that such massive data collection will require automation and the use of quantitative reproducible criteria for selecting particles. In the past, several attempts have been made to automate particle selection, but at that time the computers were too slow and contained insufficient memory to make any of these approaches practical.

The recent surge in the applications of single-particle reconstructions is reflected in many new attempts to accelerate and automate particle selection. Further indications for this trend are the organization of a separate workshop (The Scripps Institute, 2003) and the appearance of a special journal issue dedicated to this subject (see appendix 5).

Earlier attempts (Lutsch et al., 1977; van Heel, 1982) used a technique based on the computation of the local variance in the neighborhood of each pixel. In the resulting "variance map" (not to be confused with the variance map that is a byproduct of averaging of an aligned image set: cf. section 4.2), particles show up as peaks of high local variance. Frank and Wagenknecht (1984) developed a method based on a cross-correlation search with an azimuthally averaged reference image that shows the molecule in a selected orientation (figure 3.6). The azimuthal averaging ensures that particles presented in any (in-plane) orientation produce the same value of cross-correlation. Moreover, the correlation function in the vicinity of a particle strongly resembles the particle's autocorrelation function and can, therefore, be used immediately for the translation-invariant rotation search. A shortcoming of both methods is that they fail to discriminate between true particles and any other mass (e.g., aggregates of stain).

Andrews et al. (1986) developed a procedure for detecting particles in a dark-field micrograph, which incorporates low-pass filtering, application of a threshold, edge detection, and determination of mass enclosed by the particle boundaries found. Indeed, molecules imaged in dark field show up with much higher contrast than in bright field, allowing them to be readily separated from the background by the application of a threshold. The procedure of Andrews and coworkers is, however, problematic for the majority of applications, where bright-field EM is used.

The advent of more powerful computers has made it possible to explore algorithms with more extensive statistical analysis. Harauz and Fong-Lochovsky (1989) proposed a sophisticated scheme comprising (i) noise suppression and edge detection, (ii) component labeling (i.e., identifying pixels connected with one another) and feature computation, and (iii) symbolic object manipulation. Lata et al. (1994) proposed a method based on standard methods of discriminant analysis (figure 3.7). Particle candidates are first identified by performing a peak search on a low-pass filtered version of the micrograph. Subsequently, the data surrounding the peaks are extracted and several statistical measures, such as variance, skewness, kurtosis, and entropy, are computed. In a training session, approximately 100 fields are visually categorized into the three groups: "good particles," "junk," and "noise." On the basis of this reference information, a discriminant analysis is then performed on any data with the same statistical characteristics, resulting in a classification of all putative particles. For instance, a typical reconstruction project might require selection of particles from 20

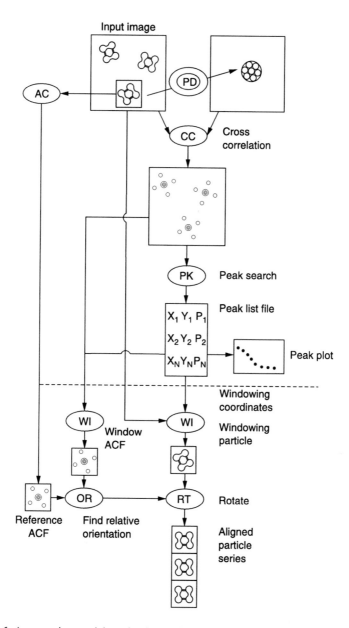

Figure 3.6 Automatic particle selection scheme of Frank and Wagenknecht (1984). A rotationally averaged version of the particle is created, and cross-correlated with the full image field. Each repeat of the particle is marked by the appearance of a version of the particle's autocorrelation function (ACF) within the cross-correlation function. The program uses the ACF subsequently to obtain a version of the particle that is rotated and correctly aligned. From Frank and Wagenknecht (1984), reproduced with permission of Elsevier.

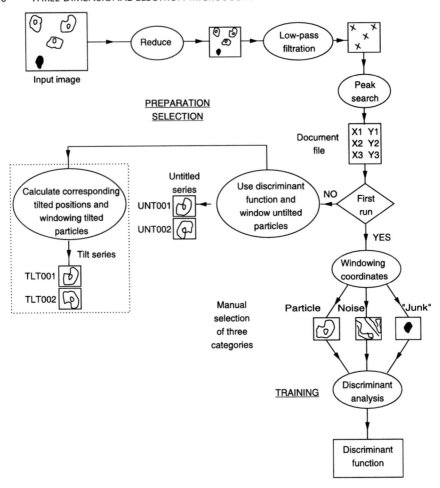

Figure 3.7 Automatic particle selection based on discriminant analysis of texture-sensitive parameters. See text for explanation. From Lata et al. (1995), reproduced with permission of Elsevier.

micrographs. We assume that these images all have the same statistical properties (i.e., originating from the same specimen, with the use of the same grid preparation method, and imaged with the same magnification, voltage, and defocus). In that case, the discriminant function for the entire data set can be set up using a single micrograph for training.

More recent attempts to obtain a fully automated system (see also the review by Nicholson and Glaeser, 2001) include works by Stoschek and Hegerl (1997a), Ogura and Sato (2001), Rath et al. (2003), Roseman (2003), and Adiga et al. (2004). Stoschek and Hegerl (1997a) enhanced the discrimination of the cross-correlation function by the use of a synthetic discriminant function combined with the MINACE (minimum noise and correlation plane energy) principle. To provide for invariance under rotation, the template is expanded into a series of circular harmonics. Ogura and Sato (2001) described the design and apparently successful performance of a neural network trained to pick particles. Roseman (2003) as well as Rath et al. (2003) found that the cross-correlation function (CCF) with local normalization gives excellent results. Computation of the local variance, required for the normalization of the CCF, can be done by fast Fourier techniques following Roseman's (2003) proposal. Adiga et al. (2004) use the principle of preprocessing the raw micrographs such that standard segmentation methods can be applied. For preprocessing, they employed a fast anisotropic nonlinear diffusion method. Zhu et al. (2004) compared 10 recently proposed particle-picking methods using a common data set (micrographs of GroEL) in a "bakeoff."

The convenience provided by techniques of automated particle selection such as the ones discussed above is often defeated by the need to adjust "fiddle parameters" or retrain the algorithm every time the data characteristics change. Even when the same specimen is used, and every experimental parameter is kept unchanged, a change in defocus (which is absolutely necessary for the purpose of covering the Fourier transform evenly; see chapter 5, section 9) will change the statistical characteristics of the data in a major way. For this reason, to date, the most widespread method of particle selection is still a combination of an auto-mated procedure—using a lenient setting of parameters so as to pass everything in the right size range as a molecule—with a visual check of the resulting gallery of particle candidates.

3. Alignment Methods

3.1. Quantitative Definitions of Alignment

Alignment is initially understood as an operation that is performed on two or more images with the aim of bringing a common motif contained in those images into register. Implicit in the term "common motif" is the concept of homogeneity of the image set: the images are deemed "essentially" the same; they differ only in the noise component, and perhaps the presence or absence of a relatively small ligand, a small change in conformation, or orientation. That the difference be small is an important stipulation; it ensures, in all alignment methods making use of the cross-correlation function (see section 3.3), that the contribution from the correlation of the main component with itself (the "autocorrelation term") is very large compared to the contribution stemming from its correlation with the difference structure. Alignment so understood is directly related to our visual concept of likeness; realizing it in the computer amounts to the challenge of making the computer perform as well as a 3-year-old child in arranging building blocks of identical shapes into a common orientation.

The introduction of dissimilar images, occurring in a heterogeneous image set, forces us to generalize the term alignment: according to this more expanded meaning, even dissimilar motifs occurring in those images are considered "aligned" when they are positioned such as to minimize a given functional. The generalized Euclidean distance (i.e., the variance of the difference image) is often used as the functional. With such a quantitative definition, the precise relative position (relating to both shift and orientation) between the motifs after digital "alignment" may or may not agree with our visual assessment, which relies on the perception of edges and marks in both images rather than a digital pixel-by-pixel comparison.

The concept of homogeneous versus heterogeneous image sets is fundamental in understanding averaging methods, their limitations and the ways these limitations can be overcome. These topics will form the bulk of the remainder of this chapter and chapter 4.

3.2. Homogeneous Versus Heterogeneous Image Sets

3.2.1. Alignment of a Homogeneous Image Set

Assume that we have a micrograph that shows N "copies" of a molecule in the same view. With the use of an interactive selection program, these molecule images are separately extracted, normally within a square "window," and stored in a stack of arrays:

$$\{p_{ij}, \ i = 1 \ldots, N; \ j = 1 \ldots, J\} \tag{3.9}$$

Within the selection window, the molecule is roughly centered, and it has normally random azimuthal "in-plane" orientations. We then seek coordinate transformations $\mathbf{T}_i$ such that their application to p_{ij} results in the precise super-imposition of all realizations of the molecule view. In such a superimposition, any pixel indexed j in the transformed arrays $p'_{ij} = \mathbf{T}_i p_{ij}$ refers to the same point in the molecule projection's coordinate system. When this goal is achieved, it is possible to form a meaningful average[4] (figures 3.8 and 3.3a,c).

$$\bar{p}_j = \frac{1}{N} \sum_{i=1}^{N} p'_{ij} \tag{3.10}$$

In contrast, the average would be meaningless if the different pixels with the same index j originated from different points of the coordinate system of the molecule projection, resulting from a failure of alignment, or from the molecules having different conformation (as the proverbial "apples and oranges"), resulting from structural or orientational heterogeneity.

If the deviations are small, however, then the resulting average will at least be similar to the ideal average that would be obtained without alignment error: the former will be a slightly blurred version of the latter. For random translational

---

[4]Throughout this book, ensemble averages are denoted by a bar over the symbol that denotes the observed quantity; for example, $\bar{p}_j$ denotes the average over multiple measurements of the pixel j. In contrast, averages of a function over its argument range will be denoted by angle brackets, as in the following example: $\langle p \rangle = \frac{1}{J} \sum_{j=1}^{J} p_j$.

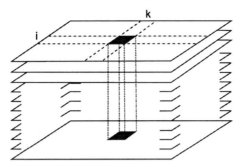

Figure 3.8 Definition of the average image. We imagine the images to be stacked up. For each pixel (indexed i, k), the column average is computed and stored in the element (i, k) of the resulting average image. At the same time, the variance of the pixel is computed and stored in the element (i, k) of the variance map. From Frank (1984b), reproduced with permission of the Electron Microscopy Foundation.

deviations, the blurring can be described in Fourier space by a Gaussian function, analog to the temperature factor of X-ray crystallography:

$$\mathfrak{F}\{\hat{p}\} = \mathfrak{F}\{\bar{p}\} \exp[-k^2/k_0^2] \qquad (3.11)$$

where $\mathfrak{F}\{\cdot\}$ stands for the Fourier transform of the term within the bracket, k is the spatial frequency, and $k_0^2 = 1/r_0^2$ is a "temperature" parameter (the Debye–Waller factor of X-ray crystallography) due to random translations characterized by the size of the r.m.s. (root mean square) deviation r_0 (see also chapter 5, section 9.3, where this subject is discussed in greater detail).

3.2.2 Alignment of a Heterogeneous Image Set

In the case of a heterogeneous image set, such as a set comprising molecules presenting different views, alignment does not have the clear and unambiguous meaning as before. Rather, it must be defined in an operational way: as a procedure that establishes a defined geometrical relationship among a set of images by minimizing a certain functional. A well-behaved alignment algorithm will have the effect that particles within a homogeneous subset are "aligned" in the same sense as defined above for homogeneous sets, while particles belonging to different subsets are brought into geometrical relationships that are consistent with one another. To put it more concretely, in the alignment of 50S ribosomal subunits falling into two views, the crown view and the kidney view, all particles in the crown view orientation will be oriented consistently, and the same will be true for all particles in the kidney view orientation. At the same time, the orientation between any of the crown-view and any of the kidney-view particles will be fixed, but the exact size of the cross-group angle will depend on the choice of the alignment algorithm. Although the size of this relative, fixed angle is irrelevant, a fixed spatial relationship is required for an objective, reproducible, and meaningful characterization of the image set by multivariate data analysis and classification. Exceptions are those methods that produce an alignment implicitly ("alignment through classification" of Dube et al., 1993; Marabini and Carazo, 1994a) or make use of invariants that lend themselves to classification (Schatz and van Heel, 1990, 1992; Schatz, 1992).

Heterogeneity may exist because molecules principally identical in structure may still be different in shape because of the different extents to which a structural component is flexed. Particles or fibers that are thin and extended may bend or deform without changing their local structure, often the true object of the study. From the point of view of studying the high-resolution structure, the diversity of overall shape may be seen as a mere obstacle and not in itself worthy of attention. Bacterial flagella present a good example of this situation. In those cases, a different approach to alignment and averaging may be possible, one in which the idealized overall particle shape is first restored by computational means. Structural homogeneity can thus be restored. The general approach to this problem is the use of curvilinear coordinate transformations. Such "unbending" methods have been introduced to straighten fibers and other linear structures in preparation for processing methods, thus far covered here, that assume rigid body behavior in all rotations and translations. The group of Alasdair Steven at the National Institutes of Health has used these methods extensively in their studies of fibrinous structures (Steven et al., 1986, 1988, 1991; Fraser et al., 1990). Geometrical unbending enables the use of helical reconstruction methods on structures whose shapes do not conform to the architecture of the ideal helix (Egelman, 1986; Steven et al., 1986; Hutchinson et al., 1990).

Yet a different concept of alignment comes to bear when an attempt is made to orient different projections with respect to one another and to a common 3D frame of reference; see section 3 in chapter 5. We will refer to this problem as the *problem of 3D projection alignment*. It is equivalent to the search for the common phase origin in the 3D reconstruction of 2D crystal sheets (Amos et al., 1982; Stewart, 1988a). An even closer analogy can be found in the common-lines methods used in the processing of images of spherical viruses (Crowther et al., 1970; Cheng et al., 1994).

3.3. Translational and Rotational Cross-Correlation

3.3.1. The Cross-Correlation Function Based on the Generalized Euclidean Distance

The cross-correlation function is the most important tool for alignment of two images. It can be derived in the following way: we seek, among all relative positions of the images (produced by rotating and translating one image with respect to the other), the one that maximizes a measure of similarity. The two images, represented by J discrete measurements on a regular grid, $\{f_1(\mathbf{r}_j),$ $j = 1 \ldots, J\}$, $\{f_2(\mathbf{r}_j), j = 1 \ldots, J\}$, may be interpreted as vectors in a J-dimensional Cartesian coordinate system (see also chapter 4, where extensive use will be made of this concept). The length of the difference vector, or the generalized Euclidean distance between the vector end points, is often used as a measure of their dissimilarity or as an inverse measure of their similarity. By introducing search parameters for the rotation and translation, $(\mathbf{R}_\alpha, \mathbf{r}')$, we obtain the expression to be minimized:

$$E_{12}^2(\mathbf{R}_\alpha, \mathbf{r}') = \sum_{j=1}^{J} [f_1(\mathbf{r}_j) - f_2(\mathbf{R}_\alpha \mathbf{r}_j + \mathbf{r}')]^2 \qquad (3.12)$$

The rotation matrix $\mathbf{R}_\alpha$ denotes a rotation of the function $f_2(\mathbf{r})$ by the angle α, while the vector $\mathbf{r}'$ denotes a shift of the rotated function. In comparing two images represented by the functions $f_1(\mathbf{r})$ and $f_2(\mathbf{r})$, we are interested in finding out whether similarity exists for *any* combination of the search parameters. This kind of comparison is similar to the comparison our eyes perform—almost instantaneously—when judging whether or not shapes presented in arbitrary orientations are identical. By writing out the expression (3.12) explicitly, we obtain

$$E_{12}^2(\mathbf{R}_\alpha, \mathbf{r}') = \sum_{j=1}^{J} [f_1(\mathbf{r}_j)]^2 + \sum_{j=1}^{J} [f_2(\mathbf{R}_\alpha \mathbf{r}_j + \mathbf{r}')]^2 - 2 \sum_{j=1}^{J} f_1(\mathbf{r}_j) f_2(\mathbf{R}_\alpha \mathbf{r}_j + \mathbf{r}') \qquad (3.13)$$

The first two terms are invariant under the coordinate transformation:

$$\mathbf{r}_j \rightarrow \mathbf{R}_\alpha \mathbf{r}_j + \mathbf{r}_k \qquad (3.14)$$

The third term is maximized, as a function of the search parameters $(\mathbf{R}_\alpha, \mathbf{r}_k)$, when the generalized Euclidean distance E_{12} assumes its minimum. This third term is called the *cross-correlation function*:

$$\Phi_{12}(\mathbf{R}_\alpha, \mathbf{r}_k) = \sum_{j=1}^{J} f_1(\mathbf{r}_j) f_2(\mathbf{R}_\alpha \mathbf{r}_j + \mathbf{r}_k) \qquad (3.15)$$

In practical application, the use of equation (3.15) is somewhat clumsy because determination of its maximum requires a 3D search (i.e., over the ranges of one rotational and two translational parameters). Functions that separately explore angular and translational space have become more important. Two additional impractical features of equation (3.15) are that (i) $\Phi_{12}(\mathbf{R}_\alpha, \mathbf{r}_k)$ is not normalized, so that it is not suitable for comparing images that originate from different experiments, and (ii) it is dependent on the size of the "bias" terms $\langle f_1 \rangle$ and $\langle f_2 \rangle$, that is, the averaged pixel values, as well as the size of the variances, which should all be irrelevant in a meaningful measure of similarity. These shortcomings are overcome by the introduction of the translational cross-correlation function and a suitable normalization, detailed in the following.

For later reference, we also introduce the *autocorrelation function* (ACF), which is simply the cross-correlation function of a function $f_1(\mathbf{r}_j)$ with itself (see appendix 1).

3.3.2. Cross-Correlation Coefficient and Translational Cross-Correlation Function

3.3.2.1. Definition of the cross-correlation coefficient The cross-correlation coefficient is a well-known measure of similarity and statistical interdependence. For two functions represented by discrete samples, $\{f_1(\mathbf{r}_j); j = 1 \ldots, J\}$ and

$\{f_2(\mathbf{r}_j); j = 1 \dots, J\}$, the cross-correlation coefficient is defined as

$$\rho_{12} = \frac{\sum_{j=1}^{J}[f_1(\mathbf{r}_j) - \langle f_1 \rangle][f_2(\mathbf{r}_j) - \langle f_2 \rangle]}{\left\{\sum_{j=1}^{J}[f_1(\mathbf{r}_j) - \langle f_1 \rangle]^2 \sum_{j=1}^{J}[f_2(\mathbf{r}_j) - \langle f_2 \rangle]^2\right\}^{1/2}} \tag{3.16}$$

where

$$\langle f_i \rangle = 1/J \sum_{j=1}^{J} f_i(\mathbf{r}_j); \quad i = 1, 2 \tag{3.17}$$

Note that $-1 \le \rho_{12} \le 1$. A high value of ρ_{12} means that the two functions are very similar for a particular choice of relative shift and orientation. However, in comparing images, it is more relevant to ask whether ρ_{12} is maximized for *any* "rigid body" coordinate transformation applied to one of the images. In the defining equation (3.16), only the numerator is sensitive to the quality of the alignment between the two functions, while the denominator contains only the variances, which are invariant under shift or rotation of the image. When a variable coordinate transformation $\mathbf{r}_j \rightarrow \mathbf{R}_\alpha \mathbf{r}_j + \mathbf{r}_k$ is applied to one of the functions, the numerator becomes identical (except for the subtraction of the averages $\langle f_1 \rangle$ and $\langle f_2 \rangle$, which only changes the constant "bias" terms) with the cross-correlation function introduced in the previous section.

3.3.2.2. Translational cross-correlation function Specifically, the translational CCF is obtained by restricting the coordinate transformation to a 2D translation $\mathbf{r}_j \rightarrow \mathbf{r}_j + \mathbf{r}_k$. However, for any $\mathbf{r}_k \ne 0$, the images no longer overlap completely, so that the summation can now be carried out only over a subset J' of the J pixels. This is indicated by the stipulation $\mathbf{r}_j + \mathbf{r}_k \in A$ under the summation symbol, meaning that only those indices j should be used, for which the sum vector $\mathbf{r}_j + \mathbf{r}_k$ is still within the boundary of the image:

$$\Phi_{12}(\mathbf{r}_k) = \frac{1}{J'} \sum_{j=1[r_j+r_k \in A]}^{J} [f_1(\mathbf{r}_j + \mathbf{r}_k) - \langle f_1 \rangle][f_1(\mathbf{r}_j + \mathbf{r}_k) - \langle f_1 \rangle] \tag{3.18}$$

By letting the "probing vector" $\mathbf{r}_k$ assume all possible positions on the grid, all relative displacements of the two functions are explored, allowing the position of best match to be found: for that particular $\mathbf{r}_k$, the CCF assumes its highest value (figure 3.9).

The formulation of the overlap contingency is somewhat awkward and indirect in the lexicographic notation, as used in equation (3.18). However, it is easy to write down a version of this formula that uses separate indexing for the x- and y-coordinates. In such a revised formula, the index boundaries become explicit. Because the practical computation of the CCF is normally done by fast

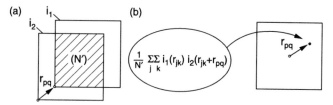

Figure 3.9 Definition of the cross-correlation function (CCF). For the computation of the CCF for a single argument $\mathbf{r}_{pq}$, (a) is shifted with respect to (b) by vector $\mathbf{r}_{pq}$. In this shifted position, the scalar product of the two images arrays is formed and put into the CCF matrix at position (p, q). For the computation of the entire function, the vector $\mathbf{r}_{pq}$ is allowed to assume all possible positions on the sampling grid. In the end, the CCF matrix has an entry in each position. From Frank (1980), reproduced with permission of Springer Verlag.

Fourier transformations (see below), we will refrain from restating the formula here. However, an issue related to the overlap will resurface later on in the context of fitting and docking in three dimensions by "globally" versus "locally" normalized cross-correlation (see chapter 6, section 6.2.2).

3.3.2.3. Computation using the fast Fourier transform   In practical application, the computation of $\Phi_{12}(\mathbf{r}_k)$ is speeded up through the use of a version of the convolution theorem (appendix 1). For notational convenience, this is restated here in the continuous form:
   The convolution product of two functions,

$$C_{12}(\mathbf{r}') = \iint f_1(\mathbf{r})f_2(\mathbf{r} - \mathbf{r}')\,d\mathbf{r} \tag{3.19}$$

is equal to the inverse Fourier transform of the product $\mathfrak{F}\{f_1\}\mathfrak{F}\{f_2\}$, *where* $\mathfrak{F}\{\cdot\}$ *denotes the Fourier transform of the function in the bracket.* The definition of the CCF differs from that of the convolution equation (3.19) only in the substitution of a plus sign for the minus in the vector argument, and this is reflected in complex conjugation when forming the product in Fourier space. The resulting recipe for fast computation of the CCF is

$$\Phi_{12}(\mathbf{r}') = \mathfrak{F}^{-1}\{\mathfrak{F}\{f_1\}\mathfrak{F}^*\{f_2\}\} \tag{3.20}$$

where $\mathfrak{F}^{-1}$ denotes the inverse Fourier transform, and $*$ denotes complex conjugation. Thus, the computation involves three discrete fast Fourier transformations and one scalar matrix multiplication.

   The usual choice of origin in the digital Fourier transform (at element [1,1] of the array) requires an adjustment: for the origin to be in a more convenient position, namely in the center of $\Phi_{12}(\mathbf{r}')$, a shift by half the image dimensions in the x- and y-directions is required. For even array dimensions (and thus for any power-of-two based algorithms) this is easily accomplished by multiplication of the Fourier product in equation (3.20) with the factor $(-1)^{(k_x+k_y)}$, where k_x, k_y are the integer grid coordinates in Fourier space.

For convenience, we will sometimes make use of the notation

$$\Phi_{12}(\mathbf{r}') = f_1 \otimes f_2 \tag{3.21}$$

in analogy to the use of

$$C_{12}(\mathbf{r}') = f_1(\mathbf{r}) \circ f_2(\mathbf{r}) \tag{3.22}$$

for the convolution product.

The computation via the Fourier transformation route implies that the 2D image array $p(l, m)$ (switching back for the moment to a 2D argument formulation for clarity) is replaced by a circulant array, with the periodic properties

$$p(l + L, m + M) = p(l, m) \tag{3.23}$$

This has the consequence that the translational CCF computed in this way, $\Phi_{12}(l', m')$ in discrete notation, contains contributions from terms $p_1(l + l' - L, m + m' - M)$ $p_2(l, m)$ when $l + l' > L$ and $m + m' > M$. We speak of a "wrap-around effect": instead of the intended overlap of image areas implied in the definition of the CCF [equation (3.18) and figure 3.9], the bottom of the first image now overlaps the top of the second, etc.

To deal with this problem, it is common practice to extend the images two-fold in both directions, by "padding" them with their respective averages (i.e., filling in the average value into the array elements unoccupied by the image):

$$\langle p_1 \rangle = \frac{1}{J} \sum_{j=1}^{J} p_1(\mathbf{r}_j); \ \langle p_2 \rangle = \frac{1}{J} \sum_{j=1}^{J} p_2(\mathbf{r}_j) \tag{3.24}$$

Alternatively, the images may be "floated," by subtraction of $\langle p_1 \rangle$ and $\langle p_2 \rangle$, respectively, then "padded" with zeros prior to the fast Fourier transform calculation, as described by DeRosier and Moore (1970). The result is the same as with the above recipe, because any additive terms have no influence on the CCF, as defined in equation (3.18). In fact, programs calculating the CCF frequently eliminate the Fourier term F_{00} at the origin in the course of the computation, rendering the outcome insensitive to the choice of padding method.

Another difference between the CCF computed by the Fourier method and the CCF computed following its real-space definition, equation 3.18, is that the normalization by $1/J'$ (J' being the varying number of terms contributing to the sum) is now replaced by $1/J$ [or, respectively, by $1/(4J)$ if twofold extension in both dimensions by padding is used].

3.3.3. Rotational Cross-Correlation Function

The rotational CCF (analogous to the *rotation function* in X-ray crystallography) is defined in a similar way as the translational CCF, but this time with a rotation as probing coordinate transformation. Here, each function is represented by samples on a polar coordinate grid defined by Δr, the radial increment, and $\Delta \phi$, the azimuthal increment:

$$\{f_i(l\Delta r, m\Delta\phi); \ l = 1\ldots, L; \ m = 1\ldots, M\}; \quad i = 1, 2 \qquad (3.25)$$

We define the discrete, weighted, rotational CCF in the following way:

$$C(k) = \sum_{l=l_1}^{l_2} w(l) \sum_{m=0}^{M} f_1(l\Delta r, \ \mathrm{mod} \ [m + k, M]\Delta\phi) f_2(l\Delta r, m\Delta\phi)\Delta\phi l\Delta r$$

$$\qquad (3.26)$$

$$= \sum_{l=l_1}^{l_2} w(l)c(l, k)l\Delta r$$

If weights $w(l) \equiv 1$ are used, the standard definition of the rotational CCF is obtained. The choice of nonuniform weights, along with the choice of a range of radii $\{l_1 \ldots l_2\}$, is sometimes useful as a means to place particular emphasis on the contribution of certain features within that range. For example, if the molecule has a peripheral protuberance that is flexible, we might want to prevent it from contributing to the rotational correlation signal.

The computation of the 1D inner sums $c(l, k)$ along rings normally takes advantage (Saxton, 1978; Frank et al., 1978a, 1986) of the *Fourier convolution theorem*, here in discrete form:

$$c(l, k) = \sum_{m'=0}^{M-1} F_1(l\Delta r, m'\Delta\phi)F_2(l\Delta r, m'\Delta\phi)\Delta\phi' \exp[2\pi i m'(\Delta\phi - \Delta\phi')] \qquad (3.27)$$

where $F_i(l\Delta r, m'\Delta\phi)$, $i = 1, 2$ are the discrete Fourier transforms of the lth ring of the functions f_i. A further gain in speed is achieved by reversing the order of the summations over rings and over Fourier terms in equations (3.26) and (3.27), as this reduces the number of inverse 1D Fourier transformations to one (Penczek et al., 1992).

3.3.4. Peak Search

The search for the precise position of a peak is a common feature of all correlation-based alignment techniques. The following rule of thumb has been found effective in safeguarding against detection of spurious peaks in aligning two particle images: not just the highest, but at least the three highest-ranking peaks are determined. Let us assume their respective values are p_1, p_2, and p_3. For a significant peak, one would intuitively expect that the ratio p_1/p_2 is well above p_2/p_3, on the assumption that the subsidiary peaks p_2 and p_3 are due to noise.

Computer programs designed for the purpose of searching peaks are straightforward: the array is scanned for the occurrence of relative peaks, that is, elements that stand out from their immediate neighbors. In a 1D search (e.g., of a rotational cross-correlation function), each element of the array is compared with its two immediate neighbors. In a 2D search (typically of a 2D translational CCF), each element is compared to its eight immediate neighbors. (In a 3D search, to be mentioned for completeness, the number of next neighbors is $8 + 9 + 9 = 26$ as in a 3D tic-tac-toe). Those elements that fulfill this criterion are put on a stack in ranking order. At the end of the scan, the stack will contain the desired list of highest peaks.

The peak position so found is given only as an integer multiple of the original sampling distance. However, the fact that the peak has finite width and originates mathematically from many independent contributions all coherently "focused" on the same spot means that the position can be found with higher accuracy by the use of some type of fitting. First, the putative peak region is defined as a normally circular region around the element with highest value found in the peak search. Elements within that region can now be used to determine an effective peak position with noninteger coordinates. Methods widely used are parabolic fitting and the computation of the center of gravity.

The accuracy in determining the peak position is determined by the SNR (see section 4.3), not by the resolution reflected in the peak width. As the linear transfer theory shows [equation (3.33) in section 3.4.1], the peak shape is determined by the (centrosymmetric) *autocorrelation* function (ACF) of the point spread function. Its precise center can be found by a fitting procedure, provided that the SNR is sufficiently large.

3.4. Reference-Based Alignment Techniques

3.4.1. Principle of Self-Detection

Reference-based alignment techniques were developed primarily for the case of homogeneous image sets, that is, images originating from particles containing identical structures and presenting the same view. In that case, all images of the set $\{p_n(\mathbf{r}); n = 1 \ldots, N\}$ have a "signal component" in common—the projection $p(\mathbf{r})$ of the structure as imaged by the instrument—while differing in the noise component $n_i(\mathbf{r})$. Any of the images can then act as reference for the rest of the image set (principle of *self-detection*; see Frank, 1975a). In formal terms:

$$p_1(\mathbf{r}) = p(\mathbf{r}) + n_1(\mathbf{r}) \tag{3.28}$$

$$p_2(\mathbf{r}) = p(\mathbf{r}) + n_2(\mathbf{r}) \tag{3.29}$$

so that, using the notation for correlation introduced in section 3.3.2,

$$\Phi_{12}(\mathbf{r}') = p_1(\mathbf{r}) \otimes p_2(\mathbf{r}) = p(\mathbf{r}) \otimes p(\mathbf{r}) + n_1(\mathbf{r}) \otimes p(\mathbf{r}) + p(\mathbf{r}) \otimes n_2(\mathbf{r}) + n_1(\mathbf{r}) \otimes n_2(\mathbf{r})$$
$$\tag{3.30}$$

The first term is the ACF of the structure common to both images, which has a sharp peak at the origin, while each of the other three terms is a CCF of two statistically unrelated functions. The shape of the peak at the center is determined solely by the ACF of the point-spread function associated with the contrast transfer function: according to equation (2.11) with $p(\mathbf{r}) = I(\mathbf{r})$ and $p_0(\mathbf{r}) = C_p(\mathbf{r})$ and, by virtue of the convolution theorem, one obtains

$$p(\mathbf{r}) \otimes p(\mathbf{r}) = [h(\mathbf{r}) \circ p_0(\mathbf{r})] \otimes [h(\mathbf{r}) \circ p_0(\mathbf{r})] = [h(\mathbf{r}) \otimes h(\mathbf{r})] \circ [p_0(\mathbf{r}) \otimes p_0(r)] \quad (3.31)$$

The term $[p_0(\mathbf{r}) \otimes p_0(\mathbf{r})]$ is, in the terminology of X-ray crystallography, the *Patterson function* of the projection of the original structure. Its most important feature in equation (3.31) is that it acts, for all practical purposes, as a delta-function, due to the appearance of the sharp peak stemming from exact overlap of the atomic object with itself for zero displacement vector; that is, in the center of the function. Since the convolution of a function with the delta-function simply acts to reproduce the function (see appendix 1), we obtain the result that equation (3.31) is essentially the same as $h(\mathbf{r}) \otimes h(\mathbf{r})$; that is, the ACF of the instrument's point spread function.[5] Its value at the origin, important for the ability to detect the peak at low SNR, is determined by the size of the "energy" integral:

$$E_B = \int_B ||H(\mathbf{k})||^2 d\mathbf{k} \quad (3.32)$$

where B is the resolution domain and $H(\mathbf{k})$ is the contrast transfer function. This follows from Parseval's theorem, which is simply a statement of the invariance of the norm of a function on transforming this function into Fourier space (see section 4.3).

In this context, it should be noted that the alignment of images of the same object taken at different defocus settings leads to a CCF peak whose shape is determined by

$$\Phi_{h_1 h_2} = h_1(\mathbf{r}) \otimes h_2(\mathbf{r}) \quad (3.33)$$

with $h_1(\mathbf{r})$ and $h_2(\mathbf{r})$ being the point-spread functions representing the action of the instrument for the two defocus settings. In this situation, the peak height is determined by the size of the "cross-energy" integral (i.e., its discrete-valued equivalent):

$$E_B = \int_B H_1(\mathbf{k}) H_2(\mathbf{k}) d\mathbf{k} \quad (3.34)$$

which is critically dependent on the relative positions of contrast transfer zones with different polarity (Frank, 1972b; Al-Ali, 1976; Al-Ali and Frank, 1980).

---

[5]This property was previously exploited by Al-Ali and Frank (1980), who proposed the use of the cross-correlation function of two micrographs of the same "stochastic" object to obtain a measure of resolution.

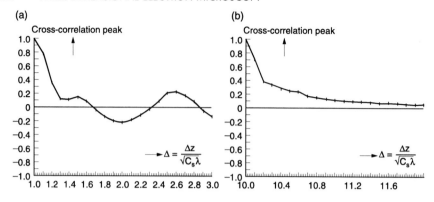

Figure 3.10 Peak value of the cross-correlation function $\Phi_{h_1 h_2}$ between two point spread functions with different defocus, plotted as a function of their defocus difference. Theoretically, the cross-correlation function of two micrographs of the same structure is obtained as a convolution product between $\Phi_{h_1 h_2}$ and the Patterson function of the structure, see equation (3.31). (a) $\Delta\hat{z} = 1$ (Scherzer focus) used as reference; (b) $\Delta\hat{z} = 10$ used a reference. It is seen that in the case of (a), as the defocus difference increases, the peak drops off to zero and changes sign twice. From Zemlin (1989b), reprinted with permission of John Wiley & Sons, Inc.

[As an aside, this integral is Linfoot's measure of *correlation quality* (Linfoot, 1964).] In fact, the value of the integral in equation (3.34) is no longer positive-definite, and unfortunate defocus combinations can result in a CCF with a peak that has inverted polarity or is so flat that it cannot be detected in the peak search as it is drowned by noise (Frank, 1980; Zemlin, 1989b; Saxton, 1994; see figure 3.10).

Saxton (1994) discussed remedies for this situation. One of them is the obvious "flipping" of transfer zones in the case where the transfer functions are known, with the aim of ensuring a positive-definite (or negative-definite, whatever the case may be) integrand (Typke et al., 1992). A more sophisticated procedure suggested by Saxton (1994) is to multiply the transform of the CCF with a factor that acts like a Wiener filter:

$$W(\mathbf{k}) = \frac{H_1(\mathbf{k})H_2(\mathbf{k})}{|H_1(\mathbf{k})|^2 |H_2(\mathbf{k})|^2 + \varepsilon} \tag{3.35}$$

where H_1 and H_2 are the known CTFs of the two images and ε is a small quantity that ensures the boundedness of $W(\mathbf{k})$, as it keeps the noise amplification in any spectral domain within reasonable margins.

Apart from the degradation of the CCF peak due to the mismatch in CTF polarities, which can be fixed by "flipping" the polarity of zones in the Fourier domain, there are other effects that diminish the size of the peak, and thus may lead to difficulties in the use of the CCF in alignment. Typke et al. (1992) identified magnification changes and local distortions of the specimen as sources

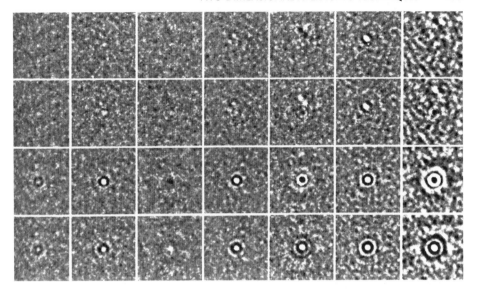

Figure 3.11 The effect of image restoration and magnification correction on the cross-correlation function (CCF) of two micrographs of a defocus series. One of eight images is cross-correlated against the remaining seven images. Only the central 128 × 128 portion of the CCFs is shown. Top row: without correction. Second row: sign reversal applied to CTFs to make the CCF Fourier integral in equation (3.34) positive-definite. Third row: micrographs corrected only for magnification changes and displacements. The CCF peak comes up strongly now, but in three cases with reversed sign. Bottom row: micrographs additionally corrected for sign reversal. All CCF peaks are now positive. From Typke et al. (1992), reproduced with permission of Elsevier.

of those problems and demonstrated that, by applying an appropriate compensation, a strong CCF peak can be restored (see figure 3.11). However, these effects come into play mainly in applications where large specimen fields must be related to one another, while they are negligible in the alignment of small (typically in the range of 64 × 64 to 150 × 150) single-particle arrays. The reason for the insensitivity in the latter case is that all local translational components are automatically accounted for by an extra shift, while the small local rotational components of the distortions (in the range of maximally 1°) affect the CCF Fourier integral underlying the CCF [equation (3.30)] only marginally in the interesting resolution range (see appendix in Frank and Wagenknecht, 1984).

3.4.2. ACF/CCF-Based Search Strategy

3.4.2.1. Rationale In order to avoid a time-consuming 3D search of the parameter space spanned by $\alpha, \Delta x, \Delta y$, translation and rotation searches are normally performed separately. One search strategy that accomplishes this separation, introduced by Langer et al. (1970), makes use of the translation invariance of the ACF (figures 3.12a–d). In fact, this method goes back to search techniques in X-ray crystallography involving the Patterson function. According

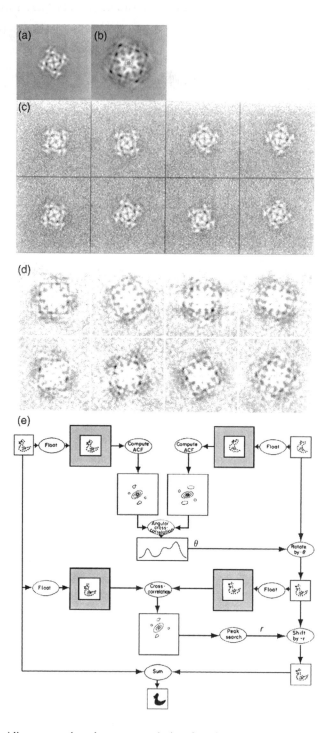

Figure 3.12 Alignment using the autocorrelation function (ACF). (a) A projection of the calcium release channel, padded into a 128×128 field. (b) ACF of (a). (c) Eight noisy realizations of (a), shifted and rotated randomly. (d) ACFs of the images in (c). Two

to this scheme (figure 3.12e), the orientation between the images is first found by determining the orientation between their ACFs, and subsequently the shift between the correctly oriented molecules is found by translational cross-correlation.

The ACF of an image $\{p(\mathbf{r}_j); j = 1 \ldots, J\}$, represented by discrete samples on a regular grid, $\mathbf{r}_j$, is obtained by simply letting

$$f_1(\mathbf{r}_j) \equiv f_2(\mathbf{r}_j) \equiv p(\mathbf{r}_j)$$

in the formula for the translational CCF [equation (3.18)]:

$$\Phi(\mathbf{r}_k) = \frac{1}{J'} \sum_{j=1(\mathbf{r}_j + \mathbf{r}_k \in A)}^{J} p(\mathbf{r}_j)p(\mathbf{r}_j + \mathbf{r}_k) \tag{3.36}$$

The property of shift invariance is immediately clear from the defining formula, since the addition of an arbitrary vector to $\mathbf{r}_j$ will not affect the outcome (except possibly for changes due to the boundary terms). Another property (not as desirable as the shift invariance; see below) is that the two functions $p(\mathbf{r}_j)$ and $p'(\mathbf{r}_j) = p(-\mathbf{r}_j)$ have the same ACF. As a result, the ACF is always centrosymmetric.

For fast computation of the ACF, the *convolution theorem* is again invoked as in the case of the CCF (section 3.3.2): the ACF is obtained by inverse Fourier transformation of the squared Fourier modulus, $|F(\mathbf{k})|^2$, where $F(\mathbf{k}) = \mathfrak{F}^{-1}\{p(r)\}$. Here, the elimination of "wrap-around" artifacts, by twofold extension and padding of the array prior to the computation of the Fourier transform, is

---

important properties of the ACF can readily be recognized: it is always centered at the origin, irrespective of the translational position of the molecule, and it has a characteristic pattern that reflects the (in-plane) orientation of the molecule in (c) from which the ACF was derived. Thus, it is seen that the ACF can be used to determine the rotation of unaligned, uncentered molecules. (e) Scheme of two-step alignment utilizing the translation invariance of the ACF. Both the reference and the image to be aligned (images on the top left and top right, respectively) are "floated," or padded into a larger field having the same average density, to avoid wrap-around artifacts. Next, the ACFs are computed and rotationally cross-correlated. The location of the peak in the rotational CCF establishes the angle θ between the ACFs. From this location it is inferred that the image on the top right has to be rotated by $-\theta$ to bring it into the same orientation as the reference. (Note, however, that this conclusion is correct only if the motif contained in the images to be aligned is centrosymmetric. Otherwise, the angle θ between the ACFs is compatible with the angle being either θ or $\theta + 180°$, which means both positions have to be tried in the following cross-correlation; see text.) Next, the correctly rotated, padded image is translationally cross-correlated with the reference, yielding the CCF. The position $\mathbf{r}$ of the peak in the CCF is found by a 2D peak search. Finally, the rotated version of the original image is shifted by $-\mathbf{r}$ to achieve complete-alignment. From Frank and Goldfarb (1980), reproduced with permission of Springer-Verlag.

particularly important, since these artifacts are peripherally located and may lead to incorrect angles in the rotation search.

We consider an idealized situation, namely two images of the same motif, which differ by a shift, a rotation α, and the addition of noise. The ACFs of these images are identical in appearance, except that they are rotated by α relative to each other. As in section 3.3.2, we now represent the ACFs in a polar coordinate system and determine the relative angle by computing their rotational CCFs:

$$C(k) = \sum_{l=l_1}^{l_2} w(l)\Phi(l\Delta r, m\Delta\phi + \alpha)\Phi(l\Delta r, \Delta\phi(m + k))\Delta\phi l\Delta r \qquad (3.37)$$

The rotational CCF will have a peak centered at the discrete position $k_{max} = \text{int}(\alpha/\Delta\phi)$ [where $\text{int}(\cdot)$ denotes the integer closest to the argument], which can be found by a maximum search over the entire range of the function.

The position may be found more accurately by using a parabolic fit or center-of-gravity determination in the vicinity of k_{max} (see section 3.3.4). The weights $w(l)$, as well as the choice of minimum and maximum radii l_1, l_2, are used to "focus" the comparison on certain ring zones of the autocorrelation function. For instance, if weights are chosen such that a radius $r = r_w$ is favored, then this has the effect that features in the molecule separated by the distance $|\mathbf{r}_1 - \mathbf{r}_2| = r_w$ will provide the strongest contributions in the computation of the rotational CCF. The weights can thus be used, for instance, to select the most stable, reliable (i.e., reproducible) distances occurring in the particle.

3.4.2.2. Ambiguity

One problem of the ACF-based orientation search is the above-mentioned symmetry, which makes the ACFs of the two functions $p_1(\mathbf{r})$ and $p_2(\mathbf{r}) = p_1(-\mathbf{r})$ indistinguishable. Consequently, the fact that the ACFs of two images optimally match for $k\Delta\Phi = \alpha$ may indicate that the images match with the relative orientations (i) α, (ii) $\alpha + 180°$, or (iii) both. (The last case would apply if the images themselves are centrosymmetric.)

Zingsheim et al. (1980) solved this problem by an "up/down cross-correlation test," in the following way (figure 3.12):

(i) Rotate image #2 by α
(ii) Compute the CCF between #1 and #2, which results in a peak maximum ρ_α at the location $\{\Delta x_\alpha, \Delta y_\alpha\}$
(iii) Rotate image #2 by $\alpha + 180°$
(iv) Compute the CCF between #1 and #2, which results in a peak maximum $\rho_{\alpha+180°}$ at $\{\Delta x_{\alpha+180°}, \Delta y_{\alpha+180°}\}$
(v) If $\rho_\alpha > \rho_{\alpha+180°}$ then shift #2 by $\{\Delta x_\alpha, \Delta y_\alpha\}$, else by $\{\Delta x_{\alpha+180°}, \Delta y_{\alpha+180°}\}$; i.e., both ways are tried, and the orientation that gives maximum cross-correlation is used

Later the introduction of multivariate data analysis and classification (see chapter 4) provided another way of distinguishing between particles lying in 180°-rotation related positions. In the multivariate data analysis of such a mixed data set, it is clear that the factor accounting for the "up" versus "down" mass redistribution will predominate, unless the particle view is close to centrosymmetric.

3.4.3. Refinement and Vectorial Addition of Alignment Parameters

Inevitably, the selection of an image for reference produces a bias in the alignment. As a remedy, an iterative refinement scheme is applied (figure 3.13), whereby the average resulting from the first alignment is used as reference in the next alignment pass, etc. (Frank et al., 1981a). The resolution of averages improves in each pass, up to a point of saturation.

However, when multiple steps of rotations and shifts are successively applied to an image, the discrete representation of images poses practical difficulties. As the image is subjected to several steps of refinement, the necessary interpolations successively degrade the resolution. As a solution to this problem, the final image is directly computed from the original image, using a rotation and shift combination that results from vectorial additions of rotations and shifts obtained in each step.

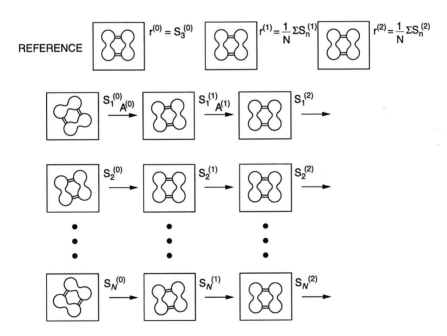

Figure 3.13 Reference-based alignment with iterative refinement. Starting with an arbitrary reference particle (picked for its "typical" appearance), the entire set of particles is aligned in the first pass. The aligned images are averaged, and the resulting average image is used as reference in the second pass of alignment. This procedure is repeated several times, until the shifts and rotation angles of the entire image set stabilize, that is, remain virtually unchanged. Adapted from Frank (1982).

For any pixel with coordinates $\mathbf{r}$ in the original image, we obtain new coordinates in the first alignment step as

$$\mathbf{r}' = \alpha_1 \mathbf{r} + \mathbf{d}_1 \tag{3.38}$$

where α_1 and $\mathbf{d}_1$ are, respectively, the 2×2 rotation matrix and translation vector found in the alignment. In the next step (i.e., the first step of refinement), we have similarly

$$\mathbf{r}'' = \alpha_2 \mathbf{r} + \mathbf{d}_2 \tag{3.39}$$

where α_1 and $\mathbf{d}_2$ are, respectively, the rotation matrix and translation vector expressing the adjustments. Instead of using the image obtained by applying the second transformation, one must apply a consolidated coordinate transformation to the original image. This consolidated transformation is

$$\mathbf{r}'' = \alpha_2(\alpha_1 \mathbf{r} + \mathbf{d}_1) + \mathbf{d}_2 = \alpha_2\alpha_1 \mathbf{r} + \alpha_2\mathbf{d}_1 + \mathbf{d}_2 = \alpha_{res}\mathbf{r} + \mathbf{d}_{res} \tag{3.40}$$

with the resulting single-step rotation $\alpha_{res} = \alpha_2\alpha_1$ and single-step translation $\mathbf{d}_{res} = \mathbf{d}_1 + \mathbf{d}_2$.

3.4.4. Multireference Methods

Multireference methods have been developed for situations in which two or more different motifs (i.e., different views, particle types, conformational states, etc.) are present in the data set (van Heel and Stöffler-Meilicke, 1985). Alignment of a set of N images with L references (i.e., prototypical images showing the different motifs) leads to an array of $L \times N$ correlation coefficients, and each image is put into one of L bins, depending on which of its L correlation coefficients is maximum. In a second round, averages are formed over the subsets so obtained, which are lower-noise realizations of the motifs and can take the place of the initial references. This refinement step produces improved values of the rotational and translational parameters, but also a migration of particles from one bin to another, on account of the now-improved discriminating power of the cross-correlation in "close-call" cases. This procedure may be repeated several times until convergence is achieved, as indicated by the fact that the parameter values stabilize and the particle migration stops. Multireference alignment, by its nature, is intertwined with classification, since the initial choice of references is already based on a presumed inventory of existing views, and the binning of particles according to highest CCF achieved in the alignment is formally a case of supervised classification (chapter 4, section 4.10).

Elaborate versions of this scheme (e.g., Harauz et al., 1988) have incorporated multivariate data analysis and classification in each step. Earlier experience of van Heel and coworkers (Boekema et al., 1986; Boekema and Böttcher, 1992; Dube et al., 1993) indicated that the multireference procedure is not necessarily stable; initial preferences may be amplified, leading to biased results, especially for small particles. The failure of this procedure is a consequence of an intrinsic problem of the reference-based alignment approach. In fact, it can be shown that the multireference alignment algorithm is closely related to the *K*-means clustering technique, sharing all of its drawbacks (P. Penczek, personal communication, 1995).

3.5. Reference-Free Alignment Techniques

3.5.1. Introduction

As we have seen, reference-based alignment methods fail to work for heterogeneous data sets when the SNR (section 4.3) drops below a certain value. The choice of the initial reference can be shown to bias the outcome (see also Penczek et al., 1992). For instance, van Heel et al. (1992a,b) and Grigorieff (2000) demonstrated that, in extreme cases, a replica of the reference emerges when the correlation averaging technique is applied to a set of images containing pure random noise. A similar observation was earlier reported by Radermacher et al. (1986b) for correlation averaging of crystals: when the SNR is reduced to 0 (i.e., only a noise component remains in the image), an average resembling the reference motif is still formed. This phenomenon is easily understood: maximum correlation between a motif and an image field occurs, as a function of translation/rotation, when the image has maximal similarity. If (as often is the case) the correlation averaging procedure employs no correlation threshold, such areas are always found in a noise field. By adding up those areas of maximum similarity, one selectively reinforces all those noise components that optimally replicate the reference pattern. This becomes quite clear from an interpretation of images as points in a multidimensional space (see chapter 4, section 1.4.2).

The emergence of the signal out of "thin air," an inverse Cheshire cat phenomenon, is in some way analogous to the outcome of an experiment in optical filtering: a mask that passes reflections on the reciprocal grid is normally used to selectively enhance Fourier components that build up the structure repeating on the crystal lattice. If one uses such a mask to filter an image that consists of pure noise, the lattice and some of the characteristics of the structure are still generated, albeit with spatially varying distortions because the assignment of phases is left to chance. Application of rigorous statistical tests would of course dismiss the spurious correlation peaks as insignificant. However, the problem with such an approach is that it requires a detailed statistical model (which varies with many factors such as object type, preparation technique, imaging conditions, and electron exposure). Such a model is usually not available.

We will see later on (chapter 5, section 8.3) that cross-validation methods can be used to safeguard against noise building up to form an average (or a 3D reconstruction) that simply reproduces the reference. These methods can also be used to detect model bias, but they do not lend themselves readily to a recipe for how to avoid it.

Attempts to overcome these problems have led to the development of reference-free methods of alignment. Three principal directions have been taken; one, proposed by Schatz and van Heel (1990, 1992; see also Schatz et al., 1990; Schatz, 1992) eliminates the need for alignment among different classes of images by the use of invariants; the second, proposed by Penzcek et al. (1992), solves the alignment problem by an iterative method; and the third, developed for larger data sets (Dube et al., 1993; Marabini and Carazo, 1994a), is based on the fact that among a set of molecules sufficiently large and presenting the same view, subsets with similar "in-plane" orientation can always be found.

In explaining these approaches, we are occasionally forced to make advance reference to the subject of multivariate data analysis and classification, which will be described in some detail in chapter 4. For a conceptual understanding of the present issues, it may be sufficient to think of classification as a "black box" procedure that is able to sort images (or any patterns derived from them) into groups or classes according to their similarities.

3.5.2. Use of Invariants: The Double-Antocorrelation Function

Certain functions that are invariant both under rotation and translation of an image, but still retain distinct structural properties, can be used to classify a heterogeneous image set. The ACF of an image is invariant under a translation of a motif contained in it. This property has been used in the reference-based alignment method originally proposed by Langer et al. (1970; see also Frank, 1975a; Frank et al., 1978a; and section 3.4.2). The *double-autocorrelation function* (DACF) has the additional property that it is invariant under rotation, as well; it is derived from the normal translational ACF by subjecting the latter to an operation of rotational autocorrelation (Schatz and van Heel, 1990).

Following Schatz and van Heel's procedure (figure 3.14), the ACF is first resampled on a polar grid to give $\hat{\Phi}(r_n, \phi_1)$. From this expression, the rotational ACF is obtained by computing 1D fast Fourier transforms (FFTs) and conjugate scalar Fourier products along rings, using the same philosophy as in the computation of the rotational CCF; see equations (3.26) or (3.37). The resulting function has the remarkable property of being invariant under *both* translation and rotation, because the addition of an arbitrary angle to ϕ_1, in the above expression $\hat{\Phi}(r_n, \phi_1)$, again does not affect the outcome.

Later, Schatz and van Heel (1992; see also van Heel et al., 1992b) introduced another function, which they termed the *double self-correlation function* (DSCF), that avoids the "dynamic" problem caused by the twofold squaring of intensities in the computation of the DACF. The modification consists of a strong suppression of the low spatial frequencies by application of a band-pass filter. Through the use of either DACF or DSCF, classification of the aligned, original images can be effectively replaced by a classification of their invariants, which are formed from the images without prior alignment. In these schemes, the problem of alignment of a heterogeneous image set is thus entirely avoided; alignment needs only to be applied separately, after classification of the invariants, to images within each homogeneous subset.

This approach of invariant classification poses a principal problem, however, which is obvious from its very definition (Schatz and van Heel, 1990; Frank et al.,

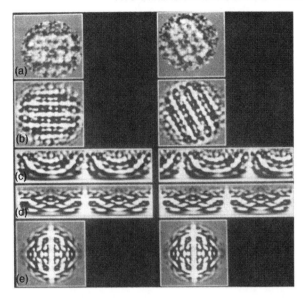

Figure 3.14 Demonstration of the invariance of the double autocorrelation function (DACF) under rotation and shift of an image. (a) Image of worm hemoglobin in side view and the same image rotated and shifted. (b) Autocorrelation functions of the images in (a). The ACF of the image on the left is seen to be identical to that on the right, but rotated by the same angle as the image. (c) The ACFs of (b) in polar coordinates (horizontal axis: azimuthal angle, from 0° to 360°; vertical axis: radius). The rotation between the ACFs is reflected by a horizontal shift of the polar coordinate representation. (d) A second, 1D autocorrelation in the horizontal direction produces identical patterns for both images: the DACF in a polar coordinate representation. (e) The DACF mapped into the Cartesian coordinate system. From Schatz (1992), reproduced with permission of Verlag Hänsel-Hohenhausen.

1992; Penczek et al., 1992): the DACFs and DSCFs are twofold degenerate representations of the initial images. Forming the ACF (or its "self-correlation" equivalent) of an image corresponds to eliminating the phase information in its Fourier transform. In the second step, in forming the DACF or DSCF from the ACF, the phases describing the rotational features in the 1D Fourier transforms along rings are also lost. Therefore, a classification of the invariants does not necessarily map into a correct classification of the images themselves. Theoretically, a very large number of patterns have the same DACF; the saving grace is that they are unlikely to be realized in the same experiment. Another problem is the reduction in SNR due to the loss of phases. Experience will tell whether the problem of ambiguity is a practical or merely an academic problem.

For completeness, another alignment approach making use of invariants, which has not gained practical importance because of inherent problems, should be mentioned. This approach is based on the use of moments (Goncharov et al., 1987; Salzman, 1990). Moments require a choice of an integration boundary and are extremely noise sensitive. They are, therefore, not well suited for alignment of raw data with very low SNRs.

3.5.3. Exhaustive Sampling of Parameter Space*

These are approaches based on the premise that groups of particles with closely related rotations can be found by applying multivariate data analysis (chapter 4) to the particle set that has been merely *translationally* aligned. If a data set is sufficiently large, then the number of particles presenting a similar appearance and falling within an angular range $\Delta\phi$ is on the average:

$$n_v = (\Delta\phi/(2\pi)) \times n_{tot} \times p_v \qquad (3.41)$$

where n_{tot} is the total number of particles, and p_v is the probability of encountering a particular view. For example, if the particle presents five views with equal probability ($p_v = 0.2$) and $n_{tot} = 10,000$, then $n_v = 60$ for $\Delta\phi = 10°$.

Any particles that have the same structure and occur in similar azimuths will then fall within the same region of factor space. In principle, corresponding averages could be derived by the following "brute force" method: the most significant subspace of factor space (which might be just 2D) is divided according to a coarse grid, and images are averaged separately according to the grid square into which they fall. The resulting statistically well-defined averages could then be related to one another by rotational correlation, yielding the azimuthal angles. The within-group alignment of particles can be obviously refined so that eventually an angle can be assigned to each particle.

The method of "alignment through classification" by Dube et al. (1993) proceeds somewhat differently from the general idea sketched out above, taking advantage of the particle's symmetry, and in fact establishing the symmetry of the molecule (in this case, the head-to-tail connector protein, or "portal protein," of bacteriophage $\phi29$) as seen in the electron micrograph. The molecule is rosette-shaped, with a symmetry that has been given variously as 12- or 13-fold by different authors. In the first step, molecules presenting the circular top view were isolated by a multireference approach. For unaligned molecules appearing in the top view with N-fold symmetry, a given pixel at the periphery of the molecule is randomly realized with high or low density across the entire population. If we proceed along a circle, to a pixel that is $360°/(2N)$ away from the one considered, the pattern of variation across the population has consistently shifted by 180°: if the pixel in a given image was dark, it is now bright, and vice versa. Dube et al. (1993) showed how these symmetry-related patterns of variation are reflected in the eigenimages produced by multivariate data analysis (chapter 4).

3.5.4. Iterative Alignment Method*

Penczek et al. (1992) introduced a method of reference-free alignment based on an iterative algorithm that also avoids singling out an image for "reference." Its rationale and implementation are described in the following.

We first go back to an earlier definition of what constitutes the alignment of an entire image set. By generalizing the alignment between two images to a set of N images, Frank et al. (1986) proposed the following definition: *a set of N images is aligned if all images are pairwise aligned.* In the notation by Penczek et al. (1992), alignment of such a set $\mathbf{P} = \{p_i; i = 1, \ldots, N)$ is achieved if the functional

$$L(\mathbf{P}, \mathbf{S}) = \int \sum_{i=1}^{N-1} \sum_{k=i+1}^{N} [p_i(\mathbf{r}; s_\alpha^i, s_x^i, s_y^i) - p_k(\mathbf{r}; s_\alpha^i, s_x^i, s_y^i)] d\mathbf{r} \qquad (3.42)$$

is minimized by appropriate choice of the set of the $3N$ parameters:

$$\mathbf{S} = \{s_\alpha^i, s_x^i, s_y^i; i = 1, \ldots, N\} \qquad (3.43)$$

The actual computation of all pair-wise cross-correlations, as would be required by the minimization of equation (3.42), would be quite impractical; it would also have the disadvantage that each term is derived from two raw images having very low SNR. Penczek and coworkers show, however, that minimization of equation (3.42) is equivalent to minimization of another functional:

$$\bar{L}(\mathbf{P}, \mathbf{S}) = \int \sum_{i=1}^{N-1} [p_i(\mathbf{r}; s_\alpha^i, s_x^i, s_y^i) - \bar{p}_i(\mathbf{r})]^2 d\mathbf{r} \qquad (3.44)$$

where

$$\bar{p}_i(\mathbf{r}) = \frac{1}{N-1} \sum_{k=1; k \neq i}^{N} p_k(\mathbf{r}; s_\alpha^i, s_x^i, s_y^i) \qquad (3.45)$$

In this reformulated expression, each image numbered i is aligned to a partial average of all images, created from the total average by subtracting the current image numbered i. This formula lends itself to the construction of an iterative algorithm involving partial sums, which have the advantage of possessing a strongly enhanced SNR when compared to the raw images.

To go into greater detail, the algorithm consists of two steps (figure 3.15). The first step, which is the random approximation of the global average, proceeds as follows (for simplicity, the argument vector $\mathbf{r}$ is dropped):

  (i) Pick two images p_i and p_k at random from the set $\mathbf{P}$ of images
 (ii) Align p_i and p_k (using any algorithm that minimizes) $\|p_i - p_k\|$)
(iii) Initialize the global average (general term to be denoted by a–m) by setting $a_2 = p_i \oplus p_k$, where the symbol $\oplus$ is used in this and the following description to denote the algebraic sum of the two images *after application of the rotation and shifts found*
 (iv) Set a counter m to 3
  (v) Pick the next image p_i from the set of $N - m + 1$ remaining images
 (vi) Align p_i and a_{m-1}
(vii) Update the global average to give $a_m = [p_i \oplus (m-1)a_{m-1}]/k$
(viii) Increase counter m by 1. If $m = N$ then stop, else go to step (v)

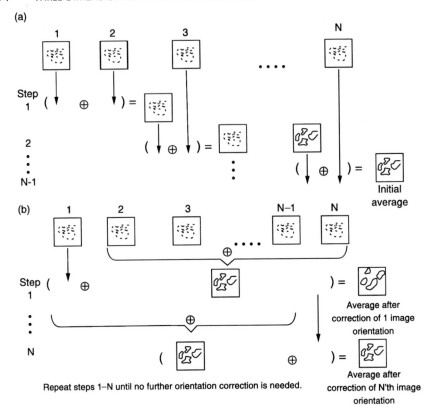

Figure 3.15 Scheme for reference-free alignment by Penczek et al. (1992). The algorithm consists of two parts (a and b); ⊕, symbol to indicate formation of the sum after the best orientation is found. For details, see text. From Penczek et al. (1992), reproduced with permission of Elsevier.

The second part of the algorithm performs an iterative refinement of the average $A = a_N$:

(i)   Set counter $m = 1$
(ii) Create the modified average A' by subtracting the current image p_k in its current position:

$$A' = (NA - p_k)/(N - 1);$$

(iii) Align p_k with A'
(iv) Update the global average A as follows:

$$A = [p_k \oplus (N - 1)A']/N;$$

(v)   Increase m by 1. If $m < N$ then go to (2)
(vi) If in step (3) any of the images changed its position significantly, then go back to step (1), else stop

The algorithm outlined has the following properties:

(i) No reference image is used, and so the result of the alignment does not depend on the choice of a reference, although there is some degree of dependency on the sequence of images being picked.
(ii) The algorithm is necessarily suboptimal since it falls short (by a wide margin) of exploring the entire parameter space.
(iii) Experience has shown that, when a heterogeneous image set is used, comprising dissimiliar subsets (e.g., relating to different particle orientations), then the images within each subset are aligned to one another on completion of the computation.

3.6. Alignment Using the Radon Transform

The 2D Radon transform (also known under the name *sinogram* [e.g., van Heel, 1987)] in its continuous form is defined for a 2D function $g(\mathbf{r})$ as

$$\hat{g}(p, \xi) = \int g(\mathbf{r})\delta(p - \xi^T \mathbf{r})d\mathbf{r} \qquad (3.45a)$$

where $\mathbf{r} = (x.y)^T$ and $\delta(p - \xi^T \mathbf{r})$ represents a line defined by the direction of the (normal) unit vector ξ. We can think of the 2D Radon transform as a systematic inventory of 1D projections of the image as a function of angle, with the angle being defined by the direction of the vector ξ. This can be seen in the relationship between the image in figure 3.16a and its Radon transform (figure 3.16c).

In this space, the extensive theory of Radon transforms and their applications in 2D and 3D image processing cannot be covered. It should merely be noted that the Radon transform yields an effective method for simultaneous translational and rotational 2D alignment of images (Lanzavecchia et al., 1996). Another brief section (chapter 5, section 7.4) is devoted to the use of the Radon transform in 3D projection matching.

4. Averaging and Global Variance Analysis*

4.1. The Statistics of Averaging

After successful alignment of a set of images representing a homogeneous population of molecules, all presenting the same view, each image element j in the coordinate system of the molecule is represented by a set of measurements:

$$\{p_i(\mathbf{r}_j); i = 1, \ldots, N\} \qquad (3.46)$$

which can be characterized by an average:

$$\bar{p}_{(N)}(\mathbf{r}_j) = \frac{1}{N}\sum_{i=1}^{N} p_i(\mathbf{r}_j) \qquad (3.47)$$

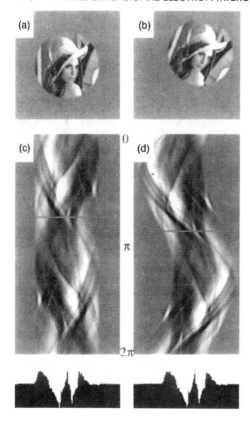

Figure 3.16 Alignment of two identical images using the Radon transform. (a) Image in reference position; (b) the same image, rotated by $\pi/8$ and shifted diagonally. (c, d) Radon transforms (sinograms) of (a, b), computed in the full interval $\{0, 2\pi\}$. Cross-correlation of the sinograms, line by line, reveals that the line marked is identical in the two Radon transforms, revealing both relative angle and shift between the two images. From Lanzavecchia et al. (1996), reproduced with permission of Oxford University Press.

and a variance:

$$v_{(N)}(\mathbf{r}_j) = s_{(N)}(\mathbf{r}_j) = \frac{1}{N-1} \sum_{i=1}^{N} [p_i(\mathbf{r}_j) - \bar{p}_{(N)}(\mathbf{r}_j)]^2 \tag{3.48}$$

Both $\{\bar{p}_{(N)}(\mathbf{r}_j); j = 1, \ldots, J\}$ and $\{V_{(N)}(\mathbf{r}_j); j = 1, \ldots, J\}$ can be represented as images, or maps, which are simply referred to as "average map" and "variance map."

The meaning of these maps depends on the statistical distribution of the pixel measurements. If we use the simple assumption of additive noise with zero-mean Gaussian statistics,

$$p_i(\mathbf{r}_j) = p(\mathbf{r}_j) + n_i(\mathbf{r}_j) \tag{3.49}$$

then $\bar{p}_{(N)}(\mathbf{r}_j)$ as defined above is an unbiased estimate of the mean and thus represents the structural motif $p(\mathbf{r})$ more and more faithfully as N is being increased. The quantity $V_{(N)}(\mathbf{r}_j) = s_{(N)}^2(\mathbf{r}_j)$ is an estimate of $\sigma^2(\mathbf{r}_j)$, the squared standard deviation of the noise.

A display of the variance map was first used in the context of image averaging in EM by Carrascosa and Steven (1979) and in single-particle averaging by Frank et al. (1981a). The variance is a function that is spatially varying, for two reasons: (i) the noise statistics varies with the electron dose, which in turn is proportional to the local image intensity, and (ii) part of the observed variations are "signal associated"; that is, they arise from components of the structure that differ from one particle to the other in density or precise location. Usually, no statistical models exist for these variations.

The variance map is particularly informative with regard to the signal-associated components as it allows regions of local inconsistency to be spotted. For instance, if half of a set of molecules were carrying a ligand, and the other half not, then a region of strong inconsistency would show up in the variance map precisely at the place of the ligand.

4.2. The Variance Map and the Analysis of Statistical Significance

We have seen in the foregoing that one of the uses of the variance map is to pinpoint image regions where the images in the set vary strongly. The possible sources of interimage variability are numerous:

(i) Presence versus absence of a molecule component, for example, in partial depletion experiments (Carazo et al., 1988)

(ii) Presence versus absence of a ligand, for example, in immunoelectron microscopy (Gogol et al., 1990; Boisset et al., 1993b), ribosome-factor binding complex (Agrawal et al., 1998; Valle et al., 2002), calcium release channel bound with a ligand (Wagenknecht et al., 1997; Sharma et al., 1998; Samsó et al., 1999, 2000; Liu et al., 2002)

(iii) Conformational change, that is, movement of a mass (Carazo et al., 1988, 1989) or of many thin flexible masses (Wagenknecht et al., 1992)

(iv) (Compositional heterogeneity

(v) Variation in orientation, for example, rocking or flip/flop variation (van Heel and Frank, 1981; Bijlholt et al., 1982)

(vi) Variation in stain depth (Frank et al., 1981a, 1982; Boisset et al., 1990a)

(vii) Variation in magnification (Bijlholt et al., 1982)

Striking examples are found in many studies of negatively stained molecules, where the variance map often reveals that the stain depth at the boundary of the molecule is the strongest varying feature. The 40S ribosomal subunit of eukaryotes shows this behavior quite clearly (Frank et al., 1981a); see figure 3.17.

An example of a study in which structural information is gleaned from the variance map is found in the paper by Wagenknecht et al. (1992) (figure 3.18). Here, a core structure (E2 cores of pyruvate dehydrogenase) is surrounded by lipoyl domains which do not show up in the single particle average because they do not appear to assume fixed positions. Their presence at the periphery of the E2 domain is nevertheless reflected in the variance map by the appearance of a strong white halo of high variance. These early findings are particularly interesting in the light of the high-resolution structure emerging now by a combination of results from cryo-EM and X-ray crystallography, and indications of the dynamical behavior of this molecule (Zhou et al., 2001c).

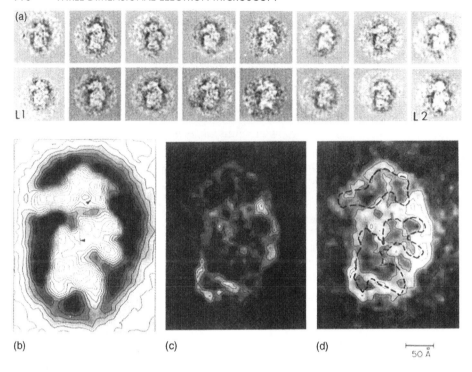

Figure 3.17 Average and variance map obtained from an aligned set of macromolecules. (a) Sixteen of a total set of 77 images showing the 40S ribosomal subunit of HeLa in L-view orientation; (b) average image; (c) variance map; and (d) standard deviation map, showing prominent variations mainly at the particle border where the amount of stain fluctuates strongly (white areas indicate high variance). From Frank et al. (1981), with permission of the American Association for the Advancement of Science.

However, this "global" variance analysis made possible by the variance map has some obvious shortcomings. While it alerts us to the presence of variations and inconsistencies among the images of a data set, and gives their location in the image field, it fails to characterize the different types of variation and to flag those images that have an outlier role. For a more specific analysis, the tools of multivariate data analysis and classification must be employed (see chapter 4).

Another important use of the variance map is the assessment of significance of local features in the average image (Frank et al., 1986), using standard methods of statistical inference (e.g., Cruickshank, 1959; Sachs, 1984): each pixel value in that image, regarded as an estimate of the mean, is accompanied by a confidence interval within which the true value of the mean is located with a given probability. In order to construct the confidence interval, we consider the random variable:

$$t = \frac{\bar{p}(\mathbf{r}_j) - p(\mathbf{r}_j)}{s(\mathbf{r}_j)} \tag{3.50}$$

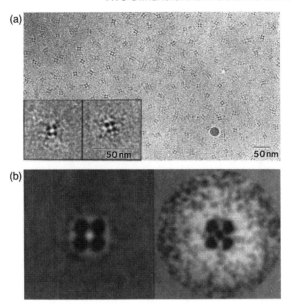

Figure 3.18 Example of the type of information contained in a variance map: visualization of a "corona" of highly flexible lipoyl domains surrounding the E2 core of puryvate dehydrogenase complex of *Escherichia coli*. (a) Electron micrograph of frozen-hydrated E2 cores presenting fourfold symmetric views. Each is surrounded by fuzzy structures believed to be lipoyl domains bound to the molecule. Due to their changing positions, the average map (b, left) depicts only the E2 core, but the variance map (b, right) shows a ring-shaped region of high variance. From Wagenknecht et al. (1992), reproduced with permission of Elsevier.

where $s(\mathbf{r}_j) = [V(\mathbf{r}_j)/N]^{1/2}$ is the standard error of the mean, which can be computed from the measured variance map expressed by equation (3.48) (see figure 3.17). The true mean lies in the interval:

$$\bar{p}_{\text{true}}(\mathbf{r}_j) = \bar{p}(\mathbf{r}_j) \pm ts(\mathbf{r}_j) \tag{3.50a}$$

the *confidence interval*, with probability P, if t satisfies

$$P = \int_{-1}^{+1} S_{N-1}(\tau)\mathrm{d}\tau \tag{3.51}$$

where $S_{N-1}(\tau)$ is the *Student distribution* with $N-1$ degrees of freedom. Beyond a number of images $N = 60$ or so, the distribution changes very little, and the confidence intervals for the most frequently used probabilities become $t = 1.96$ ($P = 95\%$), $t = 2.58$ ($P = 99\%$), and $t = 3.29$ ($P = 99.9\%$).

It must be noted, however, that the use of the variance map to assess the significance of local features in a difference map should be reserved for regions where no signal-related inconsistencies exist. In fact, the statistical analysis outlined here is strictly meaningful only for a homogeneous image set, in the sense discussed in section 3.2. Another caveat stems from the fact that the test described is valid under the assumption that the pixels are independent. This is not the case in a strict sense, because the images are normally aligned to a common reference.

Let us now discuss an application of statistical hypothesis testing to the comparison between two averaged pixels, $\bar{p}_1(\mathbf{r}_j)$ and $\bar{p}_2(\mathbf{r}_j)$, which are obtained by averaging over N_1 and N_2 realizations, respectively (Frank et al., 1988a). This comparison covers two different situations: in one, two pixels $j \neq k$ from the *same* image are being compared, and the question to be resolved is if the difference between the values of these two pixels, separated by the distance $|\mathbf{r}_j - \mathbf{r}_k|$, is significant. In the other situation, the values of the same pixel $j = k$ are compared as realized in two averages resulting from different experiments. A typical example might be the detection of extra mass at the binding site of the antibody in an immunolabeled molecule when compared with a control. Here, the question posed is whether the density difference detected at the putative binding site is statistically significant.

In both these situations, the standard error of the difference between the two averaged pixels is given by

$$s_d[p_1(\mathbf{r}_j), p_2(\mathbf{r}_j)] = [V_1(\mathbf{r}_j)/N_1 + V_2(\mathbf{r}_j)/N_2]^{1/2} \qquad (3.52)$$

Differences between two averaged image elements are deemed significant if they exceed the standard error by at least a factor of three. This choice of factor corresponds to a significance level of 0.2% [i.e., $P = 98\%$ in equation (3.52)]. In single-particle analysis, this type of significance analysis was first done by Zingsheim et al. (1982), who determined the binding site of bungarotoxin on the projection map of the nicotinic acetylcholine receptor molecule of *Torpedo marmorata*. Another example was the determination, by Wagenknecht et al. (1988a), of the anticodon binding site of P-site tRNA on the 30S ribosomal subunit as it appears in projection. An example of a t-map showing the sites where an undecagold cluster is localized in a 2D difference map is found in the work of Crum et al. (1994). Figure 3.19 shows the difference mapping of a GFP (green fluorescent protein) insert in the ryanodine receptor/calcium release channel, and a map of statistically significant regions, displayed at the 99.9% level of confidence.

The same kind of test is of course even more important in three dimensions, when reconstructions are compared that show a molecule with and without a ligand, or in different conformational states. In single-particle reconstruction, variance analysis in three dimensions is handicapped by the fact that an ensemble of 3D reconstructions does not exist; hence, a 3D variance map cannot be computed in a straightforward way following expressions that would correspond to equations (3.47) and (3.48). However, other, approximate routes can be taken, as will be detailed in chapter 6, section 2.

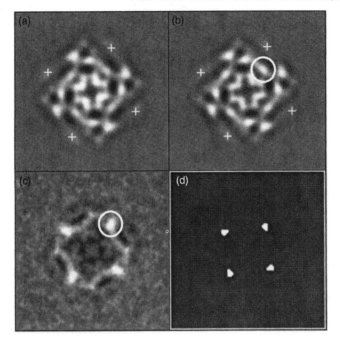

Figure 3.19 Localization of a GFP insert in the ryanodine receptor (RyR) by difference mapping and significance analysis. (a) Average of RyR wild-type ($N = 269$ particles); (b) average of RyR·GFP ($N = 250$ particles); (c) difference map b−a. The largest difference is located in the region encircled, and in the three regions related to it by fourfold symmetry. (d) Map of statistically significant regions, in the difference map, obtained by the t-test. The map is displayed at a $>99.9\%$ level of confidence. From Liu et al. (2002), reproduced with permission of The American Society for Biochemistry and Molecular Biology, Inc.

4.3. Signal-to-Noise Ratio

4.3.1. Concept and Definition

The signal-to-noise ratio (SNR) is the ratio between the variance of the signal and the variance of the noise in a given image.[6] This measure is extremely useful in assessing the quality of experimental data, as well as the power of 2D and 3D averaging methods. Unfortunately, the definition varies among different fields, with the SNR often being the square root of the above definition, or with the numerator being measured peak to peak. In the following we will use the ratio of variances, which is the most widely used definition in digital signal processing.

---

[6]Note that in the field of electrical engineering, the signal-to-noise ratio is defined as the ratio of the signal power to the noise power. This ratio is identical with the variance ratio only if the means of both signal and noise are zero.

The sample variance of an image indexed i, $\{p_{ij}; j = 1, \ldots, J\}$ is defined as

$$\text{var}(p_i) = \frac{1}{J-1} \sum_{j=1}^{J} [p_{ij} - \langle p_i \rangle]^2 \tag{3.53}$$

with the sample mean

$$\langle p \rangle = \frac{1}{J} \sum_{j=1}^{J} p_{ij} \tag{3.54}$$

According to Parseval's theorem, the variance of a band-limited function can be expressed as an integral (or its discrete equivalent) over its squared Fourier transform:

$$\text{var}(p_i) = \int_{\mathbf{B}_\square} |P_i(\mathbf{k})| d\mathbf{k} \tag{3.55}$$

where $\mathbf{B}_\square$ denotes a modified version of the *resolution domain* $\mathbf{B}$, that is, the bounded domain in Fourier space representing the signal information. The modification symbolized by the $\square$ subscript symbol is that the integration exempts the term $|P(0)|^2$ at the origin. (Parseval's theorem expresses the fact that the norm in Hilbert space is conserved on switching from one set to another set of orthonormalized basis functions.)

When we apply this theorem to both signal and additive noise portions of the image, $p(\mathbf{r}) = o(\mathbf{r}) + n(\mathbf{r})$, we obtain

$$\text{SNR} = \frac{\text{var}(o)}{\text{var}(n)} = \frac{\int_{\mathbf{B}_\square} |O(\mathbf{k})|^2 |H(\mathbf{k})|^2 d\mathbf{k}}{\int_{\mathbf{B}_\square} |N(\mathbf{k})|^2 d\mathbf{k}} \tag{3.56}$$

Often the domain, $\mathbf{B}'$, within which the signal portion of the image possesses appreciable values, is considerably smaller than $\mathbf{B}$. In that case, it is obvious that low-pass filtration of the image to band limit $\mathbf{B}'$ leads to an increased SNR without signal being sacrificed. For uniform spectral density of the noise power up to the boundary of $\mathbf{B}$, the gain in SNR on low-pass filtration to the true band limit $\mathbf{B}'$ is, according to Parseval's theorem, equation (3.55), equal to the ratio area$\{\mathbf{B}\}$/area$\{\mathbf{B}'\}$. Similarly, it often happens that the noise power spectrum $|N(\mathbf{k})|^2$ is uniform, whereas the transform of the signal transferred by the instrument, $|O(\mathbf{k})|^2 |H(\mathbf{k})|^2$, falls off radially. In that case, the elimination, through low-pass filtration, of a high-frequency band may boost the SNR considerably without affecting the interpretable resolution.

4.3.2. Measurement of the Signal-to-Noise Ratio

Generally, the unknown signal is mixed with noise, so the measurement of the SNR of an experimental image is not straightforward. Two ways of measuring the SNR of "raw" image data have been put forth: one is based on the dependence of the sample variance [equation (3.53)] of the average image on N (= the number of images averaged) and the other on the cross-correlation of two realizations of the image.

4.3.2.1. N-dependence of sample variance

We assume that the noise is additive, uncorrelated, stationary (i.e., possessing shift-independent statistics), and Gaussian; and further, that it is uncorrelated with the signal (denoted by p). In that case, the variance of a "raw" image p_i is, independently of i,

$$\text{var}(p_i) = \text{var}(p) + \text{var}(n) \qquad (3.57)$$

The variance of the average $\bar{p}_{[N]}$ of N images is

$$\text{var}(\bar{p}_{[N]}) = \text{var}(p) + \frac{1}{N}\text{var}(n) \qquad (3.58)$$

that is, with increasing N, the proportion of the noise variance in the variance of the average is reduced. This formula suggests the use of a plot of $\text{var}(\bar{p}_{[N]})$ versus $1/N$ as a means to obtain the unknown quantities $\text{var}(p)$ and $\text{var}(n)$ (Hänicke, 1981; Frank et al., 1981a; Hänicke et al., 1984). If the assumptions made at the beginning are correct, the measured values of $\text{var}(\bar{p}_{[N]})$ should lie on a straight line whose slope is the desired quantity $\text{var}(n)$ and whose intersection with the $\text{var}(p)$ axis (obtained by extrapolating it to $1/N = 0$) gives the desired quantity $\text{var}(p)$. Figure 3.20 shows such a variance plot obtained for a set of 81 images of the negatively stained 40S ribosomal subunit of HeLa cells (Frank et al., 1981a). It is seen that the linear dependency predicted by equation (3.58) is indeed a good approximation for this type of data, especially for large N.

4.3.2.2. Measurement by cross-correlation

Another approach to the measurement of the SNR makes use of the definition of the cross-correlation coefficient (CCC). The CCC of two realizations of a noisy image, p_{ij} and p_{kj}, is defined as [see equation (3.16)]

$$\rho_{12} = \frac{\sum_{j=1}^{J} [p_{ij} - \langle p_i \rangle][p_{kj} - \langle p_k \rangle]}{\left\{ \sum_{j=1}^{J} [p_{ij} - \langle p_i \rangle]^2 \sum_{j=1}^{J} [p_{kj} - \langle p_k \rangle]^2 \right\}^{1/2}} \qquad (3.59)$$

where $\langle p_i \rangle$ and $\langle p_k \rangle$ again are the sample means defined in the previous section.

When we substitute $p_{ij} = p_j + n_{ij}, p_{kj} = p_k + n_{kj}$, and observe that according to the assumptions both noise functions have the same variance $\text{var}(n)$, we obtain the simple result (Frank and Al-Ali, 1975):

$$\rho_{12} = \frac{\alpha}{1 + \alpha} \qquad (3.60)$$

from which the SNR is obtained as

$$\alpha = \frac{\rho_{12}}{1 - \rho_{12}} \qquad (3.61)$$

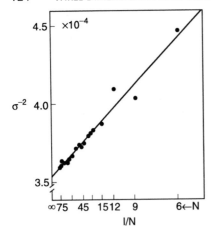

Figure 3.20 Decrease in the variance of the average image as a function of the number of images averaged, N, or (linear) increase in the variance with $1/N$. Extrapolation to $1/N = 0$ allows the variance of the signal to be measured. Averaged were $N = 81$ images of the 40S ribosomal subunit from HeLa cells negatively stained and showing the so-called L-view. From Frank et al. (1981a), reproduced with permission of The American Association for the Advancement of Science.

Thus, the recipe for estimating the SNR is quite simple: choose two images from the experimental image set, align them and filter the images to the resolution deemed relevant for the SNR measurement. Then compute the CCC and use equation (3.61) to obtain the SNR estimate. Because of the inevitable fluctuations of the results, it is advisable to repeat this procedure with several randomly picked image pairs and use the average of the different SNR measurements as a more reliable figure.

In practice, the measurement of the SNR in the image (as opposed to the SNR of the *digitized* image) is somewhat more complicated since the microdensitometer measurement will introduce another noise process which, unchecked, would lead to overly pessimistic SNR estimates. To account for this additional contribution, Frank and Al-Ali (1975) used a control experiment, in which two scans of the same micrograph were evaluated with the same method, and a corresponding SNR for the microdensitometer α_d was found. The true SNR is then obtained as

$$\alpha_{\text{true}} = \frac{1}{(1 + (1/\alpha))/(1 + (1/\alpha_d)) - 1} \tag{3.62}$$

where α is the SNR deduced from the uncorrected experiment, following equation (3.61).

5. Resolution

5.1. The Concept of Resolution

Optical definitions of resolution are based on the ability of an instrument to resolve two points separated by a given distance, d. When the two points get closer and closer, their initially separate images merge into a single one, first enclosing a "valley" between them, which gradually fills, until it entirely disappears. The definition usually stipulates that the ratio between valley and peak should be of a certain minimum size for the points to be resolved.

Two criticisms can be raised against this definition; one is quite general, the other relates to its application in EM. The first criticism (Di Francia, 1955) concerns an information-theoretical aspect of the experiment: if it is known *a priori* that the object consists of two points, then measurement of their mutual distance in the image is essentially a pattern-recognition problem, which is limited by noise, not by the size of d. (For instance, if the image of a single point, i.e., the point spread function, is a circularly symmetric function, the image of two points with nonzero distance will no longer be rotationally symmetric, and the exact distance could be simply obtained by fitting, even when the distance is much less than the diameter of the point-spread function.) The other criticism is that the resolution criterion formulated above does not lend itself to a suitable experiment when applied to high-resolution EM. On the nanometer scale, any test object, as well as the support it must be placed on, reveals its atomic makeup. Both considerations suggest that the definition and measurement of resolution require a fundamentally different approach. In fact, the most useful definitions and criteria all refer to Fourier representations of images.

In both crystallography and statistical optics, it is common to define resolution by the orders of (object-related) Fourier components available for the Fourier synthesis of the image. This so-called crystallographic resolution R_c and Raleigh's point-to-point resolution distance d for an instrument, which is diffraction limited to R_c, are related by

$$d = 0.61/R_c \qquad (3.63)$$

In electron crystallography, the signal-related Fourier components of the image are distinguished from noise-related components by the fact that the former are concentrated on the points of a regular lattice, the *reciprocal lattice*, while the latter form a continuous background. Thus, resolution can be specified by the radius of the highest diffraction orders that stand out from the background. What "stand out" means can be quantified by relating the amplitude of the peak to the mean of the background surrounding it (see below).

In single-particle averaging, on the other hand, there is no distinction in the Fourier transform between the appearance of signal and noise, and resolution estimation must take a different route. There are two categories of resolution tests; one is based on the comparison of two independent averages in the Fourier domain (*cross-resolution*), while the other is based on the multiple comparison of the Fourier transforms of all images participating in the average. The differential phase residual (Frank et al., 1981a), Fourier ring correlation (Saxton and Baumeister, 1982; van Heel et al., 1982c), and the method of Young's fringes (Frank et al., 1970; Frank, 1972a) fall into the first category, while the spectral SNR (Unser et al., 1987, 1989) and the Q-factor (van Heel and Hollenberg, 1980; Kessel et al., 1985) fall into the second.

Averaging makes it possible to recover a signal that is present in the image in very small amounts. Its distribution in Fourier space, as shown by the signal power spectrum, usually shows a steep falloff. For negatively stained specimens, this falloff is due to the imperfectness of the staining: on the molecular scale, the stain salt forms crystals and defines the boundary of the molecule only within a margin of error. In addition, the process of air-drying causes the molecule to change shape, and one must assume that this shape change is variable as it depends on stain depth, orientation, and properties of the supporting carbon film. For specimens embedded in ice, gross shape changes of the specimen are avoided, but residual variability (including genuine conformational variability) as well as instabilities of recording (drift, charging, etc.) are responsible for the decline of the power spectrum. Some of these resolution-limiting factors will be discussed below (section 5.4); for the moment, it is important to realize that there is a practical limit beyond which the signal power makes only marginal contributions to the image.

However, all resolution criteria listed below in this section have in common that they ignore the falloff of the signal in Fourier space. Thus, a resolution of $1/20\,\text{Å}^{-1}$ might be found by a consistency test, even when the signal power is very low beyond $1/30\,\text{Å}^{-1}$. An example of this kind of discrepancy was given by van Heel and Stöffler-Meilicke (1985), who studied the 30S ribosomal subunits from two eubacterial species by 2D averaging: they found a resolution of $1/17\,\text{Å}^{-1}$ by the Fourier ring correlation method, even though the power spectrum indicated the presence of minimal signal contributions beyond $1/20\,\text{Å}^{-1}$, or even beyond $1/25\,\text{Å}^{-1}$, when a more conservative assessment is used. The lesson to be learned from these observations is that, for a meaningful statement about the information actually present in the recovered molecule projection, a resolution assessment ideally should be accompanied by an assessment of the range and falloff of the power spectrum.

The same concern about the meaning of "resolution" in a situation of diminishing diffraction power has arisen in electron crystallography. Henderson et al. (1986) invented a quality factor (IQ, for *image quality*) that expresses the SNR of each crystal diffraction spot and applied a rigorous cutoff in the Fourier synthesis depending on the averaged size of this factor, with the averaging being carried out over rings in Fourier space. Glaeser and Downing (1992) have demonstrated the effect of including higher diffraction orders in the synthesis of a projection image of bacteriorhodopsin. It is seen that, with increasing spatial frequency, as the fraction of the total diffraction power drops to 15%, the actual improvement in the definition of the image (e.g., in the sharpness of peaks representing projected alpha-helices) becomes marginal.

5.2. Resolution Criteria

It should be noted that the treatment in this section applies to both 2D and 3D data: the formation of 2D averages, as well as 3D reconstruction. To avoid duplication, we will talk about both types of data below, and refer back to this section when the subject of resolution comes up again in chapter 5.

5.2.1. Definition of Region to be Tested

The resolution criteria to be detailed below all make use of the discrete Fourier transform of the images to be analyzed. It is of crucial importance for a meaningful application of these measures that no correlations are unwittingly introduced when the images are prepared for the resolution test. It is tempting to use a mask to narrowly define the region in the image where the signal—the averaged molecule image—resides, as the inclusion of surrounding material with larger inconsistency might lead to overly pessimistic results. However, imposition of a binary mask, applied to both images that are being compared, would produce correlation extending to the highest resolution. This is on account of the sharp boundary of a binary mask, with a 1-pixel falloff, which requires Fourier terms out to the Nyquist limit to be utilized in the representation. Hence, the resolution found in any of the tests described below would be falsely reported as the highest possible—corresponding to the Nyquist spatial frequency. To avoid this effect, one has to use a "soft" mask whose falloff at the edges is so slow that it introduces correlations at low spatial frequencies only. Gaussian-shaped masks are optimal for this purpose since all derivatives of a Gaussian are again Gaussian, and thus continuous.

Fourier-based resolution criteria are, therefore, governed by a type of uncertainty relationship: precise localization of features for which resolution is determined makes the resolution indeterminate; and, on the other hand, precise measurement of resolution is possible only when the notion of localizing the features is entirely abandoned.

5.2.2. Comparison of Two Subsets Versus Analysis of the Whole Data Set

5.2.2.1. Criteria based on two equally large subsets   Many criteria introduced in the following test the reproducibility of a map obtained by averaging (or, as we will later see, by 3D reconstruction) when based on two randomly drawn subsets of equal size. For example, the use of even- and odd-numbered images of the image set normally avoids any systematic trends such as related to the origin in different micrographs or different areas of the specimen. Each subset is averaged, leading to the average images $\bar{p}_1(\mathbf{r}), \bar{p}_2(\mathbf{r})$ ("subset averages").

Let $F_1(\mathbf{k})$ and $F_2(\mathbf{k})$ be the discrete Fourier transforms of the two subset averages, with the spatial frequency $\mathbf{k}$ assuming all values on the regular Fourier grid (k_x, k_y) within the Nyquist range. The Fourier transforms are now compared, and a measure of discrepancy is computed, which is averaged over rings of width Δk and radius $k = |\mathbf{k}| = [k_x^2 + k_y^2]^{1/2}$. The result is then plotted as a function of the ring radius. This curve characterizes the discrepancy between the subset averages over the entire spatial frequency range.

In principle, a normalized version of the generalized Euclidean distance, $|F_1(\mathbf{k}) - F_2(\mathbf{k})|$ could be used to serve as a measure of discrepancy, but two other measures, the *differential phase residual* and the *Fourier ring correlation*, have gained practical importance. These will be introduced in sections 5.2.3 and 5.2.4. Closely related is the criterion based on Young's fringes (section 5.2.5), which will be covered mainly because of its historical role and the insights it provides.

Even though it is clear that the information given by a curve cannot be condensed into a single number, it is nevertheless common practice to derive a single "resolution figure" for expediency.

5.2.2.2. Truly independent versus partially dependent subsets. The iterative refinement procedures that are part of the reference-based alignment (section 3.4) as well as the reference-free alignment (section 3.5) complicate the resolution estimation by halfset criteria since they inevitably introduce a statistical interdependency between the two halfset averages being compared. This problem, which is closely related to the problem of model bias, will resurface in the treatment of angular refinement of reconstructions by 3D projection matching (chapter 5, section 7.2). In the 3D case, the problem has been extensively discussed (Grigorieff, 2000; Penczek, 2002a; Yang et al., 2003). Grigorieff's (2000) suggestion, if applied to the case of 2D averaging, would call for the use of two markedly different 2D references at the outset, and for a total separation of the two randomly selected image subsets throughout the averaging procedure. The idea is if the data are solid and self-consistent, then they can be expected to converge to an average that is closely reproducible. If, on the other hand, the noise is so large that the reference leaves a strong imprint on the average, the resulting deviation between the subset averages will be reported as a lack of reproducible resolution. We come back to the idea of using independent subset averages in the discussion of the spectral signal-to-noise ratio (SSNR), since it leads to the formulation of a straightforward relationship between the SSNR and the Fourier ring correlation.

5.2.2.3. Criteria based on an evaluation of the whole data set   The criteria of the first kind have the disadvantage of large statistical uncertainty, and this is why criteria based on a statistical evaluation of the total set are principally superior, although the advantage diminishes as the numbers of particles increases, which is now often in the tens of thousands. The Q-factor and the SSNR are treated in sections 5.2.7 and 5.2.8, respectively.

5.2.3. Differential Phase Residual

The differential phase residual measures the root mean square (r.m.s.) deviation of the phase difference betweeen the two Fourier transforms, weighted by the average Fourier amplitude. If $\Delta\phi(\mathbf{k})$ is the phase difference between the two Fourier transforms for each discrete spatial frequency $\mathbf{k}$, then the differential phase residual (DPR) is defined as follows:

$$\Delta\bar{\phi}(k, \Delta k) = \left| \frac{\sum\limits_{[k, \Delta k]} [\Delta\phi(\mathbf{k})]^2 [|F_1(\mathbf{k})| + |F_2(\mathbf{k})|]}{\sum\limits_{[k, \Delta k]} [|F_1(\mathbf{k})| + |F_2(\mathbf{k})|]} \right|^{1/2} \tag{3.64}$$

The sums are computed over Fourier components falling within rings defined by spatial frequency radii $k \pm \Delta k$; $k = |\mathbf{k}|$ and plotted as a function of k (figure 3.21). In principle, as in the case of the Fourier ring correlation to be introduced below, the entire curve is needed to characterize the degree of consistency between the two averages. However, it is convenient to use a single

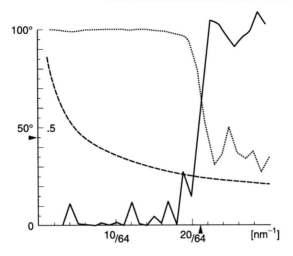

Figure 3.21 Resolution assessment by comparison of two subaverages (calcium release channel of skeletal fast twitch muscle) in Fourier space using two different criteria: differential phase residual (DPR, solid line, angular scale 0° ... 100°) and Fourier ring correlation (FRC, dashed line, scale 0 ... 1). The scale on the x-axis is in Fourier units, denoting the radius of the rings over which the expressions for DPR or FRC were evaluated. The DPR resolution limit ($\Delta\bar{\Phi} = 45°$; see arrowhead on y-axis) is $1/30\,\text{Å}^{-1}$ (arrowhead on x-axis). For FRC resolution analysis, the FRC curve was compared with twice the FRC for pure noise (dotted curve). In the current example, the two curves do not intersect within the Fourier band sampled, indicating an FRC resolution of better than $1/20\,\text{Å}^{-1}$. From Radermacher et al. (1992a), reproduced with permission of the Biophysical Society.

figure, k_{45}, the spatial frequency for which $\Delta\bar{\phi}(k, \Delta k) = 45°$. As a conceptual justification for the choice of this value, one can consider the effect of superimposing two sine waves differing by $\Delta\bar{\phi}$. If $\Delta\bar{\phi}$ is less than 45°, the waves tend to enforce each other, whereas for any $\Delta\bar{\phi} > 45°$, the maximum of one wave already tends to fall in the vicinity of the zero of the other, and destructive interference starts to occur.

It is of crucial importance in the application to EM that the phase residual (and any other Fourier-based measures of consistency, to be described in the following) be computed "differentially," over successive rings or shells, rather than globally, over the entire Fourier domain with a circle of radius k. Such global computation is often used, for instance, to align particles with helical symmetry (Unwin and Klug, 1974). Since $|F(\mathbf{k})|$ falls off rapidly as the phase difference $\Delta\phi$ increases, the figure k_{45} obtained with the global measure would not be very meaningful in our application; for instance, excellent agreement in the lower spatial frequency range can make up for poor agreement in the higher range and thus produce an overoptimistic value for k_{45}. The differential form of the phase residual, equation (3.64), was first used by Crowther (1971) to assess the preservation of icosahedral symmetry as a function of spatial frequency. It was first used in the context of single-particle averaging by Frank et al. (1981a).

One can easily verify from equation (3.64) that the DPR is sensitive to changes in scaling between the two Fourier transforms. In computational implementations, equation (3.64) is therefore replaced by an expression in which $|F_2(\mathbf{k})|$ is dynamically scaled, that is, replaced by $s|F_2(\mathbf{k})|$, where the scale factor s is allowed to run through a range from a value below 1 to a value above 1, through a range large enough to include the minimum of the function. The desired DPR then is the minimum of the curve formed by the computed residuals. One of the advantages of the DPR is that it relates to the measure frequently used in electron and X-ray crystallography to assess reproducibility and the preservation of symmetry.

5.2.4. Fourier Ring Correlation

The Fourier ring correlation (FRC) (Saxton and Baumeister, 1982; van Heel et al., 1982) is similar in concept to the DPR as it is based on a comparison of the two Fourier transforms over rings:

$$\text{FRC}(k, \Delta k) = \frac{\text{Re}\{ \sum\limits_{[k, \Delta k]} F_1(\mathbf{k})F_2^*(\mathbf{k})\}}{\{ \sum\limits_{[k, \Delta k]} |F_1(\mathbf{k})|^2 \sum\limits_{[k, \Delta k]} |F_2(\mathbf{k})|^2 \}^{1/2}} \tag{3.65}$$

Again, as in the definition of the DPR in the previous section, the notation under the sum refers to the terms that fall into a ring of certain radius and width. The FRC curve (see figure 3.21) starts with a value of 1 at low spatial frequencies, indicating perfect correlation, then falls off more or less gradually, toward a fluctuating flat region of the curve with values that originate from chance correlation.

Here, the resolution criterion is derived in two different ways: (i) by comparison of the FRC measured with the FRC expected for pure noise, $\text{FRC}_{\text{noise}} = 1/(N_{[k, \Delta k]})^{1/2}$, where $N_{[k, \Delta k]}$ denotes the number of samples in the Fourier ring zone with radius k and width Δk, or (ii) by comparison of the FRC measured with an empirical threshold value. The term $\text{FRC}_{\text{noise}}$ is often denoted as "σ," even though it does not have the meaning of a standard deviation. Using this notation, the criteria based on the noise comparison that have been variably used are 2σ (van Heel and Stöffler-Meilicke, 1985), 3σ (e.g., Orlova et al., 1997), or 5σ (Radermacher et al., 1988, 2001). As empirical threshold, in the second group of FRC-based criteria, the value 0.5 (Böttcher et al., 1997) is most frequently used.

The $\text{FRC} = 3\sigma$ criterion invariably gives much higher numerical resolution values than $\text{FRC} = 0.5$. A critical comparison of the 0.5 and the 3σ criteria, and a refutation of some of the comments in Orlova et al. (1997) are found in Penczek's appendix to Malhotra et al. (1998). A data set is considered that is split into halves, and the FRC is calculated by comparing the halves. It is shown here that $\text{FRC} = 0.5$ corresponds to $\text{SNR} = 1$. In other words, in the corresponding shell, the noise is already as strong as the signal. Therefore, the inclusion of data beyond this point appears risky. More about the application of resolution criteria to 3D reconstructions, as opposed to 2D averages considered here, will be found in chapter 5.

DPR Versus FRC   Regarding the relationship between FRC and DPR, experience has generally shown that the FRC $= 3\sigma$ gives consistently a more optimistic answer than DPR $= 45°$. In order to avoid confusion in comparisons of resolution figures, some authors have used both measures in their publications. Some light on the relationship between DPR and FRC has been shed by Unser et al. (1987), who introduced another resolution measure, the SSNR (see section 5.2.8). Their theoretical analysis confirmed the observation that always $k_{FRC[3\sigma]} > k_{DPR[45°]}$ for the same model data set. Further illumination of the relative sensitivity of these two measures was provided by Radermacher (1988) who showed, by means of a numerical test, that an FRC $= 2 \times 1/N^{1/2}$ cutoff (the criterion initially proposed, before a factor of 3 was adopted) (Orlova et al., 1997) is equivalent to a SNR of 0.2, whereas the DPR 45° cutoff is equivalent to SNR $= 1$. Thus, the FRC cutoff, and even the DPR cutoff with its fivefold increased SNR seem quite optimistic; on the other hand, for well-behaved data the DPR curve is normally quite steep, so that even a small increase in the FRC cutoff will often lead to a rapid increase in SNR.

Another observation by Radermacher (1988), later confirmed by de la Fraga et al. (1995), was that the FRC cutoff of FRC $= 2\sigma$ corresponds to a DPR cutoff of 85°.

5.2.5. Young's Fringes

Because of its close relationship with the DPR and the FRC, the method of Young's fringes (Frank et al., 1970; Frank, 1972a, 1976) should be mentioned here, even though this method is nowadays rarely used to measure the resolution of computed image averages. However, Young's fringes shed an interesting light on the relationship between common information and correlation. The method is based on the result of an optical diffraction experiment: two micrographs of the same specimen (e.g., a thin carbon film) are first brought into precise superimposition, and then a small relative translation Δx is applied. The diffraction pattern (figure 3.22) shows the following intensity distribution:

$$
\begin{aligned}
I(\mathbf{k}) &= |F_1(\mathbf{k}) + F_2(\mathbf{k}) \exp(2\pi i k_x \Delta x)|^2 \\
&= |F_1(\mathbf{k})|^2 + |F_2(\mathbf{k})|^2 \\
&\quad + 2|F_1(\mathbf{k})||F_2(\mathbf{k})| \cos[2\pi i k_x \Delta x + \phi_1(\mathbf{k}) - \phi_2(\mathbf{k})]
\end{aligned}
\tag{3.66}
$$

The third term, the Young's fringes term proper, is modulated by a cosine pattern whose wavelength is inversely proportional to the size of the image shift, and whose direction is in the direction of the shift. Since a fixed phase relationship holds only within the domain where the Fourier transform is dominated by the signal common to both superimposed images, while the relationship is random outside of that domain, the cosine fringe pattern induced by the shift can be used to visualize the extent of the resolution domain.

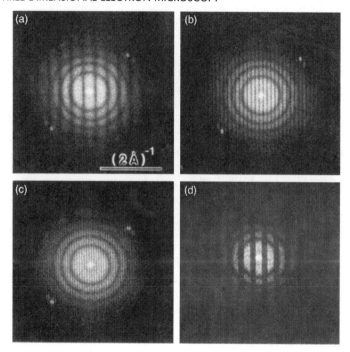

Figure 3.22 Computed diffraction patterns showing Young's fringes. The patterns are obtained by adding two different electron micrographs of the same specimen area and computing the power spectrum of the resulting image. (a–c) The micrographs are added with different horizontal displacements (corresponding to 10, 30, and 50 Å). The intensity of the resulting pattern follows a cosine function. (d) A pattern with approximately rectangular profile is obtained by linear superposition of the patterns (a–c) using appropriate weights. From Zemlin and Weiss (1993), reproduced with permission of Elsevier.

Moreover, the position of the Young's fringes is sensitive to the phase difference between the two transforms. When two images of an object are obtained with the same defocus setting, the phases are the same. Consistent shifts affecting an entire region of Fourier space show up as shifts of the fringe system. When two images of an object are obtained with different defocus settings, then the fringe system shifts by 180° wherever the contrast transfer functions differ in polarity (Frank, 1972a).

Using digital Fourier processing, the waveform of the fringes can be freely designed, by superposing normal cosine-modulated patterns with different frequencies (Zemlin and Weiss, 1993). The most sensitive detection of the band limit is achieved when the waveform is rectangular. Zemlin and Weiss (1993) obtained such a pattern of modulation experimentally (figure 3.22d).

5.2.6. Statistical Limitations of Halfset Criteria

Criteria based on splitting the data set in half in a single partition (as opposed to doing it repeatedly for a large number of permutations) are inferior because of large statistical fluctuations. The results of an evaluation by de la Fraga et al.

(1995) are interesting in this context. By numerical trial computations, these authors established confidence limits for DPR and FRC resolution tests applied to a data set of 300 experimental images of DnaB helicase. The data set was divided into halfsets using many permutations, and corresponding averages were formed in each case. The resulting DPR and FRC curves were statistically evaluated. The confidence limits for both the DPR determination of 10 Fourier units and the FRC determination of 13 units were found to be ±1 unit. This means that even with 300 images, the resolution estimates obtained by DPR_{45} or $FRC_{0.5}$ may be as much as 10% off their asymptotic value. Fortunately, though, the number of particles in typical projects nowadays go into the thousands and tens of thousands, so that the error will be much smaller.

Another obvious flaw has not been mentioned yet: the splitting of the data set in half worsens the statistics, and will inevitably underreport resolution. Since the dependence of the measured resolution on the number of particles N entering the reconstruction is unknown, there is no easy way to correct the reported figure. Morgan et al. (2000) chose to determine this relationship empirically, by extrapolating from the trend of a curve of $FSC_{0.5}$ as a function of $\log(N)$, which the authors obtained by making multiple reconstructions with increasing N.

5.2.7. Q-Factor

This and the following section deal with criteria based on an evaluation of the whole data set. The Q-factor (van Heel and Hollenberg, 1980; Kessel et al., 1985) is easily explained by reference to a vector diagram (figure 3.23b) depicting the summation of equally-indexed (i.e., relating to the same discrete spatial frequency $\mathbf{k}$) Fourier components $P_i(\mathbf{k})$ in the complex plane, which takes place when an image set is being averaged. Because of the presence of noise (figure 3.23a), the vectors associated with the individual images zigzag in the approximate direction of the common signal. The Q-factor is simply the ratio between the length of the sum vector and the length of the total pathway of the vectors contributing to it:

$$Q(\mathbf{k}) = \frac{|\sum\limits_{i=1}^{N} F_i(\mathbf{k})|}{\sum\limits_{i=1}^{N} |F_i(\mathbf{k})|} \qquad (3.67)$$

Obviously, from its definition, $0 \leq Q \leq 1$. For pure noise, $Q(\mathbf{k}) = 1/\sqrt{N}$, since this situation is equivalent to the random wandering of a particle in a plane under Brownian motion (Einstein equation).

The Q-factor is quite sensitive as an indicator for the presence of a signal component, because the normalization is specific for each Fourier coefficient. A map of $Q(\mathbf{k})$ (first used by Kessel et al., 1985; see figure 3.23c) readily shows weak signal components at high spatial frequencies standing out from the background and thus enables the ultimate limit of resolution recoverable (*potential resolution*) to be established. Again, a quantitative statement can be obtained by averaging

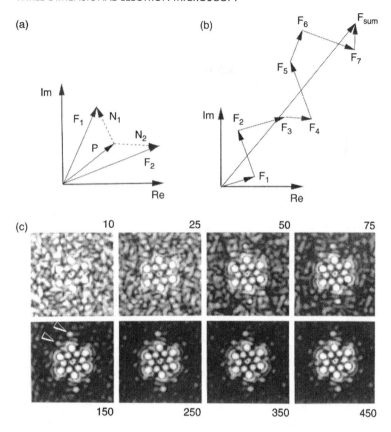

Figure 3.23 Averaging of signals in the complex plane. (a) Corresponding Fourier components (i.e., same-$\mathbf{k}$) F_1, F_2 of two aligned images that represent the same signal (P) but differ by the additive noise components N_1 and N_2. (b) Definition of the Q-factor: it relates to the addition of vectors (in this case $N = 7$) representing same-$\mathbf{k}$ Fourier components of seven images in the complex plane. F_{sum}/N is the Fourier component of the average image. The Q-factor is now defined as the ratio between the length of F_{sum} and the sum of the lengths of the contributing vectors F_i. Only in the absence of noise, can the maximum $Q = 1$ be reached. (c) Q-factor obtained in the course of averaging over an increasing number of repeats of a bacterial cell wall. As N increases, the Fourier components belonging to the noise background perform a random walk, while the signal-containing Fourier components all add up in the same direction, as illustrated in (b). As a result, the signal-related Fourier components stand out in the Q-factor map when N is sufficiently high. From Kessel et al. (1985), reproduced with permission of Blackwell Science Ltd.

this measure over rings in the spatial frequency domain, and plotting the result, $Q'(k)$, as a function of the ring radius $k = |\mathbf{k}|$. The stipulation that $Q'(k)$ should be equal to or larger than $3/\sqrt{N_{[k, \Delta k]}}$ can be used as a resolution criterion. For some additional considerations regarding the statistics of the Q-factor, see Grigorieff (1998).

Sass et al. (1989) introduced a variant of the Q-factor, which they called the S-factor:

$$S(\mathbf{k}) = \frac{|\frac{1}{N}\sum\limits_{i=1}^{N} F_i(\mathbf{k})|^2}{\frac{1}{N}\sum\limits_{i=1}^{N} |F_i(\mathbf{k})|^2} \tag{3.68}$$

This factor relates to the structural content (or, in the parlance of signal processing, the *energy*) of the images being averaged. The expectation value of the S-factor for Fourier coefficients of pure noise is $1/N_{[k,\Delta k]}$. Thus, a resolution criterion can be formulated by stipulating that the ring zone-averaged value of $S(\mathbf{k}) \geq 3/N_{[k,\Delta k]}$.

5.2.8. Spectral Signal-to-Noise Ratio

The spectral signal-to-noise ratio (SSNR) was introduced by Unser et al. (1987; see also Unser et al., 1989, and corresponding 3D forms introduced by Grigorieff, 2000, and Penczek, 2002a) as an alternative measure of resolution. It is based on a measurement of the SNR as a function of spatial frequency and has the advantage of having better statistical performance than DPR and FRC. Unser et al. (1987) also pointed out that the SSNR relates directly to the Fourier-based resolution criteria commonly used in crystallography. An additional advantage is that it allows the *improvement* in resolution to be assessed that can be expected when the data set is expanded, provided that it follows the same statistics as the initial set. This prediction can be extended to the asymptotic resolution reached in the case of an infinitely large data set. Finally, as we will show below, the SSNR provides a way to relate the DPR and the FRC to each other in a meaningful way.

Following Unser's treatment, the individual images q_i, which represent single molecules, are modeled assuming a common signal component $[p(\mathbf{r}_j); j = 1 \ldots, J]$ and zero-mean, additive noise:

$$q_i(\mathbf{r}_j) = p(\mathbf{r}_j) + n_i(\mathbf{r}_j) \qquad (i = 1 \ldots, N) \tag{3.69}$$

An equivalent relationship holds in Fourier space:

$$Q_i(\mathbf{k}_l) = P(\mathbf{k}_l) + N_i(\mathbf{k}_l) \qquad (i = 1 \ldots, N) \tag{3.70}$$

where $\mathbf{k}_l$ are the spatial frequencies on a discrete 2D grid indexed with l. Both in real and Fourier space, the signal can be estimated by averaging:

$$\bar{p}(\mathbf{r}_j) = \sum_{i=1}^{N} q_i(\mathbf{r}_j); \quad \bar{P}(\mathbf{k}_l) = \sum_{i=1}^{N} Q_i(\mathbf{k}_l) \tag{3.71}$$

The definition of the SSNR, $\alpha_\mathbf{B}$, is based on an estimate of the SNR in a local region $\mathbf{B}$ of Fourier space. It is given by

$$\alpha_\mathbf{B} = \frac{\sigma_{s\mathbf{B}}^2}{\sigma_{n\mathbf{B}}^2/N} - 1 \tag{3.72}$$

The numerator, $\sigma_{s\mathbf{B}}^2$, is the local signal variance, which can be estimated as

$$\sigma_{s\mathbf{B}}^2 = \frac{1}{n_\mathbf{B}} \sum_{l \in \mathbf{B}} |\bar{P}(\mathbf{k}_l)|^2 \tag{3.73}$$

where $n_\mathbf{B}$ is the number of Fourier components in the region $\mathbf{B}$. The denominator in equation (3.72), $\sigma_{n\mathbf{B}}^2$, is the noise variance, which can be estimated as

$$\sigma_{s\mathbf{B}}^2 = \frac{\sum_{l \in \mathbf{B}} \sum_{i=1}^{N} |P_i(\mathbf{k}_l) - \bar{P}(\mathbf{k}_l)|^2}{(N-1)n_\mathbf{B}} \tag{3.74}$$

By taking the regions $\mathbf{B}$ in successive computations of $\alpha_\mathbf{B}$ to be concentric rings of equal width in Fourier space, the spatial frequency dependence of $\alpha_\mathbf{B}$ can be found, and we obtain the curve $\alpha(\mathbf{k})$. Generally, the SSNR decreases with increasing spatial frequency. The resolution limit is taken to be the point where $\alpha(\mathbf{k})$ falls below the value $\alpha(\mathbf{k}) = 4$ (figure 3.24). Consideration of a numerical model has shown that this limit is roughly equivalent to DPR $= 45°$. The statistical analysis given by Unser et al. (1987) also allows upper and lower confidence intervals for the $\alpha = 4$ resolution limit to be established. For $N = 30$ images of the herpes simplex virus particles used as a test data set, the resolution was estimated as $1/29\,\text{Å}^{-1}$ ($\alpha = 4$), but the confidence intervals ranged from $1/27$ to $1/30\,\text{Å}^{-1}$. This indicates that for such small data sets, resolution estimates obtained with any of the measures discussed should be used with caution.

These authors also address the question of "How much is enough?," referring to the number of noisy realizations that must be averaged to obtain a satisfactory result. From the SSNR curve obtained for a given data set, the resolution improvement available by increasing N to N' can be estimated by shifting the threshold from $\alpha_N = 4$ to $\alpha_{N'} = 4N/N'$ (see figure 3.24).

The statistical properties of the SSNR are summarized in Penczek (2002a). For an FRC derived from independently processed halfsets (see section 5.2.2), the following simple relationships can be shown to hold:

$$FRC = \frac{SSNR}{SSNR + 1} \tag{3.74a}$$

and, conversely,

$$SSNR = \frac{FRC}{1 - FRC} \tag{3.74b}$$

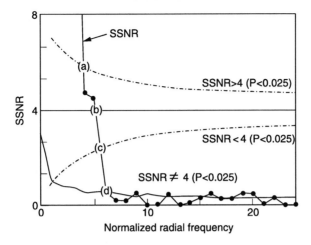

Figure 3.24 Experimental spectral signal-to-noise ratio (SSNR) curve obtained for a set of 30 images of the herpes virus Type B capsomer. With increasing spatial frequency, the curve drops rapidly from an initially high value (>8) to values close to 0. The SSNR resolution limit (here $1/29\,\text{Å}^{-1}$) is given by the spatial frequency where the experimental curve (b) intersects $SSNR = 4$. The dashed lines represent the (a) upper ($+2\sigma$) and (c) lower (-2σ) confidence limits for a measurement of $SSNR = 4$. The solid line at the bottom (d) represents the upper (2σ) confidence limit for a measurement of $SSNR = 0$. From Unser et al. (1987), reproduced with permission of Elsevier.

which needs to be modified into

$$SSNR = \frac{2FRC}{1 - FRC} \tag{3.74c}$$

taking into account that the FRC here reflects the statistics of the halfset. We see that the relationship between the two measures is the same as the one between the CCF and the SNR (section 4.3; Frank and Al-Ali, 1975).

5.3. Resolution and Cross-Resolution

At this point it will be useful to show how Fourier-based measures of resolution can be used to measure "cross-resolution." Although not rigorously defined, the concept originates with attempts to cross-validate results that come from different experiments or even different techniques of imaging. For instance, a structure might be solved both by X-ray crystallography and cryo-EM, and then the question can be asked to what extent, and to which resolution, the latter reproduces the former. It is immediately evident that this question can be answered by applying any of the *halfset criteria* (sections 5.2.3, 5.2.4, and 5.2.5) to the two different maps in question, since these will report the spatial frequency up to which structural information common to both experiments extends. An example for such an assessment was provided by Penczek et al. (1999), who compared density maps obtained from cryo-EM of ribosomes with those obtained by X-ray crystallography (Ban et al., 1999) and by Roseman et al. (2001), who compared a cryo-EM map of GroEL with a density representation derived from the X-ray structure. We will return to this point in chapter 6, section 3.3.

5.4. Resolution-Limiting Factors

Although the different resolution criteria emphasize different aspects of the signal, it is possible to characterize the resolution limit qualitatively as the limit, in Fourier space, beyond which the signal becomes so small that it "drowns" in noise. Beyond a certain limit, no amount of averaging will retrieve a trace of the signal. This "practical resolution limit" is the result of many adversary effects that diminish the contrast at high spatial frequencies. The effects of various factors on the contrast of specimens embedded in ice have been discussed in detail (Henderson and Glaeser, 1985; Henderson, 1992). Useful in this context is the so-called relative *Wilson plot*, defined in regions where $CTF \neq 0$ by

$$w(k) = \frac{\text{image Fourier amplitudes}}{\text{electron diffraction amplitudes} \times CTF} \tag{3.75}$$

where both the expressions in numerator and denominator are derived by averaging over a ring in Fourier space with radius $k = |\mathbf{k}|$. This plot is useful because it allows those effects to be gauged that are *not* caused by the CTF but nevertheless affect the image only, not the diffraction pattern of the same specimen recorded by the same instrument. In the parlance of crystallography, these would be called *phase effects* since they come to bear only because diffracted rays are recombined in the image, so that their phase relationship is important. The most important effects in this category are specimen drift, stage instability, and charging. [Here, we consider effects expressed by classical envelope terms associated with finite illumination convergence and finite energy spread (see chapter 2, section 3.3.2) as part of the CTF.] For instance, drift will leave the electron diffraction pattern unchanged, while it leads to serious deterioration in the quality of the image since it causes an integration over images with different shifts during the exposure time.

Therefore, an ideal image would have a Wilson plot of $w(k) = \text{const}$. Instead, a plot of $w(k)$ for tobacco mosaic virus, as an example of a widely used biological test specimen, shows a falloff in amplitude by more than an order of magnitude as we go from $1/\infty$ to $1/10\,\text{Å}^{-1}$. This cannot be accounted for by illumination divergence (Frank, 1973a) or energy spread (Hanszen, 1971; Wade and Frank, 1977), but must be due to other factors. Henderson (1992) considered the relative importance of contributions from four physical effects: radiation damage, inelastic scattering, specimen movement, and charging. Of these, only the last two, which are difficult to control experimentally, were thought to be most important. Both effects are locally varying, and might cause some of the spatial variation in the power spectrum across the micrograph described by Gao et al. (2002) and Sanders et al. (2003a).

Concern about beam-induced specimen movement led to the development of spot scanning (Downing and Glaeser, 1986; Bullough and Henderson, 1987; see section 3.2 in chapter 2). Charging is more difficult to control, especially at temperatures close to liquid helium where carbon becomes an insulator. Following Miyazawa et al. (1999), objective apertures coated with gold are used as a remedy, since their high yield of back-scattered electrons neutralizes the predominantly positive charging of the specimen.

In addition to the factors discussed by Henderson (1992), we have to consider the effects of conformational variability that is intrinsically larger in single molecules than in those bound together in a crystal. Such variability reduces the resolution not only directly, by washing out variable features in a 2D average or 3D reconstruction, but it also decreases the accuracy of alignment since it leads to reduced similarity among same-view images, causing the correlation signal to drop (see Fourier computation of the CCF, section 3.3.2).

5.5. Statistical Requirements following the Physics of Scattering

So far, we have looked at signal and noise in the image of a molecule, and the resolution achievable without regard to the physics of image formation. If we had the luxury of being able to use arbitrarily high doses, as is the case for many materials science applications of EM, then we could adjust the SNR to the needs of the numerical evaluation. In contrast, biological specimens are extremely radiation sensitive. Resolution is limited both on the sides of high and low doses. For high doses, radiation damage produces a structural deterioration that is by its nature stochastic and irreproducible; hence, the average of many damaged molecules is not a high-resolution rendition of a damaged molecule, but rather a low-resolution, uninformative image that cannot be sharpened by "deblurring." For low doses, the resulting statistical fluctuations in the image limit the accuracy of alignment; hence, the average is a blurred version of the signal.

The size of the molecule is a factor critical for the ability to align noisy images of macromolecules (Saxton and Frank, 1977) and bring single-particle reconstruction to fruition. The reason for this is that for a given resolution, the size determines the amount of structural information that builds up in the correlation peak ("structural content" in terms of Linfoot's criteria; see section 3.3 in chapter 2). Simply put, a particle with 200 Å diameter imaged at 5 Å resolution covers $200^2/5^2 = 1600$ resolution elements, while a particle with 100 Å diameter imaged at the same resolution covers only 400.

By comparing the size of the expected CCF peak with the fluctuations in the background of the CCF for images with Poisson noise, Saxton and Frank (1977) arrived at a formula that links the particle diameter D to the contrast c, the critical dose p_{crit}, and the sampling distance d. Accordingly, the minimum particle diameter D_{min} allowing significant detection by cross-correlation is

$$D = \frac{3}{c^2 d p_{crit}}$$

Henderson (1995) investigated the limitations of structure determination using single-particle methods in EM of unstained molecules, taking into account the yield ratio of inelastic to elastic scattering, which determines the damage sustained by the molecule, versus the amount of useful information per scattering event in bright-field microscopy. He asked the question "What is the smallest size of molecule for which it is possible to determine from images of unstained molecules the five parameters needed to define accurately its orientation (three parameters) and position (two parameters) so that averaging can be performed?" The answer he obtained was that it should be possible, in principle, to reconstruct unstained protein molecules with molecular mass in the region of 100 kDa to ~3 Å resolution.

However, by a long shot, this theoretical limit has not been reached in practice, for two main reasons: there is likely a portion of noise not covered in these calculations, and secondly, the signal amplitude in Fourier space falls off quite rapidly on account of experimental factors not considered in the contrast transfer theory, a discrepancy observed and documented earlier on (see Henderson, 1992). Experience has shown that proteins should be ~400 kDa or above in molecular mass for reconstructions at resolutions of 10 Å to be obtained. The spliceosomal U1 snRP (molecular mass of ~200 kD) reconstructed by Stark et al. (2001) is one of the smallest particles investigated with single-particle methods without the help of stain, but it does have increased contrast over protein on account of its RNA.

Another conclusion of the Henderson study that may be hard to reconcile with experiments is that the number of images required for a reconstruction at 3 Å resolution is ~19,000, independent of particle size. The reconstruction of the ribosome to 11.5 Å already required 73,000 images (Gabashvili et al., 2000), though later, due to improvements in EM imaging and image processing, the same resolution could be achieved with half that number (e.g., Valle et al., 2003a). Still, this means that 19,000 images is just half the number necessary to achieve a resolution that is lower by a factor of 3 than predicted. Only when the problems of specimen and stage instability and charging are under better control will the estimations by Henderson (1995) match with the reality of the experiment.

5.6. Noise Filtering

The reproducible resolution determined by using one of the methods described in section 3.2 of this chapter is a guide for the decision what part of the Fourier transform contains significant, signal-related information to be kept, and what part contains largely noise to be discarded. Since some signal contributions are found even beyond the nominal resolution boundary, low-pass filtration (or any nonlinear noise filtering techniques) should normally not be applied until the very end of a project, when a final 2D average or 3D reconstruction has been obtained. The reason is that low-pass filtration at any intermediate step would remove the opportunity to utilize the more spurious parts of the signal, and nonlinear filtering would prohibit the use of any linear Fourier-based processing afterwards.

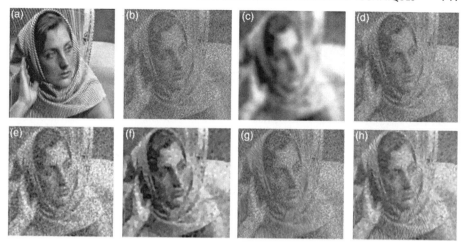

Figure 3.25 Demonstration of low-pass filtration and other means of noise reduction. (a) Test image (256×256); (b) test image with additive "white" noise ($SNR = 0.8$); (c) Gaussian filter; (d) median filter (5×5 stencil); (e) adaptive Wiener filter; (f) edge enhancing diffusion; (g) coherence enhancing diffusion; (h) hybrid diffusion. From Frangakis and Hegerl (2001), reproduced with permission of Elsevier.

In the following, the profiles of three linear filters for low-pass filtration are given. Other, nonlinear filters are discussed by Frangakis and Hegerl (2001). A demonstration of both types of filters, reproduced from that work, is given in figure 3.25.

Low-pass filters must be designed such that transitions in Fourier space are smooth, to avoid "ringing" or "overshoot" artifacts (also known as *Heaviside phenomenon*) that appear in real space along edges of objects. In those terms, a filter with a top-head profile would have the worst performance. A Gaussian filter, defined as

$$H(k) = \exp[-k^2/2k_0^2] \qquad (3.75a)$$

is ideal in this regard, but it has the disadvantage of attenuating, as a function of spatial frequency, too soon while coming to a complete blocking too late relative to the desired spatial frequency radius.

Two other Fourier filters are more frequently used as they offer a choice of width for the transition zone: the Fermi filter (Frank et al., 1985) and the Butterworth filter (Gonzales and Woods, 1993).

The Fermi filter follows the distribution of particles following the Fermi statistics at a given temperature; hence, the width of the transition zone is specified by the "temperature" parameter T:

$$H(k) = \frac{1}{1 + \exp[(k - k_0)/T]} \qquad (3.75b)$$

The Butterworth filter has the following profile:

$$H(k) = \frac{1}{1 + |(k/k_0)|^{2\kappa}} \tag{3.75c}$$

where κ is a parameter that determines the order of the filter.

Profiles of some of the filter functions, along with the associated point-spread functions, and examples for their application, are contained in appendix 2.

6. Validation of the Average Image

Questions of significance and reproducibility of features have been addressed above in this chapter (section 4.2). The problem with such treatments is that they must be based on assumptions regarding the statistical distribution of the noise. In the following we will discuss a technique that requires no such assumptions.

The question of significance of features in the average image can be addressed by a method of multiple comparison that makes no assumptions regarding the statistics: the rank sum method (Hänicke, 1981; Hänicke et al., 1984). Given a set of N images, each represented by J pixels:

$$p_{ij} = \{p_i(\mathbf{r}_j); \; j = 1 \ldots, J; \; i = 1 \ldots, N\} \tag{3.76}$$

The nonparametric statistical test is designed as follows: each pixel p_{ij} of the ith image is ranked according to its value among the entire set of pixels $\{p_{ij}; j = 1 \ldots, J\}$: the pixel with the lowest value is assigned rank 1, the second lowest, rank 2, and so forth. Finally, the pixel with the largest value receives rank J. In the case of value ties, all pixels within the equal-value group are assigned an average rank. In the end, the image is represented by a set of rank samples $\{r_{i1}, r_{i2}, \ldots, r_{iJ}\}$, which is a permutation of the set of ordered integers $\{1, 2, \ldots, J\}$ (except for those rank samples that result from ties). After forming such rank representations for each image, we form a *rank sum* for each pixel:

$$R_j = \sum_{i=1}^{N} r_{ij}, \qquad 1 \leq j \leq J \tag{3.77}$$

In order to determine whether the difference between two pixels, indexed j and k, of the average image,

$$\bar{p}_j = 1/N \sum_{i=1}^{N} p_{ij}, \qquad \bar{p}_k = 1/N \sum_{i=1}^{N} p_{ik} \tag{3.78}$$

is statistically significant, on a specified significance level α, we can use the test statistic

$$c_{jk} = |R_j - R_k| \tag{3.79}$$

Now the following rule holds: *whenever c_{jk} is greater than a critical value $D(\alpha, J, N)$, the difference between the two pixels is deemed significant.* This critical value has been tabulated by Hollander and Wolfe (1973) and, for large N, by Hänicke (1981).

In addition to answering questions about the significance of density differences, the rank sum analysis also gives information about the spatial resolution. Obviously, the smallest distance (*critical distance*) between two pixels that fulfills

$$c_{jk} > D(\alpha, J, N) \tag{3.80}$$

is closely related to the local resolution in the region around those pixels. The nature of this relationship and a way to derive a global resolution measure from this are discussed by Hänicke et al. (1984).

4

Multivariate Data Analysis and Classification of Images

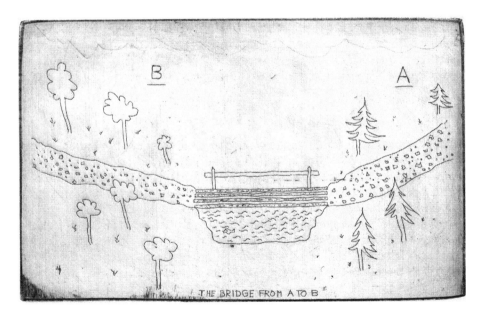

Figure 4.1 The bridge from A and B. (Etching, J. Frank.)

1. Introduction

Classification, the division of a set of images into subsets with similar features (figure 4.1), is the overarching theme of this chapter. In the reconstruction of highly ordered biological objects, which form two-dimensional (2D) crystals, or helical or icosahedral assemblies, we could say that classification has already taken place, by virtue of the equivalence of intermolecular chemical forces, which assures that only building blocks that are alike are assembled. By contrast, in the reconstruction of a molecule from single particles, classification of images according to orientation, conformation, and state of ligand binding must be done computationally, at the level of the image data.

1.1. Heterogeneity of Image Sets

The need for a quantitative analysis of variability present in the data set has become evident in chapter 3, in the treatment of 2D image averaging. Any variations in the orientation of the molecule on the support will obviously result in variations of the projected image. (In the case of one-sided negative staining, see below, these variations are even further amplified by meniscus effects.) Furthermore, macromolecules may exist in a variety of conformational states. Many of the associated image variations are subtle, too small to be observed directly in the electron micrograph, but they are revealed under close scrutiny when entire sets of images are compared. Nonstoichiometric ligand binding not only leads to inconsistent appearance of the ligand mass, but also often to variability in the conformation of the target molecule. Finally, for negatively stained, air-dried specimen preparation, the forces acting on the molecule often produce strong deformations. The molecule is effectively "flattened" in the direction perpendicular to the support, an effect that critically depends on the orientation of the molecule relative to the support film.

1.1.1. Signal Variability Versus Noise

What is important in understanding the aim of classification is the conceptual difference between variations in the image entirely *unrelated to the object* (which we might call "true noise") and those variations that are due to something that happens to the object ("signal-related variations"). Since the ultimate goal of our study is to obtain meaningful three-dimensional (3D) images of the macromolecule, we should merge only those data that originate from particles showing the macromolecule in the same state (i.e., shape, conformational state, degree of flattening, etc.). On the other hand, there is no practical way, for an unknown object, to unambiguously distinguish signal variations from noise in the data, except perhaps that the noise is not localized, but normally spread over the entire image field.

1.1.2. Homogeneous Versus Heterogeneous Data Sets

The terms "homogeneous" and "heterogeneous" are obviously relative. For instance, a certain range of orientation or variation is acceptable as long as the effect of these variations is much smaller than the resolution distance.

To consider the magnitude of this tolerance, we might take the 50S ribosomal subunit, whose base is roughly 250 Å wide. A tilt of the subunit by 1° would lead to a foreshortening of the particle base to $250 \times \cos(1°) = 249.96$ Å, which leads to a very small change of shape compared to the size of the smallest resolved features (10 Å). From this consideration, we could infer that orientation changes by 1° will not change the appearance of the projection significantly. In reality, however, the ability to differentiate among different patterns is a small function of the amount of noise. In the absence of noise, even a change that is a fraction of the resolution distance will be detectable. A general criterion for homogeneity could be formulated as follows (Frank, 1990): "a group of images is considered homogeneous if the intra-group variations are small compared with the smallest resolved structural detail." Boisset et al. (1990b) used such a test to determine homogeneous subgroups in a continuously varying molecule set.

Two ways of scrutinizing the image set based on a statistical analysis have already been introduced in chapter 3: the variance map and the rank sum analysis. Neither of these methods, however, is capable of dealing with a completely heterogeneous data set, that is, a data set consisting of two or more different groups of images. For such data sets, new tools are required, which are able to group similar images with one another, and discern clusters. Various classification techniques exist, but these are most efficient, given the low signal-to-noise ratio (SNR), when a *reduced representation* of the data is available. Much of the following is concerned with the introduction of such reduced representations. Once these are established, we then move on to the subject of classification techniques.

1.2. Images as a Set of Multivariate Data

We assume that a set comprising N images is given, $\{pi(\mathbf{r}); i = 1, \ldots, N\}$. Each of the images is represented by a set of discrete measurements on a regular Cartesian grid:

$$\{p_{ij}; j = 1, \ldots, J\} \tag{4.1}$$

where the image elements are lexicographically ordered as before (see section 2.2 in chapter 3).

We further assume that, by using one of the procedures outlined in chapter 3, the image set has already been aligned; that is, any given pixel j has the same "meaning" across the entire image set. In other words, if we were dealing with projections of "copies" of the same molecule lying in the same orientation, that pixel would refer to the same projection ray in a 3D coordinate system affixed to the molecule. In principle, the images of the set can vary in any pixels: they form a multivariate data set.

Variations among the images are in general due to a linear combination of variations affecting groups of pixels, rather than to variations of a single pixel. In the latter case, only a single pixel would change while all other pixels would

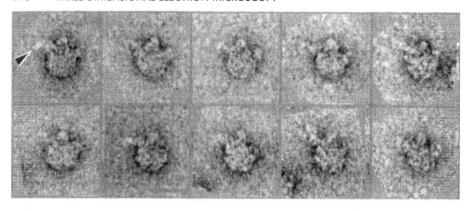

Figure 4.2 A specimen used to demonstrate the need for classification: although the micrograph shows the same view of the molecule (the crown view of the 50S ribosomal subunit from *E. coli*), there are differences due to variations in staining and in the position of flexible components such as the "stalk" (indicated by the arrow-head). From Frank et al. (1985), reproduced with permission of Van Nostrand-Reinhold.

remain exactly the same. Such a behavior is extremely unlikely as the outcome of an experiment, for two reasons: for one, there is never precise constancy over any extended area of the image, because of the presence of noise in each experiment. Second, unless the images are undersampled or critically sampled—a condition normally avoided (see section 2.2 in chapter 3)—pixels will not vary in isolation but in concert with surrounding pixels. For example, a flexible protuberance may assume different positions among the molecules represented in the image set. This means that an entire set of pixels undergoes the same coordinate transformation; for example, a rotation around a point of flexure (see figure 4.2). Another obvious source of correlation among neighboring image elements is the instrument itself: each object point, formally described by a delta-function, is imaged as an extended disk (the *point-spread function* of the instrument; see section 3.3 in chapter 2). Because of the instrument's imperfections, image points within the area of that disk are correlated; that is, they will not vary independently.

1.3. The Principle of Making Patterns Emerge from Data

"Making patterns emerge from data" is the title of a groundbreaking paper by Benzecri (1969a), introducing *correspondence analysis* to the field of taxonomy. Benzecri's treatment (see also Benzecri, 1969b) marks a departure from a statistical, model-oriented analysis in the direction of a purely descriptive analysis of multidimensional data. This new approach toward data analysis opened up the exploration of multivariate data for which no statistical models or at best sketchy models exist, for example in anthropology and laboratory medicine. Correspondence analysis is a technique closely related to *principle component analysis* (PCA), and both methods will be described in detail below.

Data displayed as patterns appeal to the eye; they are easily captured "at a glance." Whenever data can be represented in visual form, their inter-relationship can be more easily assessed than from numerical tables (see the fascinating book on this subject by Tufte, 1983). In a way, Benzecri's phrase "patterns emerging from data" anticipated the new age of supercomputers and fast workstations that allow complex numerical relationships and vast numerical fields to be presented in 3D form on the computer screen. In recent years, the power of this human interaction with the computer has been recognized, and it has become acceptable to study the behavior of complex systems from visual displays of numerical simulations. In fact, looking at such visual displays may be the only way for an observer to grasp the properties of a solution of a nonlinear system of equations. (A good example was provided by the July 1994 issue of *Institute of Electrical and Electronics Engineers Transactions on Visualization and Computer Graphics*, which specialized in visualization and featured on its cover a parametric solution to the fourth-degree Fermat equation.)

1.4. Multivariate Data Analysis: Principal Component Analysis Versus Correspondence Analysis

1.4.1. Introduction

The methods we employ for multivariate data analysis[1] (see Lebart et al., 1977, 1984) fall into the category of eigenvector methods of ordination. What these methods have in common, in the application to image data, is that they are based on a decomposition of the total interimage variance into mutually orthogonal components that are ordered according to decreasing magnitude. This decomposition is tailored specifically to the data set analyzed.

The lexicographical notation we introduced earlier preconceives the idea of the image as a vector in a multidimensional space $\mathbf{R}^J$ (sometimes referred to as *hyperspace*), where $J = L \times M$ is the total number of pixels. Alternatively, we can think of the image as a point (namely, a vector end point) in that space. As an exercise in the use of this concept, note that the points representing all images that differ in the value of a single pixel at the point with coordinate (l, m) but are constant otherwise lie on a line parallel to the coordinate axis numbered $j = (m - 1) \times L + l$.

This example shows, at the same time, that the original coordinate axes in $\mathbf{R}^J$ are not useful for analyzing variations among images, since such variations always involve simultaneous changes of numerous pixels, whereas a movement along any of the coordinate expresses the change of one pixel only, with the rest of the pixels staying constant.

---

[1] We will use the term *multivariate data analysis* (MDA) for the direct application of PCA and correspondence analysis (CA) to the images. The term *multivariate statistical analysis* (MSA), often used in this context, is too broad in its current usage in statistics (I thank Pawel Penczek and Jose-Maria Carazo for a clarification of this point). By contrast, "multivariate data analysis" draws attention to Benzecri's concept of "making patterns emerge from data" without the use of statistical models.

1.4.2. The Image Set Represented as a Cloud in Hyperspace

If a set of images is given, the vector end points form a "cloud" in "hyperspace." For an introduction to this concept, let us consider a set of images generated from the noise-free projection of a macromolecule by the addition of random noise. In the space $\mathbf{R}^J$, any image of this set would be represented by addition of a vector with random orientation and random length (as governed by the statistical distribution) to the fixed vector representing the image of the molecule. The resulting vector end points representing the entire set of molecule images would form a multidimensional cloud, whose center lies in the point that represents the average of all noisy images. The more noisy images we have, the better will their average approximate the original projection. In hyperspace, this means the point representing the average and the point representing the original image lie closely together.

1.4.3. Definition of Factors in Principal Component Analysis

Our aim is to find the vectors that define the directions of the principal extensions of the data cloud, since these vectors are naturally suited to describe the variations in the data set. These principal directions are constructed as follows: (i) find the maximum extension of the cloud; (ii) find the vector, perpendicular to the first, that points in the direction of the next-largest extension of the cloud; (iii) find the vector, perpendicular to both the first and the second, etc. The different successive "size measurements," which describe the shape of the data cloud with increasing accuracy, are components of the total interimage variance, and the method of finding these measurements along with the new, data-adapted coordinate system is called *eigenanalysis* in physics, and *principal component analysis* (PCA) in statistics. Another term for this method, which is prevalent in image and signal analysis, is *Karhunen–Loeve transformation* (see Rosenfeld and Kak, 1982). In the following, the mathematical principles of PCA are outlined before *correspondence analysis* (CA) is introduced.

The simplest way to introduce the concepts of PCA and the classification of images by considering a 2D scatter diagram of two variables, for example, elevation of a geographic site and its average temperature. Such a plot may reveal a thin stretched "cloud," whose trend suggests an anticorrelated behavior; that is, higher elevations on the average go along with lower temperatures. This trend can be represented by a straight line, thus reducing the 2D to a one-dimensional (1D) relationship. The line represents a 1D subspace of the initially 2D space. The simplest *classification* problem would arise if the points on the scatter diagram were to fall into two clusters. This would call for a judgment on which points belong to which of the two clusters. In both cases, the act of representing the data in a systematic way, by assigning a spatial coordinate to each variable in a scatter diagram, is of great help in visualizing the underlying statistical pattern.

The 2D diagram in which the data are represented has the shortcoming that it would be suitable only for representing "images" having no more that two pixels. We could for instance assign the value of pixel 1 to the horizontal and the value

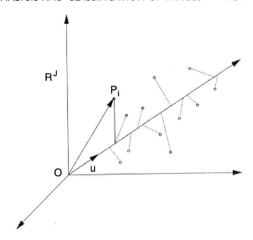

Figure 4.3 Definition of a principal axis in $\mathbf{R}^J$ resulting from the analysis of a point cloud P_i. For a given vector $\mathbf{u}$, we can construct the projections of all the vectors $\mathbf{x}_i = \bar{O}\bar{P}_i$. Among all possible choices of vectors $\mathbf{u}$, we select the one for which the sum of the squared projection is maximum. In that case, $\mathbf{u}$ points in the direction of the cloud's principal extension (after Lebart et al., 1977).

of pixel 2 to the vertical axis of the scatter diagram. Obviously, any pattern that we would normally call an "image" contains many more pixels and thus involves an initial representation in a space with much higher dimensionality. Images consisting of four pixels would already require a 4D space, beyond our capacity to visualize. Even though the dimension of the space is higher, the goal of the analysis—namely finding a low-dimensional *subspace* in which the variations can be expressed with minimum loss of information—remains the same.

In a least-squares sense, the "direction of maximum extension" of the data cloud, the term used above in describing the purpose of MDA, can be defined as follows (figure 4.3): we seek a vector $\mathbf{u}$ in $\mathbf{R}^J$ such that the sum of squared *projections[2] of the N image vectors $\mathbf{x}_i$ onto $\mathbf{u}$ is a maximum. [Before proceeding, we observe that this problem can be alternatively formulated in the N-dimensional subspace $\mathbf{R}^N$, if the number of images is less than the number of pixels, that is, $N < J$.] The above least squares condition thus leads to the equation:

$$\sum_{i=1}^{N} \bar{O}\bar{P}_i^2 = \sum_{i=1}^{N} (x_i\mathbf{u})^2 = (X\mathbf{u})'X\mathbf{u} = \mathbf{u}'X'X\mathbf{u} \rightarrow \max \qquad (4.2)$$

with the constraint:

$$u'u = 1 \qquad (4.3)$$

[2]Here and in the following, the term *projection* has two different meanings: first, the ordinary meaning, employed so far, as an image of a 3D object resulting from a projection operation; i.e., by adding up all density values of the object which lie along defined "rays of projection." The second meaning refers to the result of a vector projection operation in the J-dimensional Euclidean space $\mathbf{R}^J$, or in a factor subspace. Obviously, electron microscopic images will keep occurring in both contexts: as the result of a type-1 projection and as an element subjected to a type-2 projection. To avoid confusion, and draw attention to this distinction, the second type will henceforth be annotated with an asterisk: "*projection."

(*orthonormalization condition*). Here, we have introduced a matrix $\mathbf{X}$ that contains the image vectors as rows:

$$\mathbf{X} = \begin{pmatrix} x_{11} & x_{12} & . & . & . & x_{1J} \\ x_{21} & x_{22} & . & . & . & x_{2J} \\ . & & & & & . \\ . & & & & & . \\ . & & & & & . \\ x_{N1} & x_{N2} & & & & x_{NJ} \end{pmatrix} \tag{4.4}$$

The problem posed by equations (4.2) and (4.3) can be tackled by posing the eigenvector-eigenvalue equation:

$$\mathbf{D}\mathbf{u} = \lambda\mathbf{u} \tag{4.5}$$

where matrix $\mathbf{D}$ is defined as

$$\mathbf{D} = (\mathbf{X} - \bar{\mathbf{X}})'(\mathbf{X} - \bar{\mathbf{X}}) \tag{4.5a}$$

$\bar{\mathbf{X}}$ represents a matrix containing the average image in each row, and λ is a multiplier; $\mathbf{D}$ is called the *covariance matrix*, since its general elements are

$$d_{ii'} = \sum_{j=1}^{J} (x_{ij} - \bar{x}_j)(x_{i'j} - \bar{x}_j) \tag{4.6}$$

with $\bar{x}_j$ being the average image. Equation (4.5), solved by diagonalizing the matrix $\mathbf{D}$, has at most q solutions $\{\mathbf{u}_1, \mathbf{u}_2, \ldots, \mathbf{u}_q\}$ where $q = \min(N, J)$. These solutions form an orthogonal system of basis vectors in $\mathbf{R}^J$ and are called *eigenvectors* of matrix $\mathbf{D}$. They are associated with a set of *eigenvalues*, discrete values of the multiplier introduced in equation (4.5):

$$\lambda = \{\lambda_1, \lambda_2, \ldots, \lambda_q\} \tag{4.6a}$$

Since $\mathbf{D}$ is symmetric and positive-definite [i.e., $\det(\mathbf{D}) > 0$], the eigenvalues and eigenvectors are real, and the eigenvectors may be ranked by the size of the eigenvalues in descending order. Each eigenvalue gives the share of total interimage variance expressed by the associated eigenvector, so that the sum of all eigenvalues is the total interimage variance, equal to the trace of $\mathbf{D}$. This relationship is expressed by

$$\mathrm{Tr}(\mathbf{D}) = \sum_{m=1}^{q} \lambda_m \tag{4.7}$$

which is invariant under a rotation of the coordinate system.

1.4.4. Correspondence Analysis

Correspondence analysis (CA) is an exploratory technique designed to analyze two-way tables that contain some measure of correspondence between the rows (here: images) and columns (here: pixels). The data in each row and column are normalized into "relative frequencies" such that they add up to 1. While in PCA the analysis is based on the computation of Euclidean distances, it is based in CA on the computation of χ^2 distances between these relative frequencies.

The reason for the prominent role of CA in electron microscopic applications is in part historical and in part due to a fortuitous match of problem with technique. What counts in favor of CA compared to PCA (as defined above, via analysis of the covariance matrix) in the context of electron microscopic applications is the fact that CA ignores multiplicative factors between different images; hence, it appears better suited for analyzing images obtained in bright-field electron microscopy (EM) and recorded on film, irrespective of their exposure. For, according to measurements, the optical density on the film is in a good approximation proportional to electron dose in the interesting range (see Reimer, 1997). Thus, particle images from different micrographs can in principle be combined without rescaling.

(On the other hand, CA has the same property of "*multiplicative equivalence*" for the pixel vectors, and here this property can be detrimental because all pixels enter the analysis with the same weight, irrespective of their average value across the image series. This means, for instance, that in the analysis of cryo-EM images, the variations among the realizations of a given pixel within the particle boundary and outside of it have the same weight. However, the use of a mask sufficiently close to the boundary of the particle prevents this property from becoming a problem.)

Correspondence analysis, unlike PCA, requires positive input data. Being the results of physical measurements (electron exposure; optical density), the raw images from the electron microscope are of course always positive. However some preprocessing procedures (e.g., Boisset et al., 1993) include a step of average subtraction, which renders the preprocessed images partially negative. It should be noted that data in this form cannot be used as input to CA.

2. Theory of Correspondence Analysis

A detailed introduction into the theory of CA is found in the books by Lebart et al. (1977, 1984) and Greenacre (1984). A brief formulation using matrix notation was given by Frank and van Heel (1982a). In the current introduction, we follow the convenient presentation of Borland and van Heel (1990), who use Lebart's notation in a more compact form.

Correspondence analysis deals with the analysis of N image vectors $(x_{ij}; j = 1, \ldots, J)$ in $\mathbf{R}^J$, and, simultaneously, with the analysis of J pixel vectors $\{x_{ij}; i = 1, \ldots, N\}$ in $\mathbf{R}^N$ (the space which is said to be *conjugate* to $\mathbf{R}^J$). These two legs of the analysis are symmetric and closely linked. The geometric interpretation of image vectors as a point cloud in $\mathbf{R}^J$, which was introduced in section 1.4, is now complemented by the simultaneous interpretation of pixel vectors as a point cloud in $\mathbf{R}^N$.

2.1. Analysis of Image Vectors in $\mathbf{R}^J$

With the choice of a χ^2 metric unique to CA, the least-squares condition derived before (in section 1.4) to seek a principal direction of the data cloud leads to the new eigenvector–eigenvalue equation:

$$\mathbf{X'NXMu} = \mathbf{u}\lambda \qquad (4.8)$$

with the orthonormalization constraint:

$$\mathbf{u'\,Mu} = 1 \qquad (4.9)$$

where $\mathbf{M}$ $(N \times N)$ and $\mathbf{N}$ $(J \times J)$ are diagonal matrices describing the N objects in "image space" $\mathbf{R}^J$ and "pixel space" $\mathbf{R}^N$, respectively. The nonzero elements of $\mathbf{M}$ and $\mathbf{N}$ are, respectively,[3]

$$m_{jj} = 1/x_{\cdot j}, \quad j = 1, \ldots, J \qquad (4.10)$$

and

$$n_{ii} = 1/x_{\cdot i}, \quad i = 1, \ldots, N \qquad (4.11)$$

with the standard notation (see Lebart et al., 1984) of $x_{i\cdot}$ and $x_{\cdot j}$ as "marginal weights":

$$x_{i\cdot} = \sum_{j=1}^{J} x_{ij} \quad x_{\cdot j} = \sum_{i=1}^{N} x_{ij} \qquad (4.12)$$

(Note that, apart from the factor N, $x_{\cdot j}$ is the average image of the entire set.) As mentioned in the introductory comments, the rescaling of each row and column has the effect that the analysis treats images that are related by a multiplicative factor as identical. For example, the pixel values in a pair of bright-field images $\{x_{1j}; j = 1, \ldots, J\}$ and $\{x_{2j}; j = 1, \ldots, J\}$ of the same object recorded with two different exposure times t_1, t_2 are related by

$$x_{2j} = x_{1j} \times t_1/t_2 \qquad (4.13)$$

provided, first, that the optical density is linearly related to the electron exposure [which holds in good approximation both for recording on film and with a

---

[3]The definitions of the elements of the metrics $\mathbf{M}$ and $\mathbf{N}$ [equations (4.9) and (4.10)] in *Three-Dimensional Electron Microscopy of Macromolecular Assemblies* (Frank, 1996) were incorrect and inconsistent with the rest of the equations in that book. (I thank Luis Comolli for pointing the error out to me.)

CCD camera and, second, that the object is sufficiently resistant to beam damage and retains its structural integrity. *In that case, the two images are treated as exactly identical by CA.*]

[Comparison with PCA as introduced above shows two differences: one is that the average image is not subtracted (i.e., $\mathbf{X}$ appears in equation (4.8), not $\mathbf{X} - \bar{\mathbf{X}}$), and the other is the appearance of the metrics $\mathbf{M}$ and $\mathbf{N}$.]

Equation (4.8) generates an entire set of solutions $\{\mathbf{u}_m, m = 1, \ldots, q\}$ with associated eigenvalues $(\lambda_m, m = 1, \ldots, q)$ The eigenvectors $\mathbf{u}_i$ are of dimension J and can again be interpreted as images. Borland and van Heel (1990) used a compact formulation of the eigenvector–eigenvalue equation:

$$\mathbf{X}'\mathbf{NXMu} = \mathbf{U}\Lambda_q \tag{4.14}$$

in which the eigenvectors $\mathbf{u}_m$ are "bundled up," so that they form columns of the eigenvector matrix $\mathbf{U}$, and the eigenvalues λ_m form the diagonal of the diagonal eigenvalue matrix Λ_q. The orthonormalization condition now becomes

$$\mathbf{U}'\mathbf{MU} = \mathbf{I}_J \tag{4.15}$$

with the diagonal unitary matrix $\mathbf{I}_J$.

2.2. Analysis of Pixel Vectors in $\mathbf{R}^N$

As before (and keeping within the same compact formulation), we seek vectors that maximize the projections of the data vectors in principal directions:

$$\mathbf{XMX}'\mathbf{NV} = \mathbf{V}\Lambda_p \tag{4.16}$$

with the diagonal eigenvalue matrix Λ_p and the orthonormalization condition:

$$\mathbf{V}'\mathbf{NV} = \mathbf{I}_N \tag{4.17}$$

Based on the properties of the matrices $\mathbf{X}'\mathbf{NXM}$ and $\mathbf{XMX}'\mathbf{N}$, the eigenvalues in equations (4.16) and (4.14) are the same, assuming that all eigenvalues are different from zero. In contrast to the eigenvectors in $\mathbf{R}^J$, which lend themselves to an interpretation as images (*eigenimages*, see below), the eigenvectors that form the columns of the eigenvector matrix $\mathbf{V}$ do not have an easily interpretable meaning.

However, these latter eigenvectors and those that form the columns of $\mathbf{U}$ are linked by the so-called *transition formulas*:

$$\mathbf{V} = \mathbf{XMU}\Lambda^{-1/2} \tag{4.18}$$

$$\mathbf{U} = \mathbf{X}'\mathbf{NV}\Lambda^{-1/2} \tag{4.19}$$

These relationships are of practical importance since they allow the eigenvector set contained in U to be computed in the conjugate space R^N, where the matrix to be inverted is of dimension $N \times N$, instead of the space R^J with the matrix D [dimension $J \times J$; see equation (4.5a)]. Depending on the number of pixels passed by the mask, versus the number of images, computation in R^N may be faster.

One of the advantages of CA is that, because the data are simultaneously normalized by rows and columns, the eigenvectors U and V are on the same scale, as are the images and pixel coordinates, so that images and pixels can be represented on the same factor map (simultaneous representation). It is legitimate (see Lebart et al., 1984) to draw conclusions based on the position of a data point with respect to all data points of the conjugate space. This means, for instance, that an image that falls in an extreme position on the map will be generally surrounded by pixels that "explain" its extreme position. However, it is not possible to interpret the proximity between individual images and individual pixels in a meaningful way. Thus far, with few exceptions (e.g., Frank et al., 1982; Borland and van Heel, 1990), the simultaneous representation has received little attention in the EM literature.

2.3. Factorial Coordinates and Factor Maps

Each image $\{x_{ij}; j = 1, \ldots, J\}$ can now be represented by a set of coordinates in the new space formed by the set of eigenvectors contained in U. These eigenvectors are commonly called *factors*; hence, the term *factorial coordinates* for the image coordinates. (In the following, the terms "factor" and "eigenvector" will be used interchangeably.) As in PCA, the new space has the property that the interimage variance (or "energy") is concentrated in a small number of factors, $q_{\text{pract}} \ll q$. However, this number of useful factors depends on the number of images and on the type of specimen, as will become clear later.

It is important to re-emphasize at this point that there are exactly as many factors as there are independent data vectors, which is potentially a large number. The aforementioned concentration of energy is the result of two principles underlying the analysis: first, the search for new, optimum directions better suited to the data in the vector space (see figure 4.3), and second, the ranking of these factors according to the size of their contribution to the total variational energy.

In practice, therefore, only some small number of q_{pract} coordinates need be computed for each image. These coordinates are obtained from the original set of images, stored in the rows of matrix X [see definition in equation (4.4)], by an operation of *projection:

$$\Psi = NXMU = NV\Lambda^{1/2} \tag{4.20}$$

Similarly, the coordinates of the pixels are obtained as

$$\Phi = MX'NU = MU\Lambda^{1/2} \tag{4.21}$$

[Note here that in the second step of these derivations, the transition formulas in equations (4.18) and (4.19) were used.] In practice, the coordinates of the images, forming the elements of $\mathbf{\Psi}$, $\{\psi_{im}; m = 1, \ldots, q_{\text{pract}}; i = 1, \ldots, N\}$, constitute the immediately most useful result of CA. *Factor maps* (e.g., see figure 4.4) are 2D maps in which each image is printed as a symbol at a position with the coordinates $\{\psi_{i_{m1}}, \psi_{i_{m2}}\}$ of two selected factors m_1 and m_2. The symbol that is printed can be (i) a simple dot if one is only interested in the distribution of the image on the map, (ii) the image "ID," normally a several-digit number identifying the image, or (iii) a class symbol, if a classification has been performed (see below) and one is interested in the distribution of the classes on the factor map, but not in the identity of the class members. The second option (image ID) is most often used, preceded by a symbol characterizing the status of the image in the analysis: "a" for active, "i" for inactive.

Active images are those that participate in the analysis from the very start, forming the rows of matrix $\mathbf{X'NXM}$ [see equation (4.8)]. *Inactive images* can be any images that have the same dimensions as the active set and have been passed through the same mask, but were not included in the matrix. Therefore, they do not participate in the spanning of the factor space, but are merely "co-*projected" along with the active images and shown on the factor maps so that their relative positions can be investigated.

What precisely is the use of a factor map? We can think of it as an inventory of the image set according to rationales whose principles are laid down in the least squares formula. To obtain an insight into the rationales (i.e., the reasons for a particular ordering of the images along two factors on the map) for a given data set requires some work; in other words, merely staring at the distribution of image symbols on the map will not necessarily help. An important tool for the interpretation of the factors will be introduced below: *reconstitution*, which is a recipe for building up an image from its factorial components, as though from individual building blocks.

2.4. Reconstitution

2.4.1. Principle

To recapitulate, the eigenvectors $\{\mathbf{u}_m; m = 1 \ldots, q\}$, where $q = \min(N, J)$ is the number of nonzero eigenvalues, form an orthonormal basis. This means that it is possible to reassemble each image uniquely by a linear combination of these new vectors (figure 4.5):

$$x_{ij} = x_{i.} \left\{ x_{.j} + \sum_{m=1}^{m_{\max}} \psi_{im} u_{mj} \right\} \qquad (4.22)$$

By letting $m_{\max} = q$, the image (here indexed with i) is restored without loss of information. Another form of this so-called *full reconstitution* makes use of the relationship in equation (4.21) and expresses the eigenvectors in terms of the pixel coordinates (Bretaudiere and Frank, 1986; van Heel, 1986a):

$$x_{ij} = x_{.j} x_{i.} \left\{ 1 + \sum_{m=1}^{m_{\max}} \psi_{im} \phi_{jm} \right\} \qquad (4.23)$$

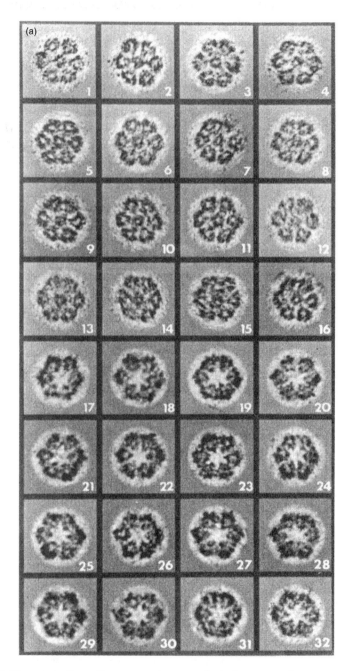

Figure 4.4 Demonstration of correspondence analysis for a set of macromolecules that vary strongly in one feature. (a) Set of worm hemoglobin molecules with (1–16) and without (17–32) extra central subunit. (b) Factor map (1 versus 2) separates images (represented by their ID numbers) unambiguously into two clusters along factor 1, the most prominent factor. Insets show images obtained by averaging over the respective clusters. From Frank and van Heel (1982b), reproduced with permission of Deutsche Gesellschaft für Elektronenmikroskopie, e.V.

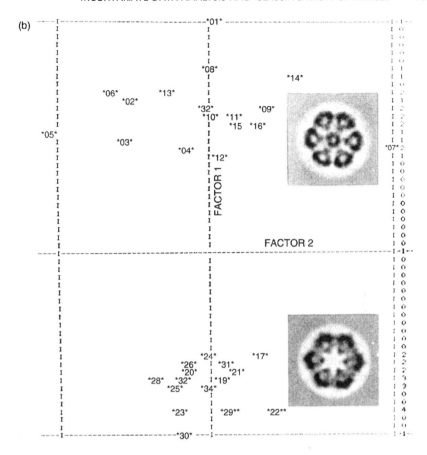

Figure 4.4 (*Continued*).

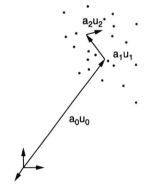

Figure 4.5 Principle of reconstitution. Each image, represented by a vector in the space $\mathbf{R}^J$, can be approximated by successive addition of vectors pointing in the direction of the eigenvectors u_1, u_2, etc., to the vector $a_0\mathbf{u}_0$ representing the average image, which is centered in the data cloud. Only a few of these additional vectors are needed to recreate the distinguishing features of the image.

A more important use of equation (4.22) is obtained when the series is broken off after a smaller number of factors, $m_{max} < q$; in that case, the expansion approximates the original image, with small, less significant components of interimage variance left out. Because of the strong condensation of interimage variance in the highest-ranking factors encountered in many experiments, the number of factors can be much smaller than q, and thus a substantial compression of information is achieved.

For example, we take 1000 images, each of which is represented by 64×64 pixels (Frank, 1990). Let us assume that the molecule is situated within a mask that has 2500 elements. Let us further assume that 10 terms are sufficient to represent the important variations among the image set. We then need 10×1000 image coordinates, plus 10×2500 pixels of the factors to be stored, or a total of 35,000 numbers. This should be compared with the original number of pixels, namely 2.5 million. Thus, the data compression factor achieved in this case is on the order of 70.

The form [equation (4.22)] of the reconstitution formula means that an image indexed i may be built up by adding to the average image the individual components:

$$x_{ijm} = x_i \cdot \psi_{im} u_{jm} \tag{4.24}$$

where x_i. is a scaling factor [the marginal weight of each image, see definition in equation (4.12)], ψ_{im} is the mth coordinate of image i, and u_{jm} is the jth element of the mth eigenvector. Depending on the value and sign of coordinate ψ_{im}, the element u_{jm} is added with different weights to (or subtracted from) the total sum. Interpreted as an image, $\{u_{jm}; j = 1 \dots, J\}$ is also known as an *eigenimage*.

2.4.2. Analogy to Fourier Analysis and Synthesis

The analogy to the discrete Fourier analysis and synthesis of an image (appendix 1) is now obvious: in that case, the eigenvectors are complex exponentials. Fourier analysis is in essence the computation of *projections of the vector that represents the image onto a fixed set of orthonormalized vectors representing the complex exponentials. The resulting set of components F_l is referred to as the *Fourier transform* of the image. In the resynthesis of the image from its Fourier transform, terms of the form $F_l \exp(2\pi i \mathbf{r}_j \mathbf{k}_l)$ are summed, which can be understood as "eigenimages" with different weights. As terms with increasing spatial frequencies $\mathbf{k}_l$ are being added, a more and more refined approximation of the image is created. As in the case of Fourier synthesis, reconstitution can be terminated by including only those terms that are deemed most important. In the case of the Fourier expansion, this termination (*low-pass filtration*; see section 4 in chapter 3 and appendix 2) has the consequence that high-resolution features and some portions of the noise are suppressed. In the case of CA or PCA, minor effects of interimage variability are suppressed by terminating the expansion at some practical index q_{pract}.

Looking at the individual terms of the expansion, we can interpret the eigenvectors as eigenimages, which are added to the average image with different weights (given by the factorial coordinates) in a closer and closer approximation to the image represented. Inspection of the eigenimages can, therefore, be quite informative in determining the meaning of individual factors (see section 3.5).

An important difference between the synthesis of an image from its Fourier components and its synthesis from its CA- or PCA-based eigenimages is that in the former case, the eigenvectors are fixed, while in the latter case, the eigenvalues are "tailored" to the variational pattern present in the entire image set. Another way of saying this is that the multidimensional coordinate system of the Fourier expansion is absolute, whereas the coordinate system used in CA or PCA is relative and data-dependent. This difference is important because it implies that the results of PCA or CA obtained for two different image sets cannot be compared, factor by factor, unless the experimental methods are precisely matched and, in addition, the pixel-defining mask and the number of images are exactly the same. Normally only a few very strong factors are found to be virtually identical; for example, in negatively stained specimens, the one that expresses the strong variation of stain level on the peripheral boundary of particles (e.g., Frank et al., 1982). With increasing factor number, the coordinate systems obtained for different image sets will increasingly differ from one another by "rotations." For this reason, reconstitution of an image requires that the image be a member of the ensemble for which the eigenvector analysis has been carried out, whereas Fourier synthesis of a single image can be done entirely without reference to such an ensemble.

Reconstitution is very useful as a means to explore the meaning of individual factors and to gauge the most important variations that exist in the image set (Bretaudiere and Frank, 1986; see section 3.5). For this purpose, depending on the circumstances, only a few selected factors, or even only a single factor, may be used in the reconstitution sum, equation (4.23).

2.5. Computational Methods

The computation of the eigenvectors and eigenvalues is done by direct matrix inversion, either of matrix $\mathbf{X'NXM}$ [equation [(4.8) or (4.14)] or of $\mathbf{XMX'N}$ equation (4.16)], depending on the number of images, N, compared with the number of pixels in the mask, J. The transition formulas, equations (4.18) and (4.19), can always be used to convert the eigenvectors from one space to the conjugate space. Often, the number of images is smaller than the number of pixels in the mask, making the inversion of $\mathbf{X'NXM}$ the fastest. For example, a set of 1000 images with a mask passing 2000 pixels requires inversion of a matrix with the size 1000×1000, which occupies 4 Mb in core memory and is easily accommodated in current workstations. If neither $N \times N$ nor $J \times J$ should fit into the memory, iterative algorithms (e.g., Lebart et al., 1984) can be used, which operate by applying the eigenvalue equation iteratively, starting with a random seed vector.

2.6. Significance Test

In electron microscopic applications with low SNR, it is important to know if the eigenvectors found by MDA relate to significant variations in the signal, or merely reflect the multidimensional fluctuations of noise. More specifically, it typically happens that eigenvectors are significant only up to a certain number, and it is important to find out that number. The answer to this question can be obtained by an analysis of noise alone. A data set consisting of pure noise has an eigenvalue spectrum whose characteristics depend on the number of images, the number of pixels, and the noise statistics (Lebart et al., 1984). For the eigenvectors to be significant, the associated eigenvalues should stand out from the noise eigenvalue spectrum. Because of the dependency of this "control" spectrum on the noise statistics, it is not possible to tabulate it in a useful way. Instead, the control analysis has to be done with direct reference to the data to be investigated (Lebart et al., 1984) by applying MDA to a noise distribution that has the same first-order statistics as the data.

For this purpose, a new dataset is created by a process of "random shuffling." Each image of the set is subjected to the following process:

(i)   Set counter to $n = 0$
(ii) Draw two numbers j_1, j_2 at random without replacement from the lexicographic range of pixel indices, $1 \ldots, J$
(iii) Exchange pixels p_{j_1} and p_{j_2}
(iv) Increment counter $n = n + 1$
(v)   If $n < J$ go to (ii) else stop

The new data set no longer has a common standard deviation since the random shuffling is independent in each image. All signal-related, systematic variations will be eliminated by this process. However, the resulting noise data set has the desired property that it possesses the same statistical distribution as the original data set and therefore acts as a true control.

An analysis of this kind was done to test the significance of results obtained when applying CA to patch-wise averaged low-dose images of the purple membrane (Frank et al., 1993).

3. Correspondence Analysis in Practice*

3.1. A Model Image Set Used for Demonstration

To understand how CA works in practice, it is most instructive to apply it to an image set with well-defined variational patterns. In the following, two image sets will be used. The first to be introduced (figure 4.6) is totally synthetic and presents a face varying in a binary way in three different features, so that eight $(= 2^3)$ different versions of the face can be distinguished (Bretaudiere and Frank, 1986). This data set will be used to exemplify the application of CA, and, in a later section, classification.

The other image set, to be introduced below (section 3.4, figure 4.10) derives from a reconstruction of the 70S ribosome of *Escherichia coli* (Gabashvili et al., 2000), which was projected into nine directions, arranged in groups of three closely related ones. In both cases, noise was added to simulate the experimental conditions.

3.2. Definition of the Image Region to Be Analyzed

3.2.1. Masking

In preparation for CA, the image set is put into a compact form and stored in a single file, one record per image. The data from each image are selectively

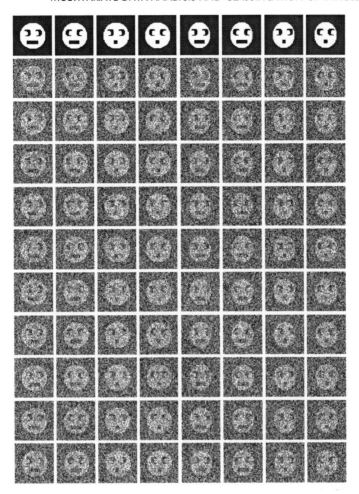

Figure 4.6 The face data set. Three different features of the face (shape, eyes, mouth) appear in two variants (round/oval, eyes left/right, mouth wide/narrow, respectively). Each of the eight noise-free prototypical images (top row) was used to generate 10 noisy versions by adding noise, resulting in the images in rows 2–11. Looking at any column, one can see how noise interferes with the recognition of the features. In some cases, small feature distinctions such as the positions of the eyes become difficult to discern. The challenge is to identify images belonging to the same prototype in a random mix of the 10 × 8 noisy versions.

copied into this master file under control of a binary mask. The mask is an array of 0s ("block") and 1s ("pass") stored in a file that has the same size as the images of the series. This mask controls which part of the image is subjected to CA. Usually, the mask is designed to pass the entire molecule, and closely follows the molecule's boundary. However, there are often applications in which the molecule presents different views, in which more than one species of molecule is present, or in which molecules with and without a bound ligand coexist. In those cases, the mask must be constructed such that it can pass all pixels of all images.

Masking off areas outside the particle ensures that only those areas in which structural information resides will participate in the analysis. Exclusion of peripheral noise increases the SNR and makes the eigenvalue spectrum more compact. In other words, the masking assures that existing variations within the molecule set are represented by the smallest number of factors. Another use of the mask is to focus on one region of the molecule while deliberately ignoring the variations in the remaining regions.

3.2.2. Tailoring a Mask to the Shape of a Molecule

Masks are rarely just geometric shapes (disk, rectangle, etc.), but are tailored specifically to the geometry of the molecule. For generating a binary mask function, two tools are needed: one that allows a binary image to be generated from a continuously valued template image, and one that allows different masks to be combined. If the image set is homogeneous, then the shape of the molecule projection is more or less described by the shape of the global average:

$$\bar{p}(\mathbf{r}) = \frac{1}{N} \sum_{i=1}^{N} p_i(\mathbf{r}) \qquad (4.25)$$

Since the average is reasonably smooth, a first approximation to a mask function would be generated by "thresholding," that is, application of a threshold such that

$$M(\mathbf{r}) = \begin{cases} 0 & \text{for} \cdot p(\mathbf{r}) < T \\ 1 & \text{elsewhere.} \end{cases} \qquad (4.26)$$

with an empirically selected threshold T. In this form, the mask function is as yet unsatisfactory, for the following reason: in the practical application, the choice of T allows little control over the size of the mask when the boundary of the particle is well defined in the average of the *majority* of the molecules. Of concern are the *minority* molecules that have features reaching outside the boundary of the average. If the mask follows the boundary of the global average too closely, then CA (or PCA) is unable to distinguish those minority molecules on the basis of their "outside" features. It is, therefore, always advisable to expand the mask produced by equation (4.26) and create a safety margin. A simple procedure for expanding (or shrinking) a binary mask function was described by Frank and Verschoor (1984): the mask function is low-pass filtered using a Gaussian function $\exp[-k^2/k_0^2]$ in Fourier space, followed by a second thresholding step (figure 4.7).

The smooth function generated by low-pass filtration has an approximate falloff halfwidth of $r_0 = 1/k_0$ near the binary edge of the original mask function. Within this distance, the choice of a new threshold T will generate a family of shapes all completely contained within one another like 2D versions of Matryoshka dolls. The expansion procedure can be repeated if necessary. A small falloff halfwith r_0 has the advantage that the expansion follows the previous boundary closely, but has the disadvantage that only a small extra margin is gained.

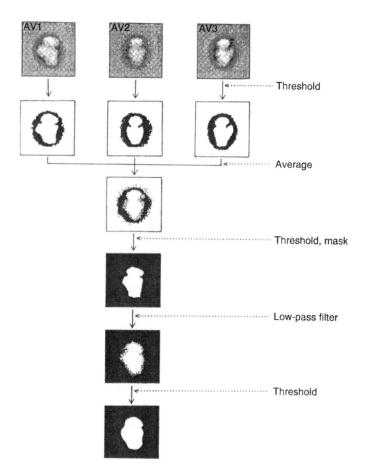

Figure 4.7 Creation of binary masks from averages. This scheme shows two concepts: the creation of an inclusive mask designed to pass several shapes, and a method for expanding (or contracting) a mask. When a binary mask is used to define the pixels to be analyzed, we wish to make sure that it will not truncate part of the distinguishing information, which might be outside of the boundary of the "majority" of the molecules. For a heterogeneous set of molecules consisting of three subsets A, B, and C, the mask has to pass the shape of molecules belonging to subset A and the shape of molecules belonging to subset B *and* the shape of molecules belonging to subset C. This can be accomplished by forming averages of subsets A, B, and C, and then going through the procedure indicated on the drawing: pass the subset averages (first row) through a threshold operation, which results in binary masks with values 0 and 1 (second row). These are added up, resulting in a mask with values 0, 1, 2, or 3 (third row). From this four-level mask, an inclusive binary mask is created by assigning 1 (=pass) to all pixels that have values 1–3 (fourth row). With appropriate choices of threshold in the first step, the inclusive binary mask will pass molecules from all three subsets. The additional, optional two steps at the bottom of this scheme can be used to change the size of the binary mask in a continuous way. The mask is first low-pass filtered (by application of box-convolution or Fourier filtration; appendix 2), then passed through another thresholding step. Depending on the choice of threshold in this step, the mask either expands or contracts. From Frank and Verschoor (1984), reproduced with permission of Elsevier.

Another operation that can be used to expand an existing binary mask is *dilation by region growth* (Russ, 1995). Along the boundary of the mask, pixels with value 1 are allowed to grow by switching neighbors to 1, until some constraint is satisfied. For an application of this technique in EM, see the procedure for separating RNA from protein in low-resolution cryo-EM density maps developed by Spahn et al. (2000).

3.2.3. Generation of a Minimal All-Inclusive Mask

A complication arises when the image set shows the molecule in dissimilar projections. In that case, the "all-inclusive" shape can be found, in principle, from the global average by choosing the threshold sufficiently low (Frank and Verschoor, 1984). However, this method breaks down, because of its sensitivity to noise, if the fraction of images in any of the subsets is small.

Alternatively, an all-inclusive mask $\hat{M}(\mathbf{r})$ can be generated from a given set of masks representing the shapes of the constituent molecule projections $M_1(\mathbf{r}), M_2(\mathbf{r}), \ldots, M_N(\mathbf{r})$ following inclusive "OR" logics. The inclusive mask is derived in the following way, going through an intermediate step that produces a nonbinary function $C(\mathbf{r})$:

$$C(\mathbf{r}) = \sum_{i=1}^{N} M_i(\mathbf{r}) \tag{4.27}$$

$$\hat{M}(\mathbf{r}) = \begin{cases} 0 & \text{for } M(\mathbf{r}) < 1 \\ 1 & \text{elsewhere} \end{cases} \tag{4.28}$$

3.3. Eigenvalue Histogram and Factor Map

3.3.1. Preliminaries

Once the data have been prepared (i.e., masked and packed into the compact form required), one needs to decide how many factors to compute. The maximum number of factors is determined by the dimensionality of the problem; that is, the minimum of the number of pixels and the number of images. This number is normally much larger than the number of factors that contribute significantly to the variational pattern of the image set. Algorithms that operate by direct matrix inversion (if the problem fits into the available computer memory) produce all factors simultaneously, so in that case it is not necessary to put any restrictions on the number of factors. For other algorithms that work iteratively a great time saving can be achieved by limiting the computation to those highest-ranking factors that are expected to be significant.

3.3.2. Number of Significant Factors

The significance of factors is assessed from the appearance of the *eigenvalue histogram*, a plot of the eigenvalues ordered by size (i.e., rank). We recall that

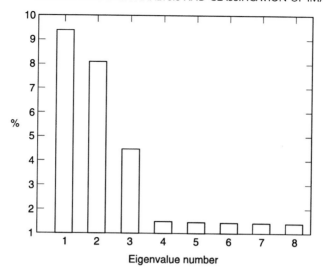

Figure 4.8 Eigenvalue histogram of the face data set. Three eigenvalues stand out clearly from the rest, indicating that the significant variations are concentrated in a three-dimensional factor space. Beyond factor 3, the histogram trails off, displaying the typical behavior noise. From this histogram, the decision would be reached that a display of factor maps 1 versus 2, 1 versus 3, and 2 versus 3 would be sufficient to gauge the distribution of the data, and the possible existence of clusters.

the eigenvalues are proportional to the share of interimage variance claimed by each factor. Typically, such a histogram shows a rapid drop as we proceed from the first to the subsequent factors (figure 4.8). This first segment of the histogram is typically followed by another segment, toward higher factors, which has a much smaller slope. The intersection between the empirical envelopes of these two segments can be taken as a guide for deciding on the factor beyond which the "signal" variation is drowned in noise.

This division into segments is not always clear-cut, however. Another way to distinguish eigenvectors attributable to signal from those attributable to noise is through the analysis of a control data set that is produced by the reshuffling process described earlier, see section 2.6.

3.3.3. Generation of Factor Maps

Having determined the number of factors, we go to the next step, the generation of factor maps. This is only useful, however, if the number of significant factors is relatively small, or if the first one, two, or three factors stand out from the rest. In that case, maps of selected pairs of factors (e.g., 1 versus 2, or 1 versus 3) (figure 4.9) can be meaningfully interpreted. In other cases, where a larger number of eigenvalues have comparable values, the analysis of the multidimensional manifold by pair-wise factor maps becomes unwieldy and incomprehensible.

A factor map is a projection of the entire point cloud onto a Cartesian plane spanned by the two selected factors. On such a map, the position of each image is

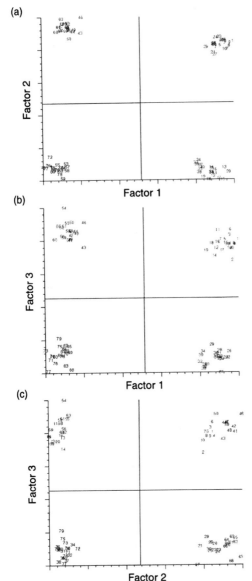

Figure 4.9 Factor maps of the face data set. Each of the factor maps, 1 versus 2 (a), 1 versus 3 (b), and 2 versus 3 (c), show very clear clustering into four clusters. This indicates that exactly eight clusters are present, sitting in the eight corners of a cubic space. A check of the numbers reveals that images are clustered according to their origin in figure 4.6.

determined by its corresponding coordinates (section 2.3). For the map of factor 1 versus factor 2, the coordinates of (a) are $\{\Psi_{i1}, \Psi_{i2}\}$.

Each image is designated by a symbol. For symbols one would normally choose the numbers identifying the images throughout the analysis (the so-called image ID), but they could also be chosen to be letters identifying entire groups of images as having originated from certain experimental batches or micrographs. They may also be chosen to indicate the membership of images in classes according to a prior classification. Finally, it may even be sufficient to use the same symbol throughout (such as a dot or asterisk) if the sole use of the map is to analyze the shape of the distribution, without the need to track down individual images. This shape may contain clues to the following questions: Are the data clustered in a nonrandom way? Is there any evidence for an ordering of the data along a multidimensional path?

In the case of the model data set, the points representing the eight groups of images evidently occupy eight separate regions in a 3D factor space. They are mostly tightly clustered. (Although the identity of the images is not shown here, one can easily verify that the division is exactly along the lines of the eight prototypes seen in figure 4.6.) Divisions are most distinct and unambiguous in the factor maps of 1 versus 2. The distribution of data across factor 3 is more scattered.

3.4. Case Study: Ribosome Images

With the tools at hand, we now move to a demonstration of the technique closer to the praxis of single-particle analysis. When the data set analyzed comprises molecules in all possible orientations (after formal alignment to a common reference; see chapter 3, section 3.4), then the interimage variability is dominated by the shape variations. This is the case for both ice-embedded and negatively stained particles. Clustering may occur if some views are preferred over others, as often happens for negatively stained molecules.

To illustrate the clustering according to shape, a data set of 90 images was constructed by projecting a density map of the ribosome (Gabashvili et al., 2000) into three entirely different directions. In each of these directions, an additional slight "rocking" was imposed, which produced a more subtle distinction into three subgroups. The nine resulting projections were "contaminated" with noise (SNR = 0.2), each in 10 different random realizations (figure 4.10).

First, we look at the eigenvalue spectrum (figure 4.11). In the analysis of EM images, the number of significant factors usually depends on the type of specimen preparation (negatively stained versus ice embedded) and on the source of variability (a data set containing a strongly asymmetric molecule in all possible orientations, versus a data set containing a molecule showing a narrow range of orientations around a preferred view, or with conformational variability).

In the current example, just two eigenvalues stand out from a slowly decaying background, indicating that two factors are all that are needed to analyze the grouping. Indeed, the 1 versus 2 factor map (figure 4.12) clearly shows a grouping of the images, but instead of the expected nine groups, only three are found. The reason for this result is that only three strongly different groups of orientations are present, while the subdivisions of each group according to a $\pm 5°$ change in orientation within each group is completely obscured by noise. It is, however, easy to verify that the division of the images into the three main clusters occurs without any misclassification.

Negatively stained molecules, as a rule, have a very strong concentration of interimage variance in a small number of factors, with a large share often claimed by the variability in the depth of the peripheral stain (see Frank et al., 1982; Boisset et al., 1993b). In contrast, when ice-embedded molecules lying in the same orientation are analyzed, they yield a much flatter, more noise-like distribution

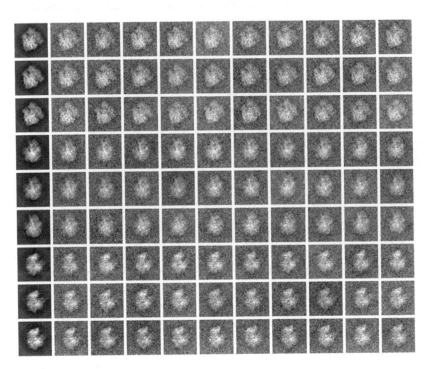

Figure 4.10 Ribosome data set. First column on the left: nine projections of the 70S ribosome density map. For each of three primary viewing directions, three projections were computed by "rocking" through the angles −5°, 0°, and +5°. Columns 2 to 11: each image was created from the image in the first column by addition of random Gaussian noise.

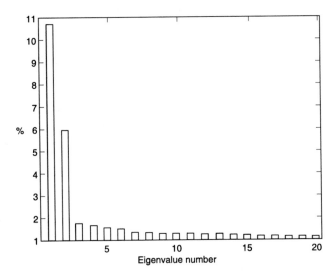

Figure 4.11 Eigenvalue histogram of the data set in figure 4.10 following application of correspondence analysis. Evidently, the main part of the variation is concentrated in two factors, which means that the distribution of data can be assessed from a single factor map, of factor 1 versus 2.

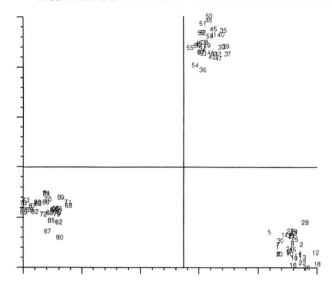

Figure 4.12 Factor map (1 versus 2) of the ribosome data set, showing clustering into the three main view groups. Further distinction into "rocking"-related subgroups is apparently drowned in noise.

of eigenvalues, which makes it difficult to recognize the jump between factors associated with signal and those with noise. If MDA is followed by classification in factor space, then it is important to know that a large number of factors (up to 50) may have to be included in the case of cryo-EM data.

3.4.1. Heterogeneity—A Relative Concept

As an aside, the result in figure 4.12 exemplifies the concepts of *heterogeneity* and *homogeneity* introduced earlier (section 1.1), and shows that they are relative. The entire data set is clearly heterogeneous as it consists of three easily distinguishable groups that must not be mixed in the computation of averages or in the input to a reconstruction. Even each group is itself heterogeneous in origin since each comes from projections with different — albeit closely related orientations. However, the presence of noise in the amount characteristic for low-dose images makes these finer distinctions difficult to detect.

A look at figure 4.12 also shows an interesting analogy between the process of imaging an object (in 3D space) with an optical instrument, and the process (represented in hyperspace) by which a motif (a 2D signal) is rendered in a noisy form. Multiple images of a motif form a noisy set that is depicted by a point cloud in hyperspace, in analogy to an object point that is depicted as an extended image—the point-spread function associated with the imaging process. We can juxtapose the concepts in the indicated in table 4.1.

Just as resolution is limited by the width of the point spread function, so the ability to classify noisy images stemming from different motifs is limited by the size of the cloud representing the noise, that is, its variance.

Table 4.1 Analogy between the Imaging Process in 3D Space with the Process of Noise Contamination as Depicted in Hyperspace

3D Space	Hyperspace
Object point	Motif
Point-spread function	Multidimensional cloud centered on motif
Distance	Dissimilarity (distance in hyperspace)
Closeness	Similarity (proximity in hyperspace)
Resolution = ability to separate images of two points	Ability to classify by separating point clouds originating from different motifs

3.5. Use of Explanatory Tools

Although, *a priori*, there is no intrinsic reason for the existence of a one-to-one relationship between factors and independent physical causes of the variation, such relationships are nevertheless found in some cases. Common examples are height of the peripheral stain meniscus (see figure 2.6), magnification changes, and azimuthal misalignment. All these are often reflected by separate factors. In that sense, we are talking about the "meaning" of a factor: what physical variational feature is associated with it?

To identify the meaning of a factor, and track down its possible physical cause, we can make use of *explanatory tools*. The following tools are available: explanatory images, local averages, eigenimages, and reconstituted images. Each will be briefly described in the following.

3.5.1. Explanatory Images

The most straightforward way of gauging the rationale underlying a factor is by displaying images that fall at the extremes of the map along the factor axis under consideration. Such "explanatory images" typify the effects of this factor. However, a shortcoming of this kind of analysis is that raw images may be quite noisy, tending to obscure the variational component expressed by the factor under consideration. Another drawback is that such images express variational components associated with other factors, as well. To avoid the second shortcoming, one could look for images that have extreme positions along the factor axis to be analyzed *and*, at the same time, have low values along the other factor axes.

3.5.2. "Local" Averages

Averaging over just a few images lying together in close proximity to one another at the extremes of the factorial axis improves the situation dramatically, since noise is reduced and the components from the other factors balance one another out. We call images generated in this way *local averages*, where the term "local" refers to a small region, in factor space, from where the images are drawn to form the average. This region could be a circumscribed region on a particular factor map (see figure 4.13), or a multidimensional region

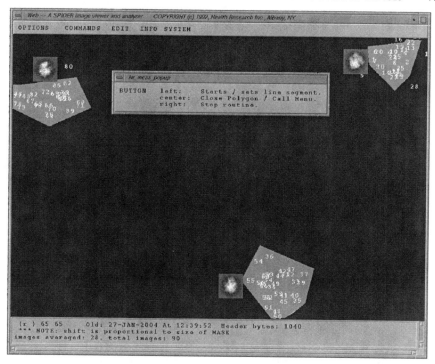

Figure 4.13 "Local" averages of ribosome images in factor space. An interactive tool (in this case, WEB; Frank et al., 1996) allows polygonal boundaries to be drawn in a factor map, to identify images to be averaged. Here, this action allows the three main views of the ribosome to be recovered (see the noise-free projections in the first column of figure 4.10).

defined by analytical boundaries, such as the selection of all images (indexed i) that fulfill

$$-a \leq \Psi_m^{(i)} \leq a; m = 1 \ldots, m_{\max} \text{ ("hypercube")} \tag{4.29}$$

or

$$\sum_{m=1}^{m_{\max}} |\Psi_m^{(i)} - \Psi_m^c|^2 \leq R^2 \text{ ("hypersphere")} \tag{4.30}$$

where $\{\Psi_m^c; m = 1 \ldots, m_{\max}\}$ are the center coordinates and R is the radius of the multidimensional sphere.

In the foregoing, attention has been exclusively on the explanation of the factors, but it should be stressed that local averages are equally useful in exploring the meaning of the grouping of images on the map: why they cluster together, and what precisely makes them special compared to the rest of the image set. In deciding on the size of the region in factor space, one must strike a balance between the need to average noise and the need to obtain a true sample of the varying features. Obviously, if the region is chosen too large, then the variations will be partially leveled.

Local averages relating to a selected factor map can be computed and displayed "on the fly" with suitable interactive software (figure 4.13). In the example shown, a polygonal boundary is drawn on the factor map, which defines a subset of images for which the average (i.e., the local average) is to be formed. This average is then instantly computed and displayed on the map right next to the polygonal boundary.

For continuous distributions of data in factor space, it is often helpful to make an exhaustive display of local averages computed on a regular grid overlaid on the factor map ("local average map"). At one glance, we are then able to see the rationale behind the particular ordering. Figure 4.14 shows such a map, made in the course of a study on the ryanodine receptor/calcium release channel (see Radermacher et al., 1994b). Evidently, the molecules are arranged primarily

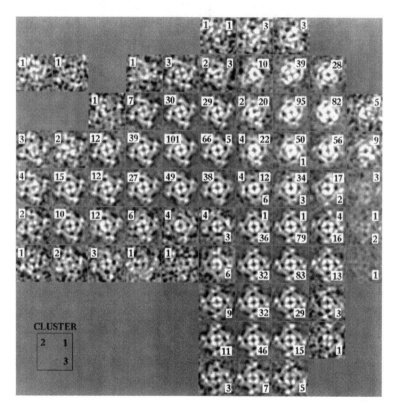

Figure 4.14 Checkerboard display of local averages (calcium release channel), each computed from images falling on a grid in factor space (factors 1 versus 2). The number on top of each average indicates the number of images falling into that grid space. Empty regions are bare of images. The distinction of main interest is between molecules lying in different orientations, related by flipping. It is seen from the peripheral pinwheel features pointing either clockwise (on the left) or counterclockwise (on the right). From Frank et al. (1996), reproduced with permission of Elsevier.

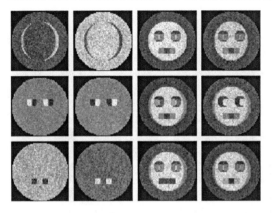

Figure 4.15 Eigenimages and their effect when added to the average image. First two columns, first through third rows: positive and negative eigenimages associated with factors 1–3. Last two columns: images obtained by adding positive or negative eigenimage to the average of all images. It is seen that the resulting two images always display the attributes expressed by the relevant factor, while displaying average attributes related to the other factors. So in the first row, the face is sharply round or oval while both mouth and eyes are blurred. In the second row, only the eyes are sharply defined (and directed) while the shapes of the face and mouth are blurred, etc.

according to their handedness (flip versus flop; horizontal direction). We will later introduce the term "inventories" for this type of systematic, exhaustive accounting (section 4.6).

3.5.3. Eigenimages

To see the effect of a factor in its purest form, we can add the corresponding eigenimage with different weights to the average image. Eigenimages have peculiar properties since their "job" is to create the associated variation (e.g., the shift of the eyes from left to right) by a correlated addition and subtraction of appropriate masses in different positions to and from the average image. In our example, the term relating to factor 2 must take away mass at two points relating to the "unwanted" position of the eyes, and add it where it is "needed." Thus, eigenimages always contain regions with positive and negative values, and are by themselves (i.e., without adding the average image) difficult to interpret. However, by adding them singly to the average image (figure 4.15), one is able to see variability they express quite clearly.

3.5.4. Reconstituted Images

Earlier we have seen that reconstitution offers a way to synthesize an image from its factorial contributions. Thus, we can gradually build up an image of interest, following equation (4.23), by adding the contribution of each factor, one at a time (figure 4.16). Each factor modifies the image in a characteristic way, and by following these changes, the meaning of the factors can be inferred.

Figure 4.16 Stepwise reconstitution of noisy images (second row of figure 4.10) and stopping after the third factor leads to virtually noise-free images that are very similar to the originals (shown for comparison in first row); second row: average image plus eigenvector #1; third row: same as second row, with eigenvector #2 added; 4th row: same as third row, with eigenvector #3 added.

4. Classification

4.1. Background

In the preceding sections, we have seen several examples of particle hetero-geneity. Data reduction techniques such as CA and PCA make it possible to quantify and visualize the multidimensional distribution of the data, and help us decide which particles "belong" to one another and form a *class*. (Without the data reduction, the grouping of images that may differ in any of their pixel values is extremely difficult.) Since structural similarity translates into proximity in factor space, clusters of points in factor space correspond to homogeneous subsets of images originating from molecules that have the same conformation and exhibit the same view. A good example is provided by figure 4.4, which shows images of a worm hemoglobin molecule with and without the central subunit, all showing the molecule in the top view. Correspondence analysis produced a sharp division of the image set, and complete segregation of the two subsets into clusters in factor space. A decision on (i) how many classes exist and (ii) which particles fall into which of the two classes is straightforward by visual inspection of the distribution on the factor map. Since the distinction in the image is so obvious, direct inspection of the molecules themselves also produces the same unambiguous answer in this case.

This section is devoted to automated or semiautomated techniques of classification, in which decisions on the number of classes and membership in these classes are made by following a quantitative analysis of the data in the computer. Often these techniques amount to a reformulation of the problem and the provision of tools that facilitate the decisions by the user. For instance, the user may need to specify the number of classes, and the algorithm will then give an unambiguous answer on which images belong to which class.

4.2. Overview over Different Approaches and Goals of Classification

4.2.1. Classification Based on Features Versus Direct Analysis of the Digital Representation

Many approaches to computer-assisted or automated classification of objects in the general literature are based on an analysis of "features"—lists of numbers, grouped into arrays (*feature vectors*), that represent measurements derived from a separate analysis of the images. For example, the width of a molecule, the angle formed by two of its domains, and the presence or absence of an antibody label would be represented by three numbers that might vary independently. We can then base the classification of a set of such molecules on a classification of their feature vectors. This approach has found little use in single-particle reconstruction. Rather, we will exclusively deal with an approach to classification that is based on a direct analysis of the digital (pixel-by-pixel) representation of the images in its entirety, or of reduced representations derived from these by MDA as outlined in sections 1–3 of this chapter.

4.2.2. Supervised Versus Unsupervised Classification

Supervised classification groups data according to the degree of their similarity to prototypes (or templates or references), while unsupervised classification groups them according to their mutual relationships without any such guidance. A typical application of supervised classification is the reference-based orientation determination by 3D projection matching, in which particle images are compared with projections of an existing reconstruction or density model (chapter 5, section 7.2). Unsupervised classification, in contrast, requires no references and is used to find any grouping of images that may have a physical origin; for example, orientational preferences of a molecule, caused by differential surface properties; presence of different molecules in a mixture; or coexistence of molecules with different conformational states or states of ligand binding.

4.2.3. Hard Versus Fuzzy Classification

The classification techniques we will be dealing with use "hard," or mutually exclusive definitions of class membership. If two classes A and B are given, then an element either falls into A or B. Fuzzy classification, by contrast, allows different degrees of membership, so in such a scheme, the image might have a degree of "A-ishness" and a simultaneously a degree of "B-ishness." The application of the latter type of classification has been explored by Carazo et al. (1990), but the technique has failed to take hold in the EM field thus far.

4.2.4. Partitional, Hierarchical Ascendant Merging, and Self-Organizing Methods of Classification

We can distinguish three different approaches to unsupervised classification that have gained importance in the field. In the first, represented by the *K-means technique* (see section 4.3), the data set is successively divided. In the second, the elements are successively merged into larger and larger groups, and a hierarchical relationship among all groups is established [*hierarchical ascendant classification* (HAC), see section 4.4]. In the third method, the relationships among all elements are represented by a self-organizing map, typically through the use of a neural network (section 4.9). We will also briefly mention a method that combines *K*-means with HAC in an effort to overcome the limitations of both methods (hybrid clustering techniques, section 4.5).

4.2.5. Continuous Classification—Charting of Trends

In all the foregoing approaches, the goal was to find discrete classes either by reference to existing templates, or on the basis of an algorithmic assessment of cohesiveness versus separation of clusters. An entirely different goal that occurs in some studies is *continuous classification*, or the charting and systematic division of data in factor space following the intrinsic geometry of the data. Approaches to this particular problem are reviewed in section 4.7 of this chapter.

4.3. *K*-Means Clustering

In the classification using the *K*-means clustering technique, the data set is successively divided into a presupposed number of classes. The way this works is described in the following: if *K* classes are expected, *K* "seeds" (i.e., individual images) are randomly drawn from the entire data set and designated as "centers of aggregation." A partition is devised based on the generalized Euclidean distances between the centers of aggregation and the elements of the data set. From the subset created by the partition, new centers of aggregation are derived by computing the centers of gravity (figure 4.17), which now take the place of the original seeds. Eventually, iteration of this algorithm produces a partition that remains more or less stable.

A great disadvantage of the *K*-means method is that the clusters tend to be (hyper) spherically shaped, as a consequence of the way they are created. Depending on the nature of the particular application, the partition created can be appropriate or totally wrong: elongated clusters, or those with more complex shapes, may not be recognized as contiguous clusters and may be split. Another disadvantage is that in many cases the final partition depends critically on the choice of initial seeds. The need to overcome this discrepancy led to the *dynamic clouds* method introduced by Diday (1971). Here, the *K*-means procedure is

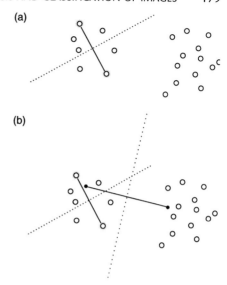

Figure 4.17 The *K*-means method of clustering, following an iterative procedure, illustrated here for the choice of $K=2$ clusters. (a) From among the images, represented by points in a high-dimensional space, two are picked at random as "seeds" (marked with concentric circles). They define a first partition of the set. (b) Next, the centers of gravity (filled dots) are computed for the two subsets created by the partition, and these centers define an improved partition, taking the place of the seeds, etc. In our example, the two clusters are already separated in the second iteration of this procedure.

applied repeatedly, each time with a different choice of seeds. A cross-tabulation of all final partitions obtained then reveals "stable clusters."

In the usual applications of *K*-means to experimental projections, those projections are first aligned to one another, using one of the methods described in chapter 3, section 3. Each of the pairwise interimage distances upon which classification is based is therefore dependent on the properties of the entire data set. To overcome this limitation, Penczek (appendix A in Penczek et al., 1996) devised a rotationally invariant *K*-means algorithm. Instead of calculating the pairwise distances from the prealigned images, he introduced the *minimum distance under rotation*. The only assumption is that all images are centered on their centers of gravity. For two images, the discrete equivalent of the following rotational squared-discrepancy function is calculated:

$$d(f, g; \alpha) = \int_{r_1}^{r_2} \int_0^{2\pi} [f(r, \phi) - g(r, \phi + \alpha)]^2 |r| \, d\phi \, dr \qquad (4.31)$$

The minimum of this function with respect to the rotation by α is then used as distance in the *K*-means algorithm. This distance must be used both between two individual images and between an image and a center of aggregation as introduced above. The center of aggregation itself needs to be redefined in Penczek's modified *K*-means: it is no longer simply the center of gravity of a cluster of images presented in their original form, but the result of reference-free alignment of those images.

This rotationally invariant *K*-means classification method was introduced in the context of the multiple common lines method, and applied to the orientational classification of ribosome images (Penczek et al., 1996). Other examples for its application are found in the work by Sewell and coworkers (Jandhyala et al., 2003; Sewell et al., 2003).

4.4. Hierarchical Ascendant Classification

4.4.1. Principle

Hierarchical ascendant classification (HAC) is a technique whereby an indexed hierarchy is constructed among the elements of the data set. Elements are successively "merged," that is, subsumed into entities or clusters comprising two or more elements, based on their distance in factor space (or in the original space $\mathbf{R}^J$, for that matter), s_{ik}, and on application of a merging rule. As distance, the generalized Euclidean distance in factor space is normally used:

$$s_{ik} = E_{ik} = \sum_{m}^{m_{max}} |\Psi_m^{(i)} - \Psi_m^{(k)}|^2 \tag{4.32}$$

However, below we will formulate the merging criteria in the most general way, referring to a distance measure s_{ik}. The choice of the merging criterion is very important since it determines the merging sequence and thereby the way the elements are combined into clusters. While distances between individual elements are defined by equation (4.32), the distance between one element and a group of elements or the distance between two groups of elements is not defined without a merging rule. (Note that HAC can be used for nonmetric data as well, and a number of merging rules have been tried out over time.)

In accordance with the merging rule, each merger yields an *index of similarity*, which can be used to construct a tree that depicts the hierarchical relationships. Such a tree is called a *dendrogram*. At this point it is instructive to look at an example of such a dendrogram (figure 4.18) since it shows the concept of successive mergers. The merging rules have the following form (Anderberg, 1973; Mezzich and Solomon, 1980): let us assume that two clusters p and q have been merged to form cluster t. The distance between the new cluster t and a given cluster r is determined by some function of the individual distances, s_{pr} and s_{qr}, between the "parent" clusters p and q, and cluster r. For the *single linkage* criterion, this function is

$$s_{tr} = \min(s_{pr}, s_{qr}) \tag{4.33}$$

It is easy to see (Anderberg, 1973) that HAC using this criterion is unable to distinguish between clusters that are so close that they practically touch each other.

Complete linkage overcomes this difficulty to some extent, by use of the criterion:

$$s_{tr} = \max(s_{pr}, s_{qr}) \tag{4.34}$$

Average linkage is obtained by computing the average of all links, and thus assumes a position somewhat intermediate between single and complete linkage (for details, see Anderberg, 1973).

The widely used criterion by Ward (1982) is based on the principle of minimum added intraclass variance. That is, the algorithm finds, at each stage, the two

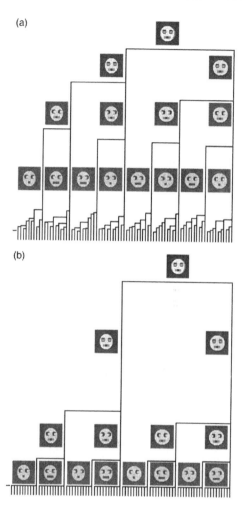

(a)

(b)

Figure 4.18 Hierachical ascendant classification (HAC) of the "face" test data set introduced in figure 4.6: (a) complete linkage; (b) Ward's criterion. Classes obtained by the different cutting levels are represented by their averages. At the very top is the average of all images, in which all three features are blurred. For the uppermost and middle cutting levels, two or one features appear blurred, respectively. Only at the lowest cutting level, all eight original classes are retrieved, as is seen from the sharp appearance of all three features in the class averages. Both HAC options perform equally well in retrieving the correct classes. However, the much higher level at which the eight classes branch off in the complete linkage case indicates more robustness against noise than in the case of Ward's criterion.

clusters whose merger results in the minimum increase ΔV_{pq} in the within-group variance. This quantity can be shown (see Anderberg, 1973) to be

$$\Delta V_{pq} = \frac{N_p N_q}{N_p + N_q} \sum_{m}^{m_{\max}} |\bar{\Psi}_m^{(p)} - \bar{\Psi}_m^{(q)}|^2 \qquad (4.35)$$

where N_p and N_q are the numbers of elements in clusters p and q, and $\bar{\Psi}_m^{(p)}$ and $\bar{\Psi}_m^{(q)}$ are the mean factorial coordinates (relating to factor m) of the pth and qth clusters:

$$\bar{\Psi}_m^{(p)} = \frac{1}{N_p} \sum_{i=1}^{N_p} \Psi_m^{(i \in P)} \; ; \; \bar{\Psi}_m^{(q)} = \frac{1}{N_q} \sum_{i=1}^{N_p} \Psi_m^{(i \in Q)} \qquad (4.36)$$

Symbols P and Q have been used to denote the index subsets of elements belonging to the two clusters. Expression (4.35) means that *the increase in intragroup variance is proportional to the squared Euclidean distance between the centroids of the merged clusters.*

4.4.2. Examples

4.4.2.1. Face data set   This data set was subjected to HAC using Ward's criterion (figure 4.18). The upside-down tree (dendrogram) shows the successive merging of elements and groups up to the highest level. The vertical lines at the bottom represent the raw images, of the type introduced earlier in figure 4.8, each obtained by adding noise to a prototypical face. This comes in eight variants relating to shape (round versus oval), direction of the eyes (left versus right) and shape of mouth (wide or pointed). Following Ward's criterion, each HAC merger (indicated by the joining of two branches) is chosen such that it corresponds to the smallest increase in intraclass variance. This increase determines the vertical placement of a branch. Hence, the scale on the vertical axis can be interpreted as a "similarity index." Depending on the choice of the dendrogram cutoff level, we obtain a classification into two classes, four classes, or the eight original classes. The dendrogram enables us to introduce the important concept of *stability of a classification*. The stability can be judged from the size of the step in the similarity index between the branching points below and above the cutoff level. The reason is that for a classification to be stable, small changes in the cutoff level should not affect the outcome. Going by that criterion, we see that the classification producing two classes (uppermost level in figure 4.18) is the most stable; the one producing four classes (middle level) is less stable but still sound, whereas the one producing eight classes is rather uncertain because a small increase in the cutoff level would have resulted in seven or six classes, while a slight increase would have given nine, ten, or even more.

4.4.2.2. Hemocyanin   Hemocyanin molecules exhibiting top views and side views will serve as a practical example for the use of HAC in the field (see figure 4.19).

4.5. Hybrid Clustering Techniques*

Some classification techniques employ a mixture of HAC and K-means, reflecting an attempt to overcome problems presented by both when used singly. We recall that K-means, from its design, tends to give clusters that are hyperspherical in shape (i.e., circular in 2D, spherical in 3D, etc.), irrespective of the actual distribution of the data, while HAC merging techniques can produce incorrect answers in cases where data are relatively unstructured, so that thin "data bridges" may be formed that persist to the very end. This problem exists even though a certain degree of control is possible by the choice of merging criterion.

In one hybrid technique, HAC is used initially, and then a "postprocessor" is employed, an algorithm that decides for each image, on a given level of classification, whether the image should be reassigned to a different class (van Heel, 1984a). The rationale for the introduction of the postprocessor is that it is able

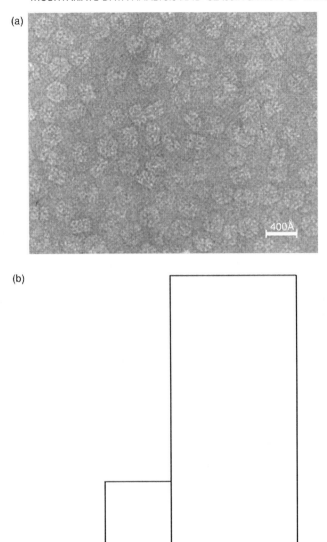

Figure 4.19 HAC with Ward's criterion applied to 2450 cryo-EM images of *Lumbricus terrestris* (earthworm) hemocyanin. (a) Appearance of a typical micrograph; (b) HAC dendrogram and class averages. The classes have (from left to right) 313, 420, 310, 311, 418, 355, and 323 images. (a) From Mouche et al. (2001), reproduced with permission of Academic Press; (b) kindly provided by N. Boisset (unpublished lecture material.)

to break premature "marriages" between images that in the course of the normal HAC scheme cannot be broken. If Ward's merging criterion is used, then the reshuffling of data amounts to an attempt to overcome the trapping in a local variance minimum. [For applications and further discussions of this modification of HAC, see van Heel and Stöffler-Meilicke (1985), Harauz et al. (1987, 1988), van Heel (1989), and Frank (1990).]

Another hybrid technique, due to Wong (1982), addresses the specific problem posed by a quasi-continuous distribution, that is, a distribution without sharp demarcations between different clusters. In such a case, the K-means techniques lead to rather arbitrary partitions depending on the choice of seeds, and the HAC techniques suffer strongly from the "early marriage" problem. (It could be argued that quasi-continuous data call for a simple grid division, treated in section 4.6, since the search for clusters will merely reveal chance fluctuations in the distribution of the data.) Wong combined Diday's (1971) *dynamic clouds* method with HAC. The dynamic clouds method works by repeated application of the K-means algorithm, in which cross-partitions that reflect the clustering history are defined. It is typical for quasi-continuous distributions that many small clusters are found in the cross-partitions. Subsequent HAC applied to the cluster centers then leads to a consolidation into larger clusters that hopefully represent "sharpened" divisions in the data set.

4.6. Inventories*

Data often vary in a continuous way within an expansive overall "cloud," that is, even though fluctuations in the multidimensional distribution are encountered, these are statistical in nature and tend to level out as more and more data are added. There is no point, in such cases, to look for clustering. However, the nature of the variability along different factors will still be of interest. The objective of techniques we will call *inventories* is to make a systematic investigation of the image variations, by imposing a grid onto factor space. By averaging over images falling into a given cell, a relatively noise-free image can be obtained that is representative of the cell's contents. Thus, by taking advantage of the ordination of information by PCA or CA, an exhaustive description of existing variability may be obtained.

This procedure can be likened to making a list of the contents of a room, by drawing a regular grid on the floor, and listing the items falling into each square. One could go a step further (stretching the analogy a bit), and describe the contents of each square by a general term, such as "furniture," "toys," or "clothes."

In the generalization into multidimensional factor space, we similarly divide the image data according to a grid overlaid upon factor space, producing square-shaped cells in two dimensions, cubes in three, or "hypercubes" in spaces with higher dimensions. By forming the average of the data falling into each cell, we will get a set of noise-reduced "samples" at regular intervals, which represent the continuously changing image.

Two-dimensional inventories are most common, and easiest to visualize (see earlier figure 4.14). Going toward higher dimensions, it is clear that the number of cells rapidly increases, making the analysis a formidable task. A simple division into four equal parts along each factor will give 64 cubes for three factors, and already 256 hypercubes for four. Two methods, listed below, overcome this problem by mapping the high-dimensional manifold into a 2D space: nonlinear mapping (section 4.8) and self-organized maps with 2D topology (section 4.9).

4.7. Analysis of Trends

In a number of interesting applications, continuous variations exist that are constrained by physical conditions along specific trajectories. We will use the term *trend* when the data appears continuous but ordered one-dimensionally along a *similarity path*. Consider for instance a molecule that has a flexible domain, which can assume different orientations related by a rotation around a hinge point. Let us assume that in the micrograph we see all the different manifestations of the molecule in the same view. In that case, all molecule images observed can be uniquely ordered according to their similarities, related to the similarity in the orientation of the domain (e.g., Walker et al., 2000; Burgess et al., 2003). Another example is a series of projections of a molecule that were obtained by tilting it around a single axis. Since the angle range is cyclic, the data path formed by the projections in factor space is closed; that is, it has inevitably the topology of a circle (Frank and van Heel, 1982b). Instead of projections generated by actively tilting the molecule, we could also have a set of projections of a barrel-shaped molecule that exists in many copies with different orientations, all related by rotation around the cylinder axis (Boekema and van Heel, 1989). Local tilt associated changes in a 2D crystal can also result in a data path that is intrinsically 1D (Frank et al., 1988c).

In all those cases, each image in factor space has an immediate precursor and an immediate successor; thus, the data are ordered along a linear path. One also speaks of *seriation* of the data. Of course, an ordering along a perfectly smooth line is never observed because of the presence of noise, which creates some variability in directions perpendicular to the data path. This is why seriated data appear distributed along a curved band in a 2D factor map, or a curved "sausage" in a 3D factor map. Just as in the case of inventories for data that are distributed in an unstructured continuum, we might want to create representative images along the data path. For this purpose, we need to create segments by cutting perpendicularly to the data path—just as we would do in cutting a curved sausage up into segments—and then average over images falling into each segment (see Frank et al., 1988c). These averages can be taken as the desired representative images that show the gradual variation.

Related to this topic are two articles by Harauz and Chiu (1991, 1993) on *event covering*, the tracking of statistical interdependencies in a suitably defined subspace.

4.8. Nonlinear Mapping*

As has become clear in the previous sections, the analysis of continuous variations—and subsequent partitioning of the data into homogeneous subsets—is based solely on visual analysis of one or two factor maps, since there exists no ready tool for automatic, data-directed partitioning. However, visual analysis becomes difficult to carry out when several factors need to be considered.

One way to overcome this difficulty is by nonlinear mapping, a way to present multidimensional variational information in two dimensions.

Formally, the goal of nonlinear mapping is to represent an N-dimensional point distribution on a 2D map in such a way that the interpoint distances are optimally preserved. The procedure is iterative and uses one of the 2D factor maps as a starting distribution. The projected positions of the points are changed such that the differences between the Euclidean distances in two dimensions, d_{ii}, and those in the m_{max}-dimensional factor subspace, d_{ii}^{max}, are minimized. As a measure of error, the following expression is used:

$$E = \frac{1}{\sum_{i<i'}^{N}[d_{ii'}^{m_{max}}]^{2-a}} \sum_{i<i'}^{N} [d_{ii'}^{m_{max}} - d_{ii}]^2 / [d_{ii'}^{m_{max}}]^a \qquad (4.37)$$

The distances are defined as follows:

$$d_{ii'} = \left[\sum_{k=1}^{2} \left(x_k^{(i)} - x_k^{(i')} \right)^2 \right]^{1/2} \qquad (4.38)$$

where $\{x_k^{(i)}, k = 1, 2\}$ are the coordinates of the ith point on the nonlinear map;

$$d_{ii'}^{m_{max}} = \left[\sum_{m=1}^{m_{max}} (\kappa_m^{(i)} - \kappa_m^{(i')})^2 \right]^{1/2} \qquad (4.39)$$

where $\{\kappa_m^{(i)}, m = 1 \ldots, m_{max}\}$ are the coordinates of the ith point in the m_{max}-dimensional factor subspace.

The parameter a $(0 \le a \le 1)$ in the denominator of equation (4.37) controls the importance of short distances $d_{ii'}^{m_{max}}$ relative to short distances. At one extreme, $a = 0$ would be used when clustering is expected, and short-range relationships within a cluster are thought to be unimportant; at the other extreme, the choice $a = 1$ would be used when importance is placed on the maintenance of short-range relationships.

Radermacher et al. (1990) tried out the technique on a set of negatively stained 50S ribosomal subunits from *E. coli* that were partially depleted of 5S rRNA. The nonlinear mapping showed continuous variability of the stalk features as well as the presence and absence of mass in the central protuberance of the subunit, attributed to 5S rRNA.

4.9. Self-Organized Maps*

Similarly as nonlinear mapping, classification by self-organized maps (SOMs) with 2D topology (Kohonen, 1990, 1997) results in a 2D ordering of images, such that immediate neighbors are most closely related. However, unlike nonlinear mapping, the ordering is not achieved by a numerical minimization, but—as the name implies—a process of self-organization that only enforces some rules about relationships among immediate neighbors, without any attempt to

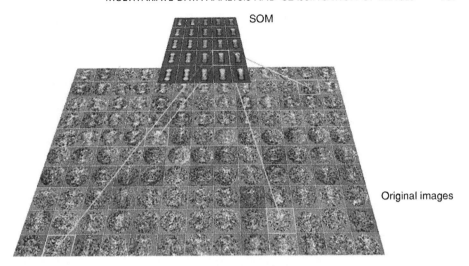

Figure 4.20 Principle of the self-organized map. Original images are in the lower layer, the "input layer." The upper layer shows "code vector images," each of which represents a number of input images most alike to each other. From Pascual-Montano et al. (2001), reproduced with permission of Elsevier.

minimize a measure of global discrepancy. SOMs are obtained by using simulated two-layered neural networks, which usually are computationally implemented on normal sequential machines. The procedure can be described as a mapping from an input layer that represents a 2D arrangement of the raw images onto an output layer that is formed by a 2D arrangement of *code vectors* (figure 4.20). For details, the reader is referred to Marabini and Carazo (1994a) and the original literature quoted there.

Marabini and Carazo (1994a) were the first to apply Kohonen's method to EM. Meanwhile the applications in this field range from the classification of single particles (Marabini and Carazo, 1994a; Pascual-Montano et al., 2001; Radermacher et al., 2001; Gomez-Lorenzo et al., 2003; Ogura et al., 2003) to the classification of the pseudo-repeats of the unit cell of a heterogeneous 2D crystal (Fernandez and Carazo, 1996), automated particle picking (Ogura and Sato, 2001, 2003), classification of rotational power spectra (Pascual-Montano et al., 2000), and classification of the pseudo-repeats in a 3D crystal imaged by electron tomography (Pascual-Montano et al., 2002). Pascual-Montano et al. (2001) developed a variant of the method ("KerDen-SOM," for kernel density SOM), in which the input–output relationships are tailored to satisfy a particular cost function. Accordingly, a set of code vectors is obtained whose probability density best resembles the probability density of the input.

The application of SOMs to large T antigen complexed with viral origin of replication (ori) may serve to illustrate the method. 4200 particles in side view were subjected to the KerDen variant of SOM. Each of the 15×10 code vector images of the output map (figure 4.21) stands for a number of "representative images" from the raw data set that are closely similar. The code vector

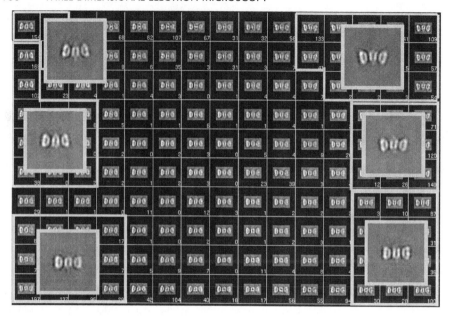

Figure 4.21 Classification of large T antigen molecule complexed with viral origin of replication (Tag–ori complex) using a self-organized map; 4200 particles presenting the side view were analyzed. Molecules represented by the nodes at the four corners are most dissimilar to one another. They are distinguished by the different degrees of bending along the longitudinal axis. The boxes marked at the four corners indicate four groups of molecules (containing 961, 904, 688, and 352 particles) selected for separate reconstructions. From Gomez-Lorenzo et al. (2003), reproduced with permission of EMBO.

images are seen to vary smoothly, as we move in different directions in the array, and these variations originate from a change in the arrangement of the density masses forming the complex, described as different degrees of "bending." The self-organized 2D ordering of the code vectors could then be used to identify groups of molecules that represent extremes of the structural rearrangement, from which separate reconstructions could be obtained.

Application, in EM, of self-organized maps with higher topology (Martinez and Schulten, 1994) have also been discussed (Zuzan et al., 1997, 1998). Such SOMs might carry promise in classifying data sets representing the entire angular range, but initial tests with real data were inconclusive (Zuzan et al., 1998).

4.10. Supervised Classification: Use of Templates

4.10.1. Principle

Supervised classification methods group images according to their similarity to existing prototypes, often called templates, or references. Correlation averaging of crystals (Frank, 1982; Saxton and Baumeister, 1982) incorporates a supervised classification of some sort: here, a reference is simply picked from the raw crystal image, in the form of a patch that is somewhat larger than a unit cell, or generated by quasi-optical Fourier filtration of the raw image. Its cross-correlation with the full crystal field produces a correlation peak wherever a match occurs. Those lattice repeats for which the correlation coefficient falls below a certain threshold are rejected. Application of such a criterion means that only those repeats that are most similar to the reference are accepted. This practice conforms with the notion that the repeats originate from the same motif, which is slightly rotated or distorted in different ways, and are corrupted by noise. Only those corresponding to the undistorted, unrotated template are picked. The dependency of the result on the choice of the template is strikingly demonstrated by the correlation averaging of an image representing a double layer (front and back) of a bacterial membrane (Kessel et al., 1985). In this example, repeats of the front layer are selected by the template obtained by quasi-optical filtering of the front layer, and those of the back layer by the back-layer template. Sosinski et al. (1990) demonstrated this selectivity by using an artificially created field of periodically repeating patterns (a hand) that differ in the position of a component (the index finger) (figure 4.22).

Supervised classification has become commonplace in reference-based orientation determination (3D projection matching). Here, the experimental data are matched with a large number of 2D templates (a case of multireference classification; see below), each of which is a projection of the same 3D map (the reference map) in a different direction of view. This subject will be treated in the context of 3D reconstruction (chapter 5).

4.10.2. Multireference Classification

Supervised classification employing multiple templates, and a decision based on the highest similarity, is also called *multireference classification* (van Heel and Stöffler-Meilicke, 1985). It is used in the alignment of a heterogeneous image set, as follows: after a preliminary reference-free alignment of the entire set followed by classification, for instance by HAC, class averages are created, which are then in turn used as multiple references. Each image of the original data set is then presented with the choice among these references, and is then classified according to its greatest similarity, as (normally) measured by the cross-correlation coefficient. Early applications in single-particle averaging are found in the works of Alasdair Steven and coworkers (Fraser et al., 1990; Trus et al., 1992). The most important application of multireference classification is 3D projection matching: the comparison of an experimental projection with projections of a 3D reference, computed for a regular angular grid. This method will be discussed at some length in chapter 5, section 7.2.

An obvious disadvantage of supervised classification, as with all reference-based schemes, is that its outcome depends strongly on the choice of the template, thus allowing, to some extent, subjective decisions to influence the course of the structural analysis.

4.11. Inference from Two to Three Dimensions

Obtaining a valid 3D reconstruction rests on the availability of a procedure that divides the macromolecules according to the appearance of their projections

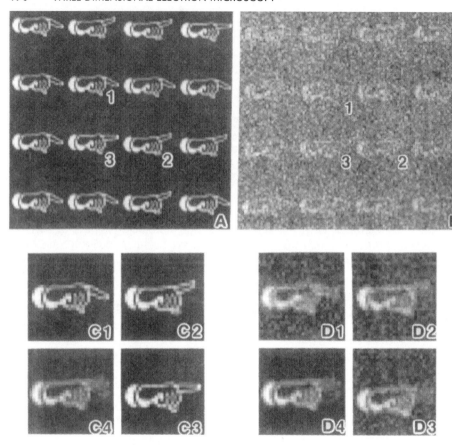

Figure 4.22 The power of supervised classification. (A) Part of a model "crystal" made by placing 240 images into a regular 2D lattice. Three versions of the right-hand motif were used, distinguished by the position of the index finger; (B) noisy version of (A); (C1–C3) averages, each computed from those 25 images in (A) that correlate best with the motif; (C4) Fourier average of the entire field in (A), which shows a blurred index finger; (D1–D3) as (C1–C3), but obtained from the noisy image (B). The averages still show the positions of the index finger, while the Fourier average (D4) is blurred as before. From Sosinsky et al. (1990), reproduced with permission of the Biophysical Society.

into homogeneous subsets. The principal problem posed by this requirement has been discussed in a previous treatment of classification in EM (Frank, 1990). This problem is contained in the ambiguity of the term "homogeneous" in this context: while our goal is to ensure that particles from which we collect data are very similar, we can go only by the similarity of their *projections*. The comparison of projections, however, is subject to a twofold uncertainty: on the one hand, two different 3D objects may have very similar projections when viewed in different directions, and, on the other hand, the observation that their projections differ from one another is not sufficient to conclude that the objects themselves are different: we may just be looking at different views of the same object.

Even though the latter situation may lead to inefficiencies in the image processing (because it will force us to treat the data separately until the very end), it is not harmful as it merely leads to pseudo-classes of particles differing in orientation only. The former situation—multiplicity of objects giving rise to a given view—is more serious, since it poses the risk of obtaining a reconstruction that is not grounded in reality. An additional discussion on this problem of ambiguity will be found later on (chapter 5).

Closely related to this issue is the issue of *handedness* (or *chirality*) ambiguity. Two enantiomorphic objects, that is, objects that are related to each other by mirroring through the origin of the 3D coordinate system, like the relationship between the left and the right hand, produce identical projections in particular orientations. Conversely, there is no way to resolve the handedness of an object by analyzing its projections alone—an additional experiment is always needed, in which the object is tilted (e.g., Leonard and Lake, 1979), unless information is available from other sources (figure 4.23). Specifically, all common-lines methods of orientation determination cannot distinguish between

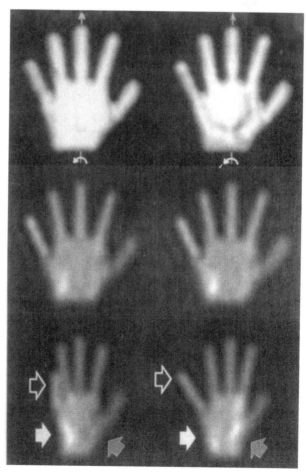

Figure 4.23 Handedness ambiguity, and its solution by a tilt experiment. Top row: models of a left hand and a right hand shown in surface representations. Middle row: in the orientations shown, the projections are identical. Bottom row: unnoticed in the views above, the little finger in the model is inclined toward the palm. Tilting the hands (here by 45° around a vertical axis) immediately reveals which hand we are dealing with—the handedness problem is solved. From Belnap et al. (1997), reproduced with permission of Elsevier.

the two solutions for the set of Eulerian angles that lead to objects with opposite handedness. [This is not true for the random-conical method (chapter 5, section 3.5), which makes use of a tilt.] Only when the resolution is reached, at about 7 Å (see Böttcher et al., 1997; Conway et al., 1997), where the handedness of the alpha-helical motif is distinguishable, can the question be solved without additional data. (For RNA-containing structures, the known hand of double-stranded helices shows already up at ~15 Å). A more recent treatment of the handedness of single particles of known symmetry is contained in the work of Rosenthal et al. (2003). Again, the handedness is resolved by evaluation of a tilt experiment.

It is interesting to consider the case of a mixed population of molecules with opposite handedness. Classification would lump projections coming from the two different species pair-wise together, since they are indistinguishable (as in figure 4.23, middle row). Fortunately for single-particle reconstruction, molecules simultaneously realized with opposite handedness do not exist in nature, so that this particular ambiguity of classification does not occur. Hands, feet, and ears are just a few examples for the fact that handedness does arise in nature, but well *above* the molecular level.

5

Three-Dimensional Reconstruction

Figure 5.1 A single projection image is plainly insufficient to infer the structure of an object. (Note, though, that TEM projections do not merely give the outline, as in this drawing, but internal features, too—the bones and internal organs of the rabbit, which we would see if the projector were to emit X-rays.) (Drawing by John O'Brien; © 1991 *The New Yorker*.)

1. Introduction

The value of a projection image is quite limited for someone trying to understand the structure of an unknown object (see figure 5.1). This limitation is illustrated by the early controversies regarding the three-dimensional (3D) model of the ribosome, which was inferred, with quite different conclusions, by visual analysis of electron micrographs (see, for instance, the juxtaposition of different models of the ribosome in Wittmann's 1983 review). Over 35 years ago, DeRosier and Klug (1968) published the first 3D reconstruction of a biological object, a phage tail with helical symmetry. Soon after that, Hoppe wrote an article (1969) that sketched out the strategy for reconstruction of a single macromolecule lacking order and symmetry [see Hoppe et al. (1974) for the first 3D reconstruction of such an object from projections]. Since then, the methodologies dealing with the two types of objects have developed more or less separately, although the existence of a common mathematical thread (Crowther et al., 1970) has been often emphasized.

This chapter is organized in the following way: first, some basic mathematical principles underlying reconstruction are laid down (section 2). Next, an overview is given over the different data collection schemes and reconstruction strategies (section 3), which answer the questions of how to maximize information, minimize radiation damage, or determine the directions of the projections, while leaving the choice of reconstruction algorithm open. The main algorithmic approaches are subsequently covered (section 4): weighted back-projection, Fourier interpolation methods, and iterative algebraic methods. This mainly theoretical part of the chapter is followed by a practical part, in which approaches to reconstructions following three strategies are outlined: random conical (section 5), common lines (also known as angular reconstitution; section 6), and reference based (section 7). It will become clear that the reference-based approach (3D projection alignment) also underlies all refinement schemes that are used in the end to refine both random-conical- and common lines-based reconstructions.

Next, the important problem of resolution assessment is covered (section 8), followed by a treatment of CTF correction that is necessary to obtain a final reconstruction faithful to the object (section 9). The chapter concludes with three sections devoted to special problems: how to compensate for large angular gaps in the data set, by a process called *restoration* (section 10), how to deal with heterogeneous data sets, that is, projections that come from two or more different molecule populations intermixed on the grid (section 11), and how to align density maps resulting from different reconstructions (section 12).

2. General Mathematical Principles

2.1. The Projection Theorem and Radon's Theorem

The projection theorem, which is of fundamental importance in attempts to recover the 3D object, is implied in the mathematical definition of the multidimensional Fourier transform (see appendix 1). In two dimensions, let us consider the Fourier representation of a function $f(x, y)$:

$$f(x, y) = \iint_B F(k_x, k_y) \exp[-2\pi i(k_x x + k_y y)] dk_x dk_y \qquad (5.1)$$

where B is the spatial frequency domain, the domain of integration. Now we form a one-dimensional (1D) projection in the y-direction. The result is

$$g(x) = \int f(x,y)dy = \int_{y=0}^{y_1} \left[\iint_B F(k_x,k_y) \exp[-2\pi i(k_x x + k_y y)]dk_x dk_y \right] dy \quad (5.2)$$

where y_1 is the extent of the object in the y-direction. Integration over y only affects the second part of the exponential term and immediately yields a delta-function:

$$g(x) = \int_{-K_x}^{K_x} \int_{-K_y}^{K_y} F(k_x,k_y) \exp[-2\pi i k_x x]\delta(k_y)dk_x dk_y = \int_{-K_x}^{K_x} F(k_x,0) \exp[-2\pi i k_x x]dk_x$$

$$(5.3)$$

Here, we have written the integration over the spatial frequency domain as two separate integrations, one over the k_x-extension $\{-K_x, K_x\}$, the other over the k_y-extension $\{-K_y, K_y\}$ of this domain. By performing the k_y-integration first, one obtains the section of the Fourier transform lying along the k_x-axis, by virtue of the properties of the delta-function (appendix 1). This means that the 1D projection of a two-dimensional (2D) function $f(x,y)$ can be obtained as the inverse 1D Fourier transform of a central section through its 2D Fourier transform $F(k_x,k_y) = \mathfrak{F}[f(x,y)]$.

The above "proof" is evidently simple when the projections in the x- and y-directions are considered, making use of the properties of the Cartesian coordinate system. Because the choice of the orientation of the coordinate system is arbitrary, the same relationship holds for any choice of projection direction; see for instance the formulation by Dover et al. (1980). An analogous relationship holds between the projection of a 3D object and the corresponding central 2D section of its 3D Fourier transform. This result immediately suggests that reconstruction can be achieved by "filling" the 3D Fourier space with data on 2D central planes that are derived from the experimentally collected projections by 2D Fourier transformation (figure 5.2).

More rigorously, the possibility of reconstructing a 3D object from its projections follows from Radon's (1917) quite general theory which has as its subject "the determination of functions through their integrals over certain manifolds." The parallel projection geometry we use to describe the image formation in the transmission electron microscope [as modeled in equation (5.2)] is a special case of Radon's theory, in that the integration is performed along parallel lines [see the integral for the 3D case, equation (5.38)]. According to Radon, "an object can be reconstructed uniquely from its line projections when all of its line projections are known."

Taken literally, this theorem is rather impractical because it does not address the questions of (i) how to reconstruct the object from a *limited* number of experimental (noisy) projections to a finite resolution and (ii) whether it also holds

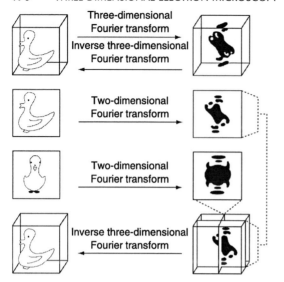

Figure 5.2 Illustration of the projection theorem and its use in 3D reconstruction. From Lake (1971), reproduced with permission of Academic Press Ltd.

for a limited number of projections, and if so, whether or not such a reconstruction would be unique. The effect of restricting the reconstruction problem to finite resolution can be understood by considering the projection theorem—the fact that each projection furnishes a central section of the object's Fourier transform—and taking into account the *boundedness* of the object (see Hoppe, 1969). First, we need to introduce the closely related concept of the *shape transform*.

2.2. Object Boundedness, Shape Transform, and Resolution

In order to explain the concept of the shape transform, we make reference to figure 5.3 (which will also later be used to explain the related fact that adjacent projections are correlated up to a resolution that depends on the angular increment and the size of the object). A *bounded object* $o(\mathbf{r})$ can be described mathematically as the product of an unbounded function $\hat{o}(\mathbf{r})$ that coincides with the object inside the object's perimeter, and has arbitrary values outside, with a *shape function* or mask $s(\mathbf{r})$; that is, a 3D function that describes the object's shape and has the value 1 within the boundary of the object and 0 outside:

$$o(\mathbf{r}) = \hat{o}(\mathbf{r})s(\mathbf{r}) \qquad (5.4)$$

The Fourier transform of $o(\mathbf{r})$, which we seek to recover from samples supplied by projections, is

$$O(\mathbf{k}) = \hat{O}(\mathbf{k}) \circ S(\mathbf{k}), \qquad (5.5)$$

that is, every point in the Fourier transform of the unlimited object can be thought of as being surrounded by the shape transform $S(\mathbf{k}) = \mathfrak{F}\{s(\mathbf{r})\}$. [See

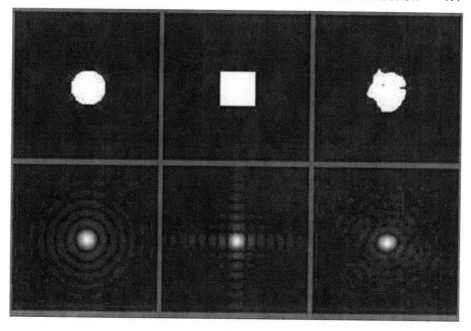

Figure 5.3 Examples for 2D shape transforms. Top row: square, disk, and shape of the ribosome in side view, as an example for an asymmetric molecule. Bottom row: corresponding shape transforms. The ribosome shape is essentially globular, reflected in transform features similar to those for the disk. But, at the same time, the presence of straight edges is reflected in features (radial spikes with cushions) that are reminiscent of the transform of the square. Note that 3D shape transforms have generally smaller side-lobes than those for 2D. In considering the "region of influence" in Fourier space, only the width of the main, central lobe of the shape transform is of importance.

examples in figure 5.3. Note that an entire catalog of shape transforms was provided in Taylor and Lipson (1964) for use in the early days of X-ray crystallography.]

The shape transform always has a central main lobe surrounded by a pattern of side ripples. The main lobe occupies a region of Fourier space whose size is $\sim 1/D$ if D is the size of the object and whose precise shape and profile depends on the properties of the shape function. The side ripples occupy patterns—and have amplitudes—that again depend on the features of the shape (see figure 5.3). If we ignore the influence of the side ripples, we can say that the Fourier transform $O(\mathbf{k})$ varies smoothly over a distance of $1/D$ or, conversely, that measurements have only to be available on a grid in Fourier space with that minimum spacing.

As a consequence (see figure 5.4), roughly

$$N = \frac{\pi D}{d} \qquad (5.6)$$

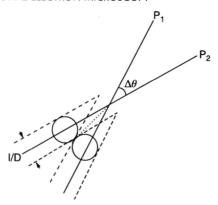

Figure 5.4 Statistical dependence of Fourier components belonging to different projections. We consider the Fourier transform along central sections P_1, P_2 representing two projections of an object with diameter D. Each Fourier component is surrounded by the shape transform, in this case a "circle of influence" with diameter $1/D$. Thus, the central section is accompanied on both sides by a margin of influence, whose boundaries are indicated by the dashed lines. The diagram explains why reconstruction is possible from a finite number of projections. It can be used to answer two interrelated questions: (1) What is the minimum number of projections with equispaced orientations that are required to reconstruct the object to a resolution R without loss of information? (2) Up to what resolution are two projections separated by $\Delta\theta$ correlated? From Frank and Radermacher (1986), reproduced with permission of Springer-Verlag.

equispaced projections need be available if one wants to reconstruct an object with diameter D to a resolution $R = 1/d$ (Bracewell and Riddle, 1967; Crowther et al., 1970). The same conclusion can be reached when one uses a least-squares approach and formulates the reconstruction problem as the problem of finding the values of the Fourier transform on a finite polar grid from a finite number of experimental projections (Klug and Crowther, 1972). (Note that in the development of that argument, the equispaced arrangement of projections angles was actually not required. However, it is easy to see that, in the presence of noise, the equations for recovering data in Fourier regions corresponding to major angular gaps are ill-determined.)

In conclusion, we can state that, *by virtue of the projection theorem and the boundedness of the object, an object can be recovered to a given resolution from a finite number of projections, provided that these projections cover the angular space evenly.* For the time being, we leave this formulation general, but the problems related to gaps in angular coverage will resurface throughout the remainder of this chapter.

2.3. Definition of Eulerian Angles, and Special Projection Geometries: Single-Axis and Conical Tilting

The purpose of this section is to define the relationship between the coordinate system of the projection and that of the molecule. Furthermore, using this formalism, we will define the two most important regular data collection geometries, *single-axis* and *conical*.

Let $\mathbf{r} = (x, y, z)^T$ be the coordinate system affixed to the molecule. By projecting the molecule along a direction $z^{(i)}$, defined by the three angles ψ_i, θ_i, and φ_i, we obtain the projection $p^{(i)} \{x^{(i)}, y^{(i)}\}$. The transformation between the vectors in the coordinate system of the molecule and those in the coordinate system of the projection indexed i is expressed by three Eulerian rotations. In the convention used by Radermacher (1991),

$$\mathbf{r}^{(i)} = R\mathbf{r} \tag{5.7}$$

with

$$R = R_{\psi_i} R_{\theta_i} R_{\phi_i} \tag{5.8}$$

where the rotation matrices performing the successive Eulerian rotations are defined as follows:

$$R_{\psi_i} = \begin{pmatrix} \cos\psi_i & \sin\psi_i & 0 \\ -\sin\psi_i & \cos\psi_i & 0 \\ 0 & 0 & 1 \end{pmatrix} \tag{5.9}$$

$$R_{\theta_i} = \begin{pmatrix} \cos\theta_i & 0 & -\sin\theta_i \\ 0 & 1 & 0 \\ \sin\theta_i & 0 & \cos\theta_i \end{pmatrix} \tag{5.10}$$

$$R_{\phi_i} = \begin{pmatrix} \cos\phi_i & \sin\phi_i & 0 \\ -\sin\phi_i & \cos\phi_i & 0 \\ 0 & 0 & 1 \end{pmatrix} \tag{5.11}$$

These rotations are defined as in classical mechanics, and they can be understood by reference to the sketch in figure 5.5: first (figure 5.5a) the molecule is rotated by the angle φ_i in the *positive* direction around its z-axis, yielding the intermediate coordinate system $\{\xi, \eta, \zeta\}$, then (figure 5.5b) by the angle θ_i in the *negative* direction around its new y-axis (η in figure 5.5a, termed the *line of nodes*) to yield the second intermediate coordinate system $\{\xi', \eta', \zeta'\}$, and finally (figure 5.5c) by the angle ψ_i in the *positive* direction around its new z-axis, yielding the final coordinate system (x', y', z').

One commonly associates the angle θ_i with the concept of "tilt," although the exact tilt direction must be first established by the size of the first "azimuthal"

(a)

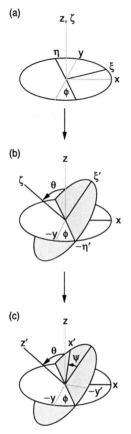

(b)

(c)

Figure 5.5 Definition of Eulerian angles. (a) Rotation by φ around the z-axis leads from the (x, y, z) coordinate system to the coordinate system (ξ, η, ζ). (b) Tilt by θ around the η axis ("line of nodes") leads to the coordinate system (ξ', η', ζ'). (c) Rotation in the tilted plane by ψ around axis ζ' yields the final coordinate system (x', y', z'). [To avoid visual confusion, the hidden axes y and η' in (b) are not shown as dashed lines, but indicated by their negative parts -y and -η', respectively. The same applies to the hidden axes y and y' in (c).] Adapted from the Ph.D. thesis of I. M. Withers, Materials Research Institute, Sheffield Hallam University (http://dl9s6.chem.unc.edu/imw/thesis/node22.html)

angle φ_i. Note that for small θ_i, the combination of Eulerian angles $\{\varphi_i, \theta_i, -\psi_i\}$ is virtually equivalent to $\{0, \theta_i, 0\}$.

The orientations of projections accessible in a given experiment are defined by technical constraints; these constraints are tied to the degrees of freedom of the tilt stage and to the way the molecules are distributed on the specimen grid. Referring to the geometry defined by these constraints, we speak of the *data collection geometry*.

The regular single-axis tilt geometry (figure 5.6a) is generated by letting

$$\psi_i = 0 \quad \text{and} \quad \varphi_i = 0 \qquad (5.12)$$

that is, the molecule is tilted by θ_i in equal increments around the y-axis and then projected along $z^{(i)}$.

The regular conical tilt geometry (figure 5.6b) is generated by

$$\varphi_i = i \times 2\pi/N, \ \theta_i = \theta_0 = \text{const., and } \psi_i = 0 \qquad (5.13)$$

The molecule is first rotated around its z-axis by the "azimuthal angle" φ_i into one of N discrete positions and then tilted by θ_0 around the y-axis.

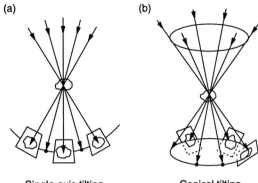

(a)   (b)

Figure 5.6 Data collection by (a) single-axis and (b) conical tilting. From Radermacher (1980).

Single-axis tilting   Conical tilting

Finally, for later reference, the random-conical geometry should be mentioned, which is equivalent to the regular conical tilt geometry, except that the azimuth is randomly distributed in the entire azimuthal range $\{0, 2\pi\}$. (Another important distinction lies in the fact that in the random-conical data collection, the different azimuths are realized by an ensemble of molecules, not by moving a single molecule.)

3. The Rationales of Data Collection: Reconstruction Schemes

3.1. Introduction

In attempting to reconstruct a macromolecule from projections to a resolution of $1/20 \, \text{Å}^{-1}$ or better, we must satisfy several mutually contradictory requirements:

(i) We need *many different* projections of the same structure to cover Fourier space as evenly as possible (this requirement often excludes the direct use of images of molecules showing preferred orientations, since the number of those is normally insufficient, and the angular coverage is far from even).

(ii) The total dose must not exceed a certain maximum tolerable dose, which is resolution-dependent (see chapter 2, section 4.1). This stipulation limits the resolution achievable by tomographic experiments of the type that Hoppe et al. (1974) introduced.

(iii) The reconstruction should be representative of the ensemble average of macromolecules in the specimen. This stipulation excludes the use of automated tomography, by collecting all projections from a single particle while keeping the total dose low (cf. Dierksen et al., 1992, 1993), unless a sizable number of such reconstructions are obtained that can be subsequently combined (see section 12) to form a statistically meaningful 3D ensemble average.

(iv) Since the stipulations (1) through (3) imply that the projections have to be obtained from different "copies" of the molecule (i.e., different realizations of the same structure), we need to establish the relative orientations of those molecules in a common frame of reference.

In other words, for a data collection and reconstruction scheme to be viable, it must be able to "index" projections with sufficient accuracy; that is, it must be able to find their 3D orientations. Of all the stipulations listed above, the last is perhaps the most difficult one to fulfill in practice. The reason that the random-conical scheme, to be described in section 3.5, has found a permanent place among several schemes proposed over the years, is that it solves the problem of finding the relative orientations of different projections unambiguously, by the use of two exposures of the same specimen field. Other methods, such as the method of angular reconstitution (Goncharov et al., 1987; van Heel, 1987b; Orlova and van Heel, 1994), have to find the angles *a posteriori*, based on common lines, but an ancillary tilt experiment or extraneous data are required to solve the handedness problem.

The three possible strategies that lead from a set of molecule projections to a statistically significant 3D image were summarized by Frank and Radermacher (1986) in a diagram (figure 5.7): following the first strategy, individual molecules are separately reconstructed from their (tomographic) tilt series, then their reconstructions are aligned and averaged in three dimensions (figure 5.7a). Following the second strategy, molecule projections found in the micrograph are aligned, classified, and averaged by class. The resulting class averages are then used in a common-lines scheme to determine their relative orientations. When a sufficient number of views are present, the molecule can be reconstructed from these class averages (figure 5.7b). The third possibility is to determine the orientations of individual molecule projections based on a known experimental geometry, for instance by the random-conical scheme, or by reference to an existing 3D map, so that an averaged 3D reconstruction can be *directly* computed from the raw data (figure 5.7c).

In the following, we will first, for the sake of completeness, outline a method of data collection and reconstruction designed for a very simplified situation, that draws from a single, averaged projection and thus does not require orientation determination (section 3.2). Next, an important issue, the question of compatibility of projections, which determines the validity of all schemes that combine data from different particles, will be discussed (section 3.3). After that, the methods of *angular reconstitution* (i.e., common-lines determination of orientation) (section 3.4) and random-conical data collection (section 3.5) will be outlined, followed by a comparison (section 3.6). For completeness, reconstruction methods are mentioned that rely on uniform angular coverage (section 3.7).

It will become clear that single-particle reconstruction, as it is now widely practiced, involves two stages: the *bootstrap* stage and the *refinement* stage. The former establishes the angular relationship among the projections of a molecule, and results in a first, crude reconstruction. The latter stage starts with this crude reconstruction and refines it through an iterative scheme.

3.2. Cylindrically Averaged Reconstruction

For some structures, the deviation from cylindrical symmetry is not recognizable at the resolution achieved (1/20 to $1/40\,\text{Å}^{-1}$), and so there is no way, except

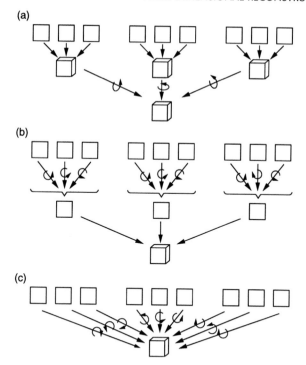

Figure 5.7 Three possible principal strategies for combining projections into a statistically meaningful 3D map. (a) Molecules are separately reconstructed from different projection sets (normally single- or double-tilt series), and then the reconstructions are merged after determining their relative orientations. (b) A "naturally" occurring projection set is divided into classes of different views, an average is obtained for each class, the relative orientations are found for all classes, and—if a sufficient number of views are available—the molecule is reconstructed. (c) The projections are directly merged into a 3D reconstruction after their relative orientations have been found. From Frank and Radermacher (1986), reproduced with permission of Springer-Verlag.

possibly by antibody labeling, to distinguish the particle orientation (with respect to its long axis, which is running parallel to the specimen grid) from the appearance of its side views. The presentation of a cylindrically averaged reconstruction may be the best one can do under these circumstances. The way from the projection to the 3D reconstruction is provided by the inversion of the *Abel transform* (Vest, 1974; Steven et al., 1984).

Let us consider the 2D case. The projection of a function $f(x, y)$ can be represented by the line integral:

$$f_L(R, \theta) = \int_{-\infty}^{\infty} \int_{-\infty}^{\infty} f(x, y) \delta(x \cos \theta + y \sin \theta - R) dx dy \qquad (5.14)$$

which, considered as a function of the variables R and θ, is the transform of $f(x, y)$. A choice of θ defines the direction of projection, and R defines the exact projection ray. Now if $f(x, y)$ is the slice of a cylindrically symmetric structure, it depends only on $r = \left(x^2 + y^2\right)^{1/2}$. In that case, equation (5.14) simplifies into the *Abel transform*:

$$f_L(R, \theta) = f_L(x) = \int_{-\infty}^{\infty} \int_{-\infty}^{\infty} f(r)\delta(x - R)dxdy = \int_{x}^{\infty} \frac{f_L(x)dx}{(x^2 - r^2)^{1/2}} \tag{5.15}$$

Equation (5.15) can be inverted and solved for the unknown profile $f(r)$ by use of the *inverse Abel transform*:

$$f_L(R, \theta) = f_L(x) = \int_{-\infty}^{\infty} \int_{-\infty}^{\infty} f(r)\delta(x - R)dxdy = \int_{x}^{\infty} \frac{f_L(x)dx}{(x^2 - r^2)^{1/2}} \tag{5.16}$$

The practical computation makes use of the fact that for a rotationally symmetric function, the Abel transform is equivalent to the Fourier transform of the Hankel transform.

This method has been used with success in the earlier investigation of flagellar basal bodies both negatively stained (Stallmeyer et al., 1989a,b) and frozen-hydrated (Sosinsky et al., 1992; Francis et al., 1994; see figure 5.8). Basal bodies are rotational motors driving bacterial flagella, which are used to propel bacteria swimming in water. More recent studies attempted to determine the 3D structure

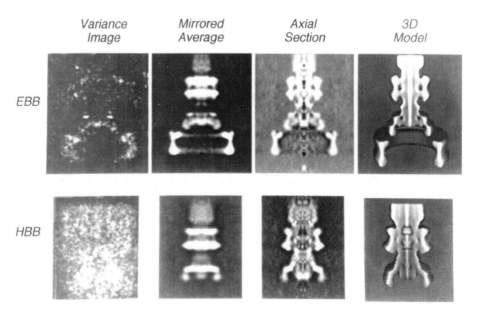

Figure 5.8 Flagellar motor (basal body) of *Salmonella*, reconstructed from single averages of frozen-hydrated specimen, assuming cylindrical symmetry. EBB, extended basal body; HBB, hook plus basal body, lacking the C-ring complex and the switch protein. From Francis et al. (1994), reproduced with permission of Elsevier.

of these motors by "true" cryo-EM reconstructions without assumption of rotational symmetry (Thomas et al., 2001; Shaikh et al., 2005).

3.3. Compatibility of Projections

When projections from different particles are combined in a 3D reconstruction, the implicit assumption is that they represent different views of the same structure. If this assumption is incorrect—that is, if the structure is differently deformed in different particles—then the reconstruction will not produce a faithful 3D image of the macromolecule. Moreover, some methods of data collection and reconstruction determine the relative directions of projections by making use of mathematical relationships among them, which are fulfilled only when the structures they originate from are identical. However, macromolecules are often deformed because of an anisotropic environment: when prepared by negative staining and air-drying on a carbon grid, they are strongly flattened, down to as much as 50% of their original z-dimension, in the direction normal to the plane of the grid. (Even ice embedment may not prevent deformation entirely, because of the forces acting on a molecule at the air–water interface.)

One important consideration in assessing the viability of different reconstruction schemes for negatively stained molecules is, therefore, whether or not they require the mixing of projections from particles lying in different orientations. If they do, then some kind of check is required to make sure that the structure is not deformed in different ways (see following). If they do not, and the molecule is deformed, then it is at least faithfully reconstructed, without resolution loss, in its unique deformed state.

It has been argued (e.g., van Heel, 1987b) that by using solely 0° projections, that is, projections perpendicular to the plane of the specimen grid, one essentially circumvents the problem of direction-dependent deformation, as this mainly affects the dimension of the particle perpendicular to the grid. Following this argument, a secondary effect could be expected to *increase* the width for each view, leading to a reconstruction that would render the macromolecule in a uniformly expanded form. As yet, this argument has not been tested. It would appear that specimens will vary widely in their behavior and that the degree of expansion may be orientation dependent as well.

Conservation of the 3D shape of the molecule on the specimen grid can be checked by an *interconversion experiment*. (Such experiments initially played an important role in interactive model building, that is, in attempts to build a physical model intuitively, by assigning angles to the different views, and shaping a malleable material so that the model complies with the observed views.) An interconversion experiment is designed to establish an angular relationship between two particles presenting different views, A and B: upon tilting, the particle appearing in view A changes its appearance into A', and the one appearing in view B into B'. The experiment tests the hypothesis that the two particles are in fact identical but lie in different orientations. In that case, it should be possible to find a tilt angle θ, around an appropriate axis, that renders A' and B identical. Inverse tilt around that axis by an angle of $-\theta$ should also render views B' and A identical.

An example for a successful interconversion experiment is the study of Stoops et al. (1991a), who found that the two prominent views of negatively stained α_2-macroglobulin, the "lip" and "padlock" views, interconvert for a tilt angle of 45° around the long axis of the molecule. Numerous interconversion experiments were also done, in an earlier phase of research into ribosome morphology, to relate the views of ribosomes and their subunits to one another (e.g., Wabl et al., 1973; Leonard and Lake, 1979) so that the 3D shape might be inferred.

Ultimately, the most conclusive test for heterogeneity of the data set underlying a reconstruction is found in the 3D *variance map* (chapter 6).

3.4. Relating Projections to One Another Using Common Lines

3.4.1. Principle of Common Lines

Methods designed to relate different projections of a structure to one another make use of the common lines (Crowther et al., 1970). These are lines along which, according to the projection theorem, the Fourier transforms of the projections should be identical in the absence of noise. The common-lines concept is important in several approaches to electron microscopic reconstruction and will be discussed again later (section 8). In particular, common lines have been used extensively in the reconstruction of icosahedral viruses (Crowther, 1971; Baker and Cheng, 1996; Fuller et al., 1996).

At the moment we make use of a simple model: let us represent two arbitrary projections of an object in 3D Fourier space by their associated central sections (figure 5.9). These intersect on a line through the origin, their *common line*. Suppose now that we do not know the relative azimuthal orientation of these projections. We can then find their common line by "brute force," by comparing every (1D) central section of one 2D Fourier transform with every central section of the other transform. The comparison is conveniently done by cross-correlation, which is in Fourier space equivalent to the forming of the summed conjugate

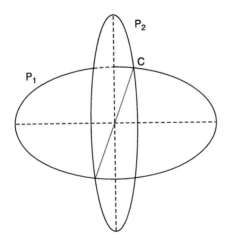

Figure 5.9 The principle of common lines. Two projections of the same object, represented by central sections of the object's 3D Fourier transform, intersect each other along a common line through the origin. Along that line, their Fourier transforms must be identical in the noise-free case.

product (see appendix 1). This product will assume a maximum when a match occurs.

Once the common line is found, the (in-plane) orientation of the two central Fourier sections (and thereby, of the corresponding projections) is fixed. In the absence of additional information, the two central sections can still move, in an unconstrained way, around the fixed common line, which thereby acts as a "hinge" of a flexible geometrical construction. Obviously, a third projection—provided that it does not share the same tilt axis shared by projections #1 and #2—will fix this movement and lead to a complete determination of relative angles among the three projections (apart from an unresolved ambiguity of handedness, because the third central section can be added in two different ways, as obvious from the geometric construction).

Starting with this system of orientations, new projections are added by combining them with pairs of projections already placed. This, in essence, is the method of angular reconstitution (Goncharov et al., 1987; van Heel, 1987b; Orlova and van Heel, 1994). In practice, the common line search is often performed in real space with the help of the so-called sinogram; this is a data table that contains in its rows the 1D projections of a 2D image (in our case, of a 2D projection) exhaustively computed for all angles on the angular grid. (Note that sinogram is another word for the discrete 2D Radon transform.) If the 2D projections originate from the same object, then there exists an angle for which their 1D projections are identical (or closely similar): thus, in real space, the equivalent to the common line is the common 1D projection. The angle is found by comparing or correlating the sinograms of the two 2D projections (Vainshtein and Goncharov, 1986; Goncharov et al., 1987; van Heel, 1987b).

3.4.2. Limitations

Although elegant in concept, the common line (or common 1D projection) method of orienting "raw data" projections, and thus the method of angular reconstitution, as proposed originally, is hampered by the low signal-to-noise ratio (SNR) of the data when applied to cryo-EM images of single molecules. However, as we know, the SNR can be dramatically improved by averaging, either over projections of molecules presenting closely related views, or over symmetry-related projections of a particle exhibiting symmetry. Examples of sinograms for class averages of a molecule are presented in figure 5.10. From these examples it is clear that the determination of the common 1D projection, and hence the determination of the relative angle between two projections represented by class averages, should be quite robust. On the other hand, the averaging of projections falling into classes that represent a range of orientations necessarily entails a resolution loss, which will be reflected by a loss in the quality of the reconstruction. Only by adding an angular refinement step, to be described in section 8 of this chapter, can the full resolution in the data be realized.

The other inherent limitation of the technique, already mentioned at the outset, is the lack of handedness information, which necessitates the use either of an ancillary experiment or of extraneous information.

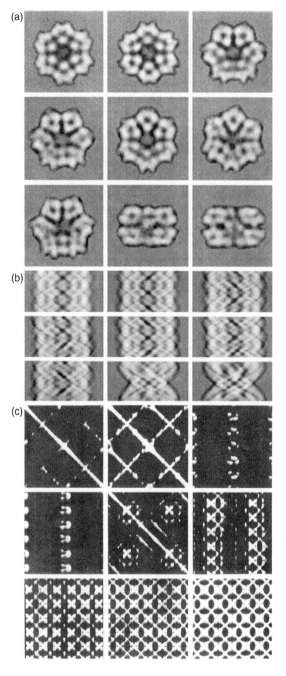

Figure 5.10 Sinograms and their use in finding relative projection angles. (a) Class averages of worm hemoglobin showing the molecule in different views; (b) corresponding sinograms. Each horizontal line in the sinogram represents a 1D projection of the corresponding molecule image in a certain direction. The lines are ordered according to increasing angle of projection, covering the full 360° range. A rotation of an image is reflected by a cyclical vertical shift of the corresponding sinogram. (c) Sinogram correlation functions (sine-corr). The sine-corr between sinograms of projections 1 and 2 is derived in the following way: the cross-correlation coefficient is computed between the first row of sinogram 1 and each row of sinogram 2, and the resulting values are placed into the first row of the sine-corr, and so on with the following rows of sinogram 1. The position of the maximum in the sine-corr indicates the angular relationship between the projections. Meaning of the panels from left to right, top to bottom: sine-corrs of 1 vs. 1, 1 vs. 2, 1 vs. 3, 1 vs. 4, 4 vs. 4, 1 vs. 9, 3 vs. 9, 9 vs. 9, and 8 vs. 8. Multiple maxima occur because of the sixfold symmetry of the molecule. From Schatz (1992), reproduced with permission of the author.

Reconstructions utilizing this concept, initially mostly applied to macromolecules with symmetries, were reported by van Heel and collaborators (Schatz, 1992; Dube et al., 1994; Schatz et al., 1994; Serysheva et al., 1995). As symmetries provide internal checks, they were useful in the starting phase of the application

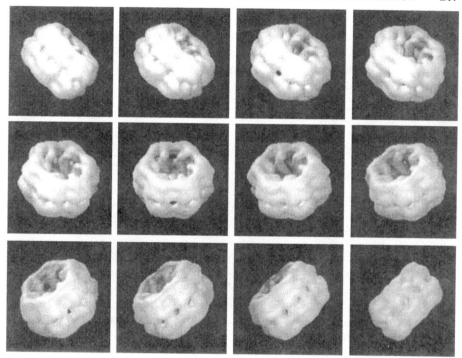

Figure 5.11 Three-dimensional reconstruction of *Lumbricus terrestris* erythrocruorin embedded in ice from class averages shown in figure 5.10a. In part, the angles were assigned based on the technique of "angular reconstitution" (see also Schatz et al., 1994). From Schatz (1992), reproduced with permission of the author.

of common-lines techniques to single particles. Figure 5.11 presents a model of worm hemoglobin obtained by Schatz (1992), partly making use of the sinogram-based angle assignments shown in figure 5.10. Increased sophistication of the technique allowed its widespread application to particles entirely without symmetry (e.g., Stark et al., 1995, 2000).

Recognizing the noise sensitivity of van Heel's and Goncharov's method in its original form (which was intended to recover relative orientations from raw data), Farrow and Ottensmeyer (1992) and Ottensmeyer and Farrow (1992) developed an extension of the technique. Solutions are found for many projection triplets using the common-lines triangulation that is at the core of the angular reconstitution technique, and the results are then reconciled using quaternion mathematics (see Harauz, 1990).

Detailed accounts on the way angular reconstitution is used for single-particle reconstruction were given by Serysheva et al. (1995) and the review by van Heel et al. (2000), as well as articles quoted there.

The simultaneous minimization technique (Penczek et al., 1996) provides a way to find a simultaneous solution for the orientation search problem, given any number (> 3) of projections multiple common lines. Concurrent processing of a large number of projections relaxes the requirement of high SNR for the input data. The method uses a discrepancy measure that accounts for the uneven distribution of common lines in Fourier space. The minimization program begins the search from an initial random assignment of angles for the projections. The authors showed that the correct solution (as judged from the result of 3D projection alignment to a merged random-conical reconstruction; Frank et al., 1995a) was found in 40% of the trials. For applications of this generalized common-lines approach, see Craighead et al. (2002), Jandhyala et al. (2003), Sewell et al. (2003), and examples in section 4.

3.5. The Random-Conical Data Collection Method

3.5.1. Principle

The principle of this data collection scheme was initially mentioned in the context of 2D averaging of molecule projections (Frank et al., 1978a) as an effective way for extending the single particle averaging into three dimensions. An explicit formulation and a discussion of the equivalent Fourier geometry was given by Frank and Goldfarb (1980). First attempts to implement the reconstruction technique led to the design of a Fourier-based computer program that proved unwieldy (W. Goldfarb and J. Frank, unpublished results, 1981). The first implementation of a reconstruction method making use of the random-conical data collection was achieved by Radermacher et al. (1986a, 1987a,b). For this implementation, numerous problems had to be solved first, including the determination of the precise tilt geometry, the practical problem of pairwise particle selection, alignment of tilted projections, relative scaling of projection data, weighting of projections in a generalized geometry, and managing the massive amount of book keeping required (Radermacher, 1988). In the following, the different solutions to these problems will be described in some detail.

The method takes advantage of the fact that single macromolecules often assume preferred orientations on the specimen grid (see section 1.5 in chapter 3). Any subset of molecules showing identical views in an untilted specimen forms a rotation series with random azimuth, φ_i. When the specimen grid is tilted by a fixed angle, θ_0 (figure 5.12a,b), the above subset will appear in the micrograph as a conical projection series with random azimuths and θ_0 as cone angle (figure 5.12c). In the actual experiment, the specimen field is recorded twice: once tilted and once untilted (in this order, to reduce the accumulated radiation damage, since only the tilted-grid data are used in the 3D reconstruction). The first micrograph is used to extract the projections for the reconstruction. The purpose of the second micrograph is twofold: (i) to separate the particles according to their views (*classification*) and (ii) within each subset (or class), to determine the relative azimuths of all particles (*alignment*). The advantages of this scheme are evident: it allows the orientations of all molecules to be readily determined, while allowing the dose to be kept to a minimum.

The instructive drawing of Lanzavecchia et al. (1993) in figure 5.13 shows the coverage of Fourier space afforded by the conical geometry. Since each projection is sampled on a square grid, its discrete Fourier transform is available

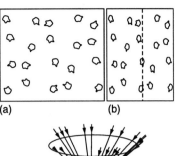

Figure 5.12 Principle of the random-conical data collection: (a) untilted; (b) tilted field with molecule attached to the support in a preferred orientation; (c) equivalent projection geometry. From Radermacher et al. (1987b), reproduced with permission of Blackwell Science Ltd.

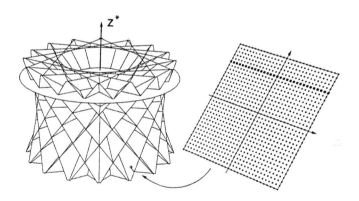

Figure 5.13 Coverage of 3D Fourier space achieved by regular conical tilting. Shown is the relationship between inclined Fourier plane (on the right), representing a single projection, and the 3D Fourier transform (on the left). For random-conical data collection, the spacings between successive planes are irregular. Adapted from Lanzavecchia et al. (1993).

within a square-shaped domain. The body formed by rotating an inclined square around its center resembles a yo-yo with a central cone spared out. Since the resolution of each projection is limited to a circular domain (unless anisotropic resolution-limiting effects such as axial astigmatism intervene, see section 3.2 in chapter 2), the coverage of the 3D Fourier transform by useful information is confined to a *sphere* contained within the perimeter of the yo-yo (not shown in figure 5.13).

We will come back to the procedural details of the random-conical data collection and reconstruction after giving a general overview over the different reconstruction algorithms.

3.5.2. Limitations

There are some obvious limitations that restrict the resolution of the reconstruction: one is because the observed "preferred orientation" in reality encompasses a finite orientation range (see section 7.5 for the likely size of the angular range). Another, related limitation, stems from the need to classify particles on the basis of their 0° appearance—a task which may have ambiguous results (see section 4.11 in chapter 4). A third limitation has to do with the fact that the azimuthal angles (as well as the subsequent classification) are determined from the images of particles (at 0°) that have already been exposed to the electron beam and may have been damaged. All three limitations are in the end mitigated by the use of a refinement method, according to which each projection is allowed to vary its orientation with respect to the reconstruction from the entire data set (see section 7.2).

3.6. Comparison of Common Lines Versus Random-Conical Methods

Several macromolecules have been reconstructed with both random-conical and common-lines methods for initial angle assignment (though both using the same 3D projection matching approach in the refinement), and a comparison shows that the results are equivalent—if one abstracts from differences in filtration, representation, and quality of the data, which are not germane to the particular methods of angular assignment used. Examples for such comparisons of contemporaneous studies are the *E. coli* 70S ribosome (Frank et al., 1995a; Stark et al., 1995), calcium release channel (Radermacher et al., 1994a,b; Serysheva et al., 1995), and a ribosomal release complex (Klaholz et al., 2003; Rawat et al., 2003).

Since the common-lines method is based on the presence of multiple views covering the whole angular space, it can be seen as complementary to the random-conical method, which exploits a different situation: the presence of a few preferred views, or even a single one.

Common-lines methods require a minimum number of views to succeed—otherwise the initial reconstruction becomes so crude that it cannot be used in the refinement, and, at any rate, not enough data are present, according to the premise, to fill the angular range as required for a meaningful reconstruction.

In contrast, the random-conical method works in both extremes of angular prevalence—with projections showing a single view, as well as a large range of different views. In the latter case, subsets of particles falling in different views can be handled one at a time, but only a few, most populous subsets need be used (see Penczek et al., 1994; Radermacher et al., 2001).

3.7. Reconstruction Schemes Based on Uniform Angular Coverage

For completeness, two reconstruction schemes that rely on a uniform coverage of the space of orientations should be mentioned. Both use spherical harmonics as a means of representing the object and its relationship to the input projections. The requirement of statistical uniformity and the choice of the rather involved mathematical representation have restricted the use of these schemes to model computations and few demonstrations with experimental data. The first scheme, proposed by Zvi Kam (1980), is based on a sophisticated statistical approach difficult to paraphrase here. The second scheme, introduced by Provencher and Vogel (1983, 1988) and Vogel and Provencher (1988), is designed to determine the relative orientations of the projections of a set of particles by a least-squares method, but it requires approximate starting orientations. Thus far, only a single reconstruction of an asymmetric particle, the 50S ribosomal subunit, has been obtained with this latter method (Vogel and Provencher, 1988).

4. Overview of Existing Reconstruction Techniques

4.1. Preliminaries

Given a data collection scheme that produces a set of projections over an angular range of sufficient size, with angle assignments for each projection, there is still a choice among different techniques for obtaining the reconstruction. Under "technique" we understand the mathematical algorithm and—closely linked to it—its numerical and computational realization. The value of a reconstruction technique can be judged by its mathematical tractability, speed, computational efficiency, stability in the presence of noise, and various other criteria (see, for instance, Sorzano et al., 2001). Competing techniques can be roughly grouped into weighted back-projection, real-space iterative techniques, and techniques based on Fourier interpolation. Weighted back-projection (section 4.2) is a real-space technique which has gained wide popularity on account of the fact that it is very fast compared to all iterative techniques (section 4.4). However, iterative techniques can be superior to weighted back-projection according to several criteria, such as sensitivity to angular gaps and smoothness (e.g., Sorzano et al., 2001). The quality of Fourier interpolation techniques (including so-called gridding techniques) (section 4.3), which match weighted back-projection in speed, depends critically on the interpolation method used.

Apart from computational efficiency, two mutually contradictory criteria that are considered important in the reconstruction of single macromolecules from electron micrographs are the *linearity* of a technique and its ability to allow *incorporation of constraints*. Here, what is meant by "linearity" is that the reconstruction technique can be considered as a black box with "input" and "output" channels, and that the output signal (the reconstruction) can be derived by linear superimposition of elementary output signals, each of which is the response of the box to a delta-shaped input signal (i.e., the projection of a single point). In analogy with the point-spread function, defined as the point response of an optical system, we could speak of the "point-spread function" of the combined system formed by the data collection and the subsequent reconstruction. This analogy is well developed in Radermacher's (1988) treatment of general weighting functions for random-conical reconstruction.

Strictly speaking, linearity holds for weighted back-projection only in the case of even angular spacing, where the weighting is by $|\mathbf{k}|$. For the more complicated geometries with randomly distributed projection angles, linearity is violated by the necessary practice of limiting the constructed weighting functions to prevent noise amplification (see section 4.2). Still, the property of linearity is maintained in an approximate sense, and it has been important in the development and practical implementation of the random conical reconstruction method because of is mathematical tractability. [Some iterative techniques such as the algebraic reconstruction technique (ART) and simultaneous iterative reconstruction technique (SIRT), which also share the property of linearity, have not been used for random conical reconstruction until more recently because of their relatively low speed.] The practical importance of linearity also lies in the fact that it allows the 3D variance distribution to be readily estimated from projection noise estimates (see section 2.2 in chapter 6).

On the other hand, the second criterion—the ability of a technique to allow incorporation of constraints—is important in connection with efforts to fill the angular gap. Weighted back-projection as well as Fourier reconstruction techniques fall into the class of linear reconstruction schemes, which make use only of the projection data and do not consider the noise explicitly. In contrast, the various iterative algebraic techniques lend themselves readily to the incorporation of constraints and to techniques that take the noise statistics explicitly into account. However, these techniques are not necessarily linear. For example, the modified SIRT technique described by Penczek et al. (1992) incorporates nonlinear constraints.

In comparing the two different approaches, one must bear in mind that one of the disadvantages of weighted back-projection—its failure to fill the missing gap—can be mitigated by subsequent application of restoration, which is, however, again a nonlinear operation. Thus, when one compares the two approaches in their entirety—weighted back-projection plus restoration versus any of the nonlinear iterative reconstruction techniques—the importance of the linearity stipulation is somewhat weakened by its eventual compromise.

4.2. Weighted Back-Projection

Back-projection is an operation that is the inverse of the projection operation: while the latter produces a 2D image of the 3D object, back-projection "smears out" a 2D image into a 3D volume ["back-projection body," see Hoppe et al. (1986), where this term was coined] by translation into the direction normal to the plane of the image (figure 5.14). The topic of modified back-projection, as it is applied to the reconstruction of single particles, has been systematically

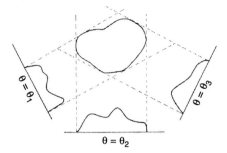

Figure 5.14 Illustration of the back-projection method of 3D reconstruction. The density distribution across a projection is "smeared out" in the original direction of projection, forming a "back-projection body." Summation of these back-projection bodies generated for all projections yields an approximation to the object. For reasons that become clear from an analysis of the problem in Fourier space, the resulting reconstruction is dominated by low-spatial frequency terms. This problem is solved by Fourier weighting of the projections prior to the back-projection step, or by weighting the 3D Fourier transform appropriately. From Frank et al. (1985), reproduced with permission of van Nostrand-Reinhold.

presented by Radermacher (1988, 1991, 1992), and some of this work will be paraphrased here.

Let us consider a set of N projections into arbitrary angles. As a notational convention, we keep track of the different 2D coordinate systems of the projections by a superscript; thus, $p_i(\mathbf{r}^{(i)})$ is the ith projection, $\mathbf{r}^{(i)} = \{x^{(i)}, y^{(i)}\}$ are the coordinates in the ith projection plane, and $z^{(i)}$ is the coordinate perpendicular to that.

With this convention, the back-projection body belonging to the ith projection is

$$b_i(\mathbf{r}^{(i)}, z^{(i)}) = p_i(\mathbf{r}'^{(i)})t(z^{(i)}) \tag{5.17}$$

where $t(z)$ is a "top hat" function:

$$t(z) = \begin{cases} 1 & \text{for } -D/2 \leq z \leq D/2 \\ 0 & \text{elsewhere} \end{cases} \tag{5.18}$$

Thus, b_i is the result of translating the projection by D (a distance that should be chosen larger than the anticipated object diameter). As more and more such back-projection bodies for different angles are added together, a crude reconstruction of the object will emerge:

$$\bar{\rho}(\mathbf{r}) = \sum_{i=1}^{N} b_i(\mathbf{r}^{(i)}, z^{(i)}) \tag{5.19}$$

with $\mathbf{r} = (x, y, z)$ being the coordinate system of the object.

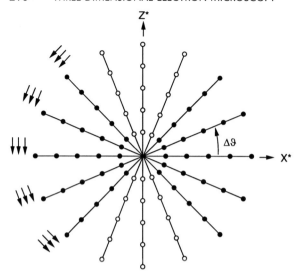

Figure 5.15 The density of sampling points in Fourier space obtained by projections decreases with increasing spatial frequency. Although this is shown here for single-axis tilting, the same is obviously true for all other data collection geometries. From Frank and Radermacher (1986), reproduced with permission of Springer-Verlag.

The reason why such a reconstruction is quite crude is found by an analysis of the back-projection summation in Fourier space: it essentially corresponds to a simple filling of Fourier space by adding the central sections associated with the projections. Analysis in Fourier space (figure 5.15) makes it immediately clear that the density of sampling points decreases with increasing spatial frequency, with the result that Fourier amplitudes at low spatial frequencies are overemphasized in comparison with those at higher spatial frequencies. The 3D image formed by back-projection therefore looks like a blurred version of the object. Intuitively, it is clear that multiplication with a suitable radius-dependent weight might restore the correct balance in Fourier space. This intuition is borne out by a rigorous analysis.

Weighted back-projection makes use of a 3D weighting function $W_3(\mathbf{k})$ in Fourier space, tailored to the angular distribution of the projections:

$$\rho(\mathbf{r}) = F^{-1}\{W_3(\mathbf{k})F\{\bar{\rho}(\mathbf{r})\}\} \tag{5.20}$$

where $\bar{\rho}(\mathbf{r})$ and $\rho(\mathbf{r})$ are the 3D density distributions before and after weighting, respectively. An equivalent result can be obtained by application of a *two-dimensional* weighting function to the projections, before carrying out the simple back-projection procedure following equations (5.17)–(5.19)

$$p'(\mathbf{r}) = F^{-1}\{W_2(\mathbf{k})F\{p(\mathbf{r})\}\} \tag{5.21}$$

The weighted back-projection technique makes it possible to design weighting functions for arbitrary projection geometries. Thus, it is possible to use weighted back-projection for the reconstruction of an object from projections of randomly oriented particles. Specifically, it allows us to deal with the random azimuths encountered in the random-conical data collection (Radermacher et al., 1986a,

1987a,b). In the following, we must distinguish between the coordinates affixed to the object (denoted by uppercase letters) and those affixed to the individual projections (lowercase letters). Similarly, we need to distinguish between the 2D Fourier space coordinates $\mathbf{k}^{(i)} = \{k_x^{(i)}, k_y^{(i)}\}$ of the ith projection and the 3D Fourier space coordinates of the object, $\mathbf{K} = \{K_x, K_y, K_z\}$. $\mathbf{R} = \{X, Y, Z\}$ are object coordinates, with Z being the coordinate perpendicular to the specimen plane. $\mathbf{R}^{(i)} = \{X^{(i)}, Y^{(i)}, Z^{(i)}\}$ are the coordinates of the ith projection body.

The transformations from the coordinate system of the object to that of the ith projection (azimuth φ_i, tilt angle θ_i) are defined as follows:

$$\mathbf{r}^{(i)} = \mathbf{R}_y(\theta_i)\mathbf{R}_z(\varphi_i)\mathbf{R} \tag{5.22}$$

where $\mathbf{R}_y, \mathbf{R}_z$ are rotation matrices representing rotations around the y- and z-axes, respectively.

The weighting function for arbitrary geometry is now derived by comparing the Fourier transform of the reconstruction resulting from back-projection, $F\{\bar{\rho}(\mathbf{R})\}$, with the Fourier transform of the original object (Radermacher, 1991):

$$F\{\bar{\rho}(\mathbf{R})\} = \sum_{i=1}^{N} F\{b_i(\mathbf{r}^{(i)}, z^{(i)})\} \tag{5.23}$$

$$= \sum_{i=1}^{N} P(\mathbf{k}^{(i)})D \operatorname{sinc}\{D\pi k_z^{(i)}\} \tag{5.24}$$

Here, $P(\mathbf{k}^{(i)})$ denotes the Fourier transform of the ith projection. The symbol $\operatorname{sinc}(x)$ represents the function $\sin(x)/x$, frequently used because it is the *shape transform* of a top-hat function and thus describes the effect, in Fourier space, of a 1D real-space limitation. Each central section associated with a projection is "smeared out" in the k_x-direction and thereby "thickened." In the discrete formulation, each Fourier coefficient of such a central section is spread out and modulated in the direction normal to the section plane. This is in direct analogy with the continuous "spikes" associated with the reflections in the transform of a thin crystal (Amos et al., 1982). However, in contrast to the sparse sampling of the Fourier transform by the reciprocal lattice that occurs in the case of the crystal, we now have a sampling that is virtually continuous.

It is at once clear that the degree of overlap between adjacent "thick central sections" is dependent on the spatial frequency radius. The exact radius beyond which there is no overlap is the resolution limit given by Crowther et al. (1970). The different degrees of overlap produce an initially imbalanced weighting, which the weighting function is supposed to overcome.

For the reconstruction of a point (represented by a delta function) from its projections,

$$P(\mathbf{k}^{(i)}) \equiv 1 \tag{5.25}$$

so that the Fourier transform of the reconstruction by back-projection becomes

$$H(\mathbf{K}) = \sum_{i=1}^{N} D \, \text{sinc}(D\pi k_z^{(i)}) \tag{5.26}$$

In line with the definition of a linear transfer function (see chapter 2, section 3.3), equation (5.26) can be regarded as the transfer function associated with the simple back-projection operation. From this expression the weighting function appropriate for the general geometry can be immediately derived as

$$W(\mathbf{K}) = 1/H(\mathbf{K}) \tag{5.27}$$

From this weighting function for general geometries, any weighting functions for special geometries can be derived (Radermacher, 1991). Specifically, the case of single-axis tilting with regular angular intervals yields the well-known "r^*-weighting" (in our nomenclature, $|\mathbf{K}|$-weighting) of Gilbert (1972).

In constructing the weighting function $W(\mathbf{K})$ according to equation (5.27), the regions where $H(\mathbf{K}) \to 0$ require special attention as they lead to singularities. To avoid artifacts, Radermacher et al. (1987a) found it sufficient to impose a threshold on the weighting function:

$$W_{\text{pract}}(\mathbf{K}) = \begin{cases} W(\mathbf{K}) & \text{for } W(\mathbf{K}) \le T \\ T & W(\mathbf{K}) > T \end{cases} \tag{5.28}$$

that is, to replace $W(\mathbf{K})$ by a value T (chosen to be $= 1.66$) wherever $1/H(\mathbf{K})$ exceeded that limit. In principle, a more accurate treatment can be conceived that would take the spectral behavior of the noise explicitly into account.

Note that $W(\mathbf{K})$ as derived here is a 3D function, and can in principle be used directly in 3D Fourier space to obtain the corrected reconstruction (Hegerl et al., 1991). In most implementations, however, its central sections $w(\mathbf{k}^{(i)})$ are used instead and applied to the 2D Fourier transforms of the projections prior to 3D reconstruction. In fact, the analysis of the discrete case (P. Penczek, personal communication) shows that the two ways of weighting are not exactly equivalent, and that only the 2D weighting gives the correct results.

Radermacher (1992) discussed reconstruction artifacts caused by approximation of equation (5.27), and recommended replacing the sinc-function by a Gaussian function in equation (5.24). This substitution is tantamount to the assumption of a "soft," Gaussian-shaped mask representing the object boundary.

We conclude this section by drawing attention to the artifacts resulting from a missing angular range. This can be illustrated by considering an object, the face of Einstein, for which a limited tilt series of projections is collected (figure 5.16). Artifacts caused by a large angular gap are much more severe than artifacts caused by large angular increments: by going from 2° to 5° increments, some streaking appears, but without affecting the important features when the tilt range is ±90°. On the other hand, the face of Einstein is no longer recognizable when the tilt range spans ±60° even with a tilt increment of 2°.

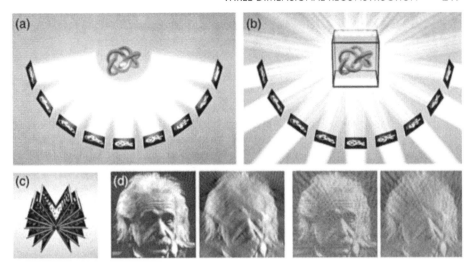

Figure 5.16 Reconstruction of a popular motif by weighted back-projection from a tilt series, using different angular collection geometries. (a, b) Principle of data collection and reconstruction by weighted back-projection. (c) Coverage in Fourier space. (d) Reconstructions with different numbers of projections. First image: $\pm 90°$, in $2°$ increments. Nearly perfect reconstruction is achieved. Second image: $\pm 60°$, $2°$ increments. The image has become distorted, to the point that the face can be hardly recognized, due to the gap in the angular range. Third and fourth images in (d): same as first and second, but the angular increment is now $5°$. Even the $\pm 90°$ reconstruction (third image) is now degraded, but still recognizable. From Baumeister et al. (1999), reproduced with permission of Elsevier.

4.3. Fourier Reconstruction Methods

Fourier approaches to reconstruction [see, for instance, Radermacher's (1992a) brief overview] utilize the projection theorem directly and regard the Fourier components of the projections as samples of the 3D transform to be determined. In most cases, the positions of these samples do not coincide with the regular 3D Fourier grid. This situation leads to a complicated interpolation problem, which can be stated as follows: given a number of measurements in Fourier space at arbitrary points not lying on the sampling grid, what set of Fourier components on the sampling grid are consistent with these measurements? The key to this problem lies in the fact that the object is of finite dimensions (Hoppe, 1969); because of this property, the given arbitrary measurements are related to those on the grid by the Whittaker–Shannon interpolation (Hoppe, 1969; Crowther et al., 1970; Radermacher, 1992a; Lanzavecchia et al., 1993; Lanzavecchia and Bellon, 1994).

Following Radermacher's account, we consider the unknown object bounded by a rectangular box with side lengths a, b, and c. In that case, its 3D Fourier transform is completely determined when all samples on the regular 3D Fourier grid with reciprocal cell size $(1/a, 1/b, 1/c)$ are known. We index them as

$$F_{hkl} = F(h/a, k/b, l/c) \qquad (5.28a)$$

For an arbitrary position (denoted by the coordinate triple x^*, y^*, z^*) not lying on this grid, the Whittaker–Shannon theorem yields the relationship:

$$F(x^*, y^*, z^*) = \sum_h \sum_k \sum_l \frac{\sin \pi(ax^* - h) \sin \pi(by^* - k) \sin \pi(cz^* - l)}{\pi(ax^* - h)\pi(by^* - k)\pi(cz^* - l)} \qquad (5.29)$$

(The three terms whose product form the coefficient of F_{hkl} are again sinc-functions.) This relationship tells us how to compute the value of the Fourier transform at an arbitrary point from those on the regular grid, but the problem we wish to solve is exactly the reverse: how to compute the values of the Fourier transform at every point h, k, l of the grid from a given set of measurements at arbitrary positions $\{x_j^*, y_j^*, z_j^*; j = 1, \ldots, J\}$ as furnished by the projections. By writing equation (5.29) for each of these measurements, we create a system of M equations for $H \times K \times L$ unknown Fourier coefficients on the regular grid. The matrix $\mathbf{C}$ representing the equation system has the general element:

$$C_{jhkl} = \frac{\sin \pi(ax_j^* - h) \sin \pi(by_j^* - k) \sin \pi(cz_j^* - l)}{\pi(ax_j^* - h)\pi(by_j^* - k)\pi(cz_j^* - l)} \qquad (5.30)$$

To solve this problem, we must solve the resulting equation system as follows:

$$F_{hkl} = \sum_{j=1}^{N} F(x_j^*, y_j^*, z_j^*) C_{hkl}^{-1} \qquad (5.31)$$

where C_{hkl}^{-1} are the elements of the matrix that is the inverse of $\mathbf{C}$. It is obvious that this approach is not practically feasible because of the large number of terms. Basically, this intractability is the result of the fact that at any point that does not coincide with a regular Fourier grid point, the Fourier transform receives contributions from sinc-functions centered on *every* single grid point of the Fourier grid within the resolution domain.

Remedies designed to make the Fourier approach numerically feasible have been discussed by Lanzavecchia and coworkers (1993, 1994). These authors use the so-called "moving window" method to curb the number of sinc-functions contributing in the interpolation, and thus obtain an overall computational speed that is considerably faster than the efficient weighted back-projection technique. A demonstration with experimental data—albeit with evenly distributed projections—indicated that the results of the two techniques were virtually identical. Other remedies are to truncate the sinc-functions or to use different (e.g., triangular) interpolation functions of finite extent.

Recently, a fast and accurate Fourier interpolation (in the class of so-called gridding-based direct Fourier inversion algorithms) method of reconstruction has been added to this arsenal (Penczek et al., 2004). Generally, gridding methods furnish samples on a regular grid by interpolation from irregularly sampled data. The method developed by Penczek and coworkers uses a regular grid on a sphere as the target grid. Test results with phantoms, with and without noise, confirm that this technique produces reconstructions superior to weighted back-projection, and comparable to SIRT, which as an iterative technique is necessarily slower.

4.4. Iterative Algebraic Reconstruction Methods

In the discrete representation of the reconstruction problem, the relationship between the object and the set of projections can be formulated by a set of algebraic equations (Crowther et al., 1970; Gordon et al., 1970). For the parallel projection geometry, the jth sample of projection i is obtained by summing the object ρ along parallel rays (indexed j) defined by the projection direction θ_i. The object is a continuous density function, represented by samples ρ_k on a regular gid, with some rule (interpolation rule) for how to obtain the values on points not falling on the grid from those lying on the grid. Now the discrete points of the object contribute to the projection rays according to weights $w_{jk}^{(i)}$ that reflect the angle of projection and the particular interpolation rule:

$$p_j^{(i)} = \sum_k w_{jk}^{(i)} \rho_k \tag{5.32}$$

With a sufficient number of projections available at different angles, equation (5.32) could be formally solved by matrix inversion (Crowther et al., 1970), but the number of unknowns is too large to make this approach feasible. An alternative is provided by a least-squares pseudo-inverse method (Klug and Crowther, 1972; Carazo, 1992), which involves the inversion of a matrix whose dimensionality is given by the number of independent measurements—still a large number but somewhat closer to being manageable for the 2D (or single-axis tilt) case (Klug and Crowther, 1972; Zhang, 1992; Schröder et al., 2002).

There is, however, an entire class of reconstruction algorithms based on an approach to estimate the solution to equation (5.32) iteratively. The strategy of these algorithms is that they start from an original estimate $\rho_k^{(0)}$ and compute its projection $\hat{p}_j^{(i)}$ following equation (5.32). The discrepancy between the actually observed projection $p_j^{(i)}$ and the "trial projection" $\hat{p}_j^{(i)}$ can now be used to modify each sample of the estimate $\rho_k^{(0)}$, yielding the new estimate $\rho_k^{(1)}$, etc. In the *algebraic reconstruction technique* [proposed by Gordon et al. (1970), but essentially identical with the Karczmarz (1937) algorithm for approximating the solutions of linear equations], the discrepancy is subtracted in each step from each voxel along the projection rays, so that perfect agreement is achieved for the particular projection direction considered. In the SIRT (proposed by Gilbert, 1972), the discrepancies of all projections are simultaneously corrected, not successively as in ART. For an exhaustive description of these and other iterative techniques, the reader is referred to the authoritative book by Herman (1980).

Iterative methods have the advantage over other approaches to 3D reconstruction in that they are quite flexible, allowing constraints and statistical considerations to be introduced into the reconstruction process (e.g., Penczek et al., 1992). They have the disadvantage of requiring a much larger computational effort than the weighted back-projection method. The use of nonlinear constraints (e.g., prescribed value range) introduces another disadvantage: the reconstrution process is no longer linear, making its characterization by a point-spread function (see section 4.2) or a 3D variance estimation by projection variance back-projection (section 2.2 in chapter 6) difficult to achieve.

5. The Random-Conical Reconstruction in Practice

5.1. Overview

The concept of the random-conical data collection was introduced in section 3.5. In the following section, all steps of the reconstruction scheme that makes use of this data collection method will be outlined. In this treatment we will follow the detailed account given by Radermacher (1988), but with the addition of the multivariate data analysis (MDA)/classification step and some modifications that reflect changes in the procedures as they have developed since 1988.[1]

To start with an overview of the procedure (figure 5.17), the particles are first selected ("windowed") simultaneously from the tilted- and the untilted-specimen field (steps 1 and 2, respectively), yielding two sets of images. Next, the untilted-specimen image set is subjected to alignment (step 3), producing a set of "aligned" images. These are then analyzed using MDA and classification (step 4), resulting in the rejection of certain particles and in the division of the data set into different classes according to particle view. From here on, the different classes are processed separately. For each class, the following steps are followed: the tilted-specimen images are sorted according to the azimuthal angle found in the alignment procedure. Once in correct order, they are aligned with respect to one another or with a common reference (step 5). After this, they may be used to obtain a 3D reconstruction (step 6). Thus, in the end, as many reconstructions are obtained as there are classes, the only requirement being that the class has to be large enough for the reconstruction to be meaningful.

5.2. Optical Diffraction Screening

The tilted-specimen micrograph covers a field with a typical defocus range of 1.4 μm (at 50,000 × magnification and 50° tilt) normal to the direction of the tilt axis. Because of the properties of the contrast transfer function (section 3.3 in chapter 2), useful imaging conditions require the defocus to be in the range of underfocus [i.e., $\Delta z < 0$ in equation (2.5) of chapter 2]. In addition, the range must be restricted, to prevent a blurring of the reconstruction on account of the defocus variation, which is equivalent to the effect of energy spread (section 3.3.2 in chapter 2). How much the range must be restricted, to

---

[1]SPIDER batch files that follow each step of processing, are accesible at the website http://www.wadsworth.org/spider_doc/spider/docs/master.html

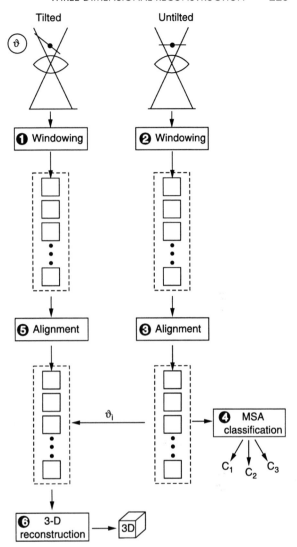

Figure 5.17 Schematic diagram of the data flow in the random-conical reconstruction. Simultaneously, particles are selected from the micrograph (1) of the tilted specimen and that (2) of the untilted specimen. Those from the untilted-specimen field are aligned (3), resulting in azimuthal angles φ_i, and classified (4), resulting in the separation into classes C_1–C_3. Separately, for each of these classes, tilted-specimen projections are now aligned (5) and passed to the 3D reconstruction program (6). From Radermacher et al. (1987b), reproduced with permission of Blackwell Science Ltd.

a practical "defocus corridor" parallel to the tilt axis, depends on the resolution expected and can be inferred by reference to the CTF characteristics (chapter 2, section 3.3) and the sharp falloff in the "energy spread" envelope produced by the defocus spread.

When the micrograph shows a field with carbon film, the useful defocus range can be found by optical diffraction analysis (see chapter 2, section 3.6). Details of this screening procedure have been described by Radermacher (1988). The selection aperture must be small enough so that the focus variation across the aperture is kept in limits. By probing different parts of the micrograph, the tilt axis direction is readily found as that direction in which the optical diffraction pattern remains unchanged. Perpendicularly to that direction, the diffraction

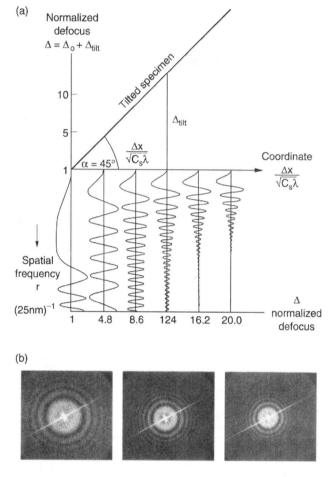

Figure 5.18 Change of contrast transfer function across the micrograph of a tilted specimen. (a) Theoretical behavior; (b) optical diffraction patterns obtained at different positions of the area-selecting aperture of the optical diffractometer along a line perpendicular to the tilt axis. From Zemlin (1989b), reproduced with permission of John Wiley & Sons, Inc.

pattern changes most strongly (see figure 5.18), and following this direction of steepest defocus change, the useful range must be established by comparison with the CTF characteristics.

Electron microscopes equipped for spot scanning allow an automatic compensation for the defocus change in the direction normal to the tilt axis (Zemlin, 1989b; Downing, 1992). Thus, for micrographs or data collected electronically from such microscopes, particles from the entire specimen field can be used for processing (Typke et al., 1992).

Another way of including all data in the processing irrespective of their defocus is by keeping track of a spatial variable, in the course of selecting and storing the individual particle windows, that gives the particle position within the micrograph field in the direction normal to the tilt axis. This variable can later be interpreted in terms of the effective local defocus, and used to make compensations or assign appropriate Fourier weighting functions in the reconstruction.

5.3. Interactive Tilted/Untilted Particle Selection

Manual selection of particles from a pair of tilted- and untilted-specimen micrographs is a tedious task. A computer program for simultaneous interactive selection was first described in the review by Radermacher (1988) and has since been upgraded. The two fields are displayed side by side on the screen of the workstation (figure 5.19) with a size reduction depending on the microscope magnification and computer screen size (in pixels). In some instances, the size reduction makes it possible to select all particles from a micrograph field at once. It has the beneficial side effect that the particles stand out with enhanced contrast, since size reduction with concurrent band limitation enhances the SNR, provided (as is the case here) that the signal spectrum falls off faster then the noise spectrum with increasing spatial frequency (see chapter 3, section 4.3).

Such a program is laid out to calculate the geometrical relationships between the coordinate systems of the two fields, as sketched out in the article by Radermacher et al. (1987b). For this purpose, the user selects a number of particle images in pairs—tilted and untilted—by going back and forth with the cursor between the two fields. After this initial manual selection, the program is prompted to compute the direction of the tilt axis and the precise coordinate transformation between the two projections. This transformation is given as (Radermacher, 1988)

$$
\begin{pmatrix} x' \\ y' \end{pmatrix} = \begin{pmatrix} \cos\beta & \sin\beta \\ -\sin\beta & \cos\beta \end{pmatrix} \times \begin{pmatrix} \cos\theta & 0 \\ 1 & 1 \end{pmatrix} \times \begin{pmatrix} \cos\alpha & -\sin\alpha \\ \sin\alpha & \cos\alpha \end{pmatrix}
$$

$$
\times \left(\begin{pmatrix} x \\ y \end{pmatrix} - \begin{pmatrix} x_0 \\ y_0 \end{pmatrix} \right) + \begin{pmatrix} x_0' \\ y_0' \end{pmatrix}
$$

(5.33)

where α is the angle between tilt axis in the untilted-specimen image and its y-axis; β is the corresponding angle for the tilted-specimen image; $\{x_0, y_0\}$ are the coordinates of an arbitrary selectable origin in the untilted-specimen image; and $\{x_0', y_0'\}$ are the corresponding origin coordinates in the tilted-specimen image. After these geometrical relationships have been established, the user can select additional particles from one field (normally the untilted-specimen field, as the particles are better recognizable there), and the program automatically

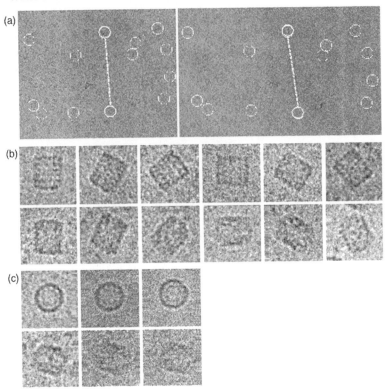

Figure 5.19 Interactive particle selection, and views of the hemocyanin molecule. (a) Two micrographs of the same field (left, tilted by 45°; right, untilted), displayed side by side. Equivalent particle images are identified and numbered in both micrographs. Shown here is the beginning phase of the program where each particle has to be tracked down in both micrographs. After the initial phase, the parameters of the underlying coordinate transformation are known, and the program is now able to identify the companion in the second micrograph for any particle selected in the first micrograph. (b) Molecules presenting the side view. Top row: 0°; bottom row: 45°. (c) Molecules presenting the top view. Top row: 0°; bottom row: 45°. (From N. Boisset, unpublished lecture material.)

finds the particle as it appears in the other field. A document file is used by the program to store all particle positions in both fields, later to be used for windowing the particles images from the raw data files.

As an alternative to interactive particle selection, various automated particle picking programs have been developed (see chapter 3, section 2.4).

5.4. Optical Density Scaling

As in electron crystallography of crystals (Amos et al., 1982), correct mutual scaling of the optical densities of the projections is necessary so that no projection will enter the analysis with undue weight, which would lead to a distorted representation of the structure.

Radermacher et al. (1987b) normalized the individual images representing tilted particles as follows: next to the particle (but staying away from the heavy stain accumulation, if the preparation is with negative staining), the average density of a small reference field, $\langle D \rangle$, is calculated. The density values measured in the particle window are then rescaled according to the formula:

$$D'_i = D_i - \langle D \rangle \tag{5.34}$$

Boisset et al. (1993a) introduced a more elaborate scaling procedure that makes explicit use of the statistical distribution of the background noise: it is assumed that the background noise surrounding the particle has the same statistical distribution throughout. Using a large portion of one of the tilted-specimen micrographs, a reference histogram is calculated. Subsequently, the density histogram of the area around each particle is compared with the reference histogram, and the parameters a and b of a linear density transformation:

$$D'_i = aD_i + b \tag{5.35}$$

are thereby estimated. This transformation is then applied to the entire particle image, with the result that in the end all images entering the reconstruction have identical noise statistics. In fact, the match in statistics among the corrected images extends beyond the first and second moment.

5.5. Processing of Untilted-Particle Images

5.5.1. Alignment and Classification

Following the random-conical scheme, the untilted-specimen projections are first aligned and then classified. The alignment furnishes the azimuthal angle φ_i of the particle, which is needed for placing the corresponding tilted-specimen projections into the conical geometry. The classification results in a division into L subsets of particles which are presumed to have different orientations and ideally should be processed separately to give L different reconstructions. In practice, however, one chooses the classification cutoff level rather low, so that many classes are initially generated. By analyzing and comparing the class averages, it is possible to gauge whether some mergers of similar classes are possible without compromising resolution. This procedure leads to some small number of $L_1 < L$ "superclasses," which are fairly homogeneous on the one hand but, on the other, contain sufficient numbers of particles to proceed with 3D reconstruction. A substantial number of particles that fall into none of the superclasses are left out at this stage, their main fault being that their view is underrepresented, with a number of particles that is insufficient for a 3D reconstruction. These particles have to "wait" until a later stage of the project when they can be merged with a well-defined reconstruction which is itself based on a merger of the superclass reconstructions (see section 6).

5.5.2. Theoretical Resolution

The resolution of a reconstruction from a conical projection set with equispaced angles was derived by Radermacher (1991). In the case of an even number of projections N, the result is

$$d = 2\pi(D/N)\sin\theta_0 \qquad (5.36)$$

where $d = 1/R$ is the "resolution distance," that is, the inverse of the resolution; D is the object diameter; and θ_0 is the tilt angle of the specimen grid. Furthermore, it should be noted that, for such a data-collection geometry, the resolution is direction-dependent, and the above formula gives only the resolution in a direction perpendicular to the direction of the electron beam. In directions that form an angle oblique to those directions, the resolution is deteriorated. In the beam direction, where the effect of the missing cone is strongest, the resolution falls off by a factor of 1.58 (for $\theta_0 = 45°$) or 1.23 (for $\theta_0 = 60°$).

5.5.3. Minimum Number of Particles Required: Angular Histogram

The number of projections required for computing a "self-standing" reconstruction is determined by the statistics of the angular coverage (Radermacher et al., 1987b; see figure 5.20). For such a reconstruction to be mathematically supported, going by the conical-reconstruction resolution formula [equation (5.36)], there should be no gap in the azimuthal distribution larger than $\Delta\theta_{min} = 360°/N = 360° \times d/(2\pi D \sin\theta_0)$, where N is the number of equidistant projections in a regular conical series, $d = 1/R$ is the resolution distance aimed for, D is the diameter of the object, and θ_0 is the cone angle. To reconstruct the 50S ribosomal subunit of E. coli ($D \approx 200$ Å) to a resolution of $R = 1/30$ Å^{-1} (i.e., $d = 30$ Å) from a 50° tilted specimen (i.e., $\theta_0 = 40°$), the minimum angular increment works out to be $\Delta\theta_{min} = 13°$. Because of the statistical fluctuations that occur in a random coverage of the azimuthal range, the required number of projections is much larger then the minimum of $N = 360/13 \approx 28$. To give an example, a set of 250 projections proved sufficient to cover the 360° range with the largest gap being 5° (Radermacher et al., 1987b). Although there is no ironclad rule, molecules in that size range (200–300 Å) and to that resolution (1/30 Å^{-1}) appear to require a minimum of 200 projections.

5.6. Processing of Tilted-Particle Images

5.6.1. Alignment

Reference to the Fourier formulation (figure 5.3) explains why an alignment of neighboring tilted-particle projections by cross-correlation is possible: it is

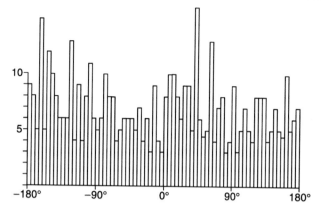

Figure 5.20 Distribution of azimuthal angles of 50S ribosomal particles extracted from five pairs of micrographs. The angles were determined by alignment of particles showing the crown view as they appear in the untilted-specimen micrographs. From Radermacher et al. (1987b), reproduced with permission of The Royal Microscopy Society.

because each Fourier component is surrounded by a "disk of influence," as a result of the boundedness of the object. It is the very reason why a reconstruction of an object from a finite number of its projections is feasible (see the early statement of this principle in Hoppe, 1969).

There are four methods of alignment of projections that have been tried at one stage or another: (i) sequential alignment "along the circle" with cosine stretching (Radermacher et al., 1987b), (ii) alignment to the corresponding 0° projection with cosine stretching (Carazo et al., 1988; Radermacher, 1988), (iii) alignment to a perspectively distorted disk (Radermacher, 1988), and (iv) alignment to a disk or "blob" (Penczek et al., 1992).

The reference to cosine stretching requires an explanation. According to Guckenberger (1982), the projection that needs to be aligned to the untilted-particle projection [i.e., method (ii) above] must first be stretched by the factor $1/\cos \theta$ in the direction perpendicular to the tilt axis, if θ is the tilt angle. When this method is applied to the alignment of two adjacent tilted-particle projections [method (i) above], both must be stretched by that factor prior to alignment. The cosine-stretching procedure appears to work well for oblate objects, that is, those objects that are more extended in the x- and y-directions than in the z-direction, as for instance the 50S ribosomal subunit in the negatively stained double-layer preparation (Radermacher et al., 1987a,b; Radermacher, 1988). The reason why it works well for such specimens is that the common signal includes the surrounding carbon film, which is flat. For globular objects, such as the 70S ribosome embedded in ice (Frank et al., 1991, 1995a,b; Penczek et al., 1992, 1994), alignment to an unstretched disk ("blob"), whose diameter is equal to the diameter of the particle, leads to better results.

The method of sequential alignment of neighboring projections is error-prone because errors are bound to accumulate in the course of several hundred alignments, as may be easily checked by computing the closure error (Radermacher et al., 1987b). Radermacher et al. (1987b) developed a fail-safe variant of the sequential method, in which neighboring projections are first averaged in groups spanning an angle of 10°; these averaged tilt projections are then aligned to one another "along the circle," and, in the end, the individual projections are aligned to their corresponding group averages. However, for cryo-specimens with their decreased SNR, sequential alignment has largely been abandoned in favor of the alignment of each tilted-particle projection with the 0° projection or, as pointed out before, with a "blob."

It may seem that the availability of 3D refinement methods (see section 8) has relaxed the requirement for precise tilted-particle projection alignment at the stage of the first reconstruction; on the other hand, it must be realized that the initial stability and speed of convergence of the refinement depends on the quality of the reference reconstruction, which is in turn critically dependent on the success of the initial alignment.

5.6.2. Screening for Consistency

Screening of tilted-particle projections is a way of circumventing the degeneracy of 0° classification (section 4.11 in chapter 4). The most important stipulation to apply is the continuity of shape in a series of angularly ordered projections. As members of a conical tilt series, by virtue of the boundedness of the particle (see section 2.2), projections must occur in a strict 1D similarity order: for instance, in an object without symmetries, the ordering of five closely spaced projections along the cone in the sequence A–B–C–D–E implies that the cross-correlation coefficients (for brevity denoted by symbol $\otimes$) must be ranked in the following way:

$$A \otimes B > A \otimes C > A \otimes D > A \otimes E \qquad (5.37)$$

If we look at projections forming an entire conical series, then the associated similarity pathway is a closed loop. The reason is that in any short segment of the loop, we find a cross-correlation ranking of the type stated above as a local property. If all projections of the conical series are closely spaced, then the shape variation from one neighbor to the next is quasi-continuous. In practice this means that a gallery of tilted-particle projections, presented in the sequence in which they are arranged on the cone, should show smooth transitions, and the last image of the series should again be similar to the first. The most sensitive test of shape continuity is a presentation of the entire projection series as a "movie": any discontinuity is immediately spotted by eye.

The correct ordering along a similarity pathway can also be monitored by multivariate data analysis (MDA). In the absence of noise, the factor map should show the projections ordered on a closed loop (Frank and van Heel, 1982b; van Heel, 1984a; Harauz and Chiu, 1991, 1993). In fact, the ordering is along a closed loop in a high-dimensional space $\mathbf{R}^J$, and this closed loop appears in many different *projected versions in the factor maps. In the presence of noise, it proves difficult to visualize the similarity pathway (and thereby spot any outliers that lie off the path). One way out of this dilemma is to form "local" averages over short angular intervals, and apply MDA (i.e., CA or PCA) to these very robust manifestations of the projections (Frank et al., 1986). A closed loop becomes then indeed visible. This method can be used, in principle, to screen the original projections according to their distance, in factor space, from the averaged pathway. The approach has not been pursued, however, having been partly replaced by the 3D projection matching (section 7.2) and 3D Radon transform (section 7.4) methods.

Flip/flop ambiguities are normally resolved by classification of the $0°$ views, on the basis of the fact that the projected molecule shapes are related by a mirror (van Heel and Frank, 1981). However, a peculiar problem emerges when the shape of the molecule in projection is symmetric. In that case, there is no distinction between flip and flop projections, unless induced by one-sidedness of staining. Lambert et al. (1994a) had to deal with this problem when they processed images of the barrel-shaped chiton hemocyanin, which sits on the grid with one of its (unequal) round faces. As these authors realized, the cylindrical shape is unique in that it allows flip and flop orientations of the molecule to be sorted by multivariate data analysis of the tilted-particle images. Perfect cylinders give, uniformly, rise to double-elliptic barrel projections, irrespective of the direction of tilt. Any asymmetry in the z distribution of mass (in this case, the existence of a "crown" on one side of the molecule) leads to a difference in appearance between molecules tilted to one side from those tilted to the other. This difference was clearly picked up by correspondence analysis, and the two different populations could be separated on this basis.

5.7. Carrying Out the Reconstruction

After the screening step, which is done to verify that the tilted-particle projections follow one another in a reasonable sequence, the particle set is ready for 3D reconstruction. Initially, weighted back-projection with generalized weights was used (Radermacher et al., 1987b), but later a variety of options including a modification of SIRT (Penczek et al., 1992) and techniques based on conjugate gradients (Penczek, unpublished; see SPIDER documentation) and Fourier interpolation (Lanzavecchia and Bellon, 1996; Grigorieff, 1998; Penczek, 2004) have been added (see section 4).

The fact that $0°$ images are available yet are not used as input to the reconstruction can be exploited for a consistency check: the $0°$ projection of the reconstruction should agree with the average of the aligned $0°$ images.

After successful reconstruction, the 3D density map exists as a set of slices (see figure 6.7 in chapter 6). Each slice can be represented by a 2D image that gives the density distribution on a given z level, which is an integer multiple of the sampling step. The representation as a gallery of successive slices (the equivalent of a cartoon if the third dimension were time) is the most comprehensive but also the most confusing way of representing 3D results. It is comprehensive since it preserves all of the interior density distribution and avoids the imposition of a choice of density threshold for the molecular boundary. However, this representation is also confusing since it does not lend itself to a visual comprehension of the 3D relationships among features seen in the slices. For this reason, *surface renderings* are now in common use, even though they hide all interior features and entail a commitment to a fixed boundary if any comparisons with other results are made. While *volume renderings* are suitable for conveying the continuous 3D density distribution within a volume in its entirety, the impression is rarely satisfactory, and highly dependent on the data. A general discussion of the various visualization options is contained in section 4 in chapter 6.

6. Common-Lines Methods (or "Angular Reconstitution") in Practice

As common-lines techniques require a high SNR, the data set must first be subjected to classification, so that classes with sufficient numbers of images can be determined, from which averages may be computed. Classification can be of the unsupervised [K-means or HAC (hierarchical ascendant classification) being the most common methods; see example A below] or supervised kind (projection matching with a known 3D reference; see example B below). In the second case, there may seem to be no point in performing orientation determination twice (i.e., once by 3D projection matching and once by common lines), but the rationale can be to eliminate any reference bias when the molecule set differs considerably from the 3D reference, as in the case of the application by Craighead et al. (2002).

The examples shown here (figures 5.21 and 5.22) make use of the simultaneous common-lines search technique of Penczek et al. (1996), which has the advantage that it does not require user interaction in the choice of the sequence in which projections enter, or the choice of "anchor projections." However, as in all common-lines based orientation search techniques, handedness information came from separate experiments.

7. Reference-Based Methods and Refinement

7.1. Introduction

To summarize the contents of the preceding part of this chapter, depending on reconstruction strategy, initial angles of particles are determined by one

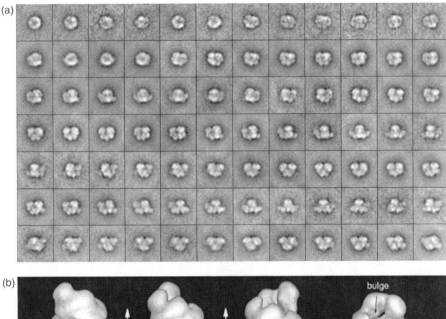

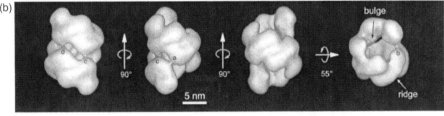

Figure 5.21 Reconstruction of cyanide-degrading nitrilase following orientation determination with common lines. (a) Gallery of 84 class averages. These were obtained from a
set of 7008 images by using the rotationally invariant K-means clustering method of
Penczek et al. (1996) followed by several cycles of multireference alignment as detailed
in Joyeux and Penczek (2002), in the course of which poorly matched images were
excluded. (b) Reconstruction of the molecule, seen from different viewing directions. From
the 84 class averages shown in (a), the 49 most populated were selected for application
of the common-lines procedure. The initial reconstruction was refined in 64 cycles of 3D
projection matching algorithm. After the 47th cycle, twofold symmetry became apparent,
and was then reinforced in the remaining 17 cycles. Several lines of evidence permitted
the choice of handedness. For example, a homolog of the protein makes a transition
from short spirals to long helices at pH 4. By shadowing the specimen in the long helix
form, Sewell and coworkers were able to prove left-handedness. From Sewell et al. (2003),
reproduced with permission of Elsevier.

of the following methods: (i) random-conical data collection; (ii) multireference alignment with projections of an existing reconstruction (a supervised classification method, in the sense defined in chapter 4, section 4.2); or
(iii) multiple common lines using class averages (also known as angular

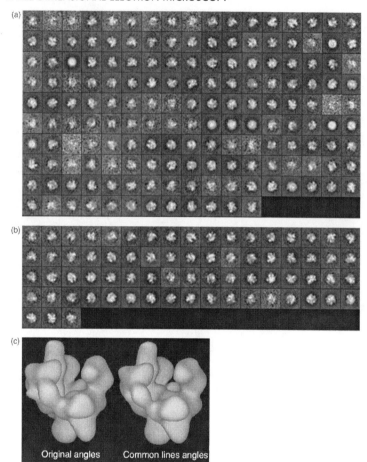

Figure 5.22 Reconstruction of the 10-subunit yeast RNA polymerase II (missing subunit Rpb4 and Rpb7) using angles found by common lines. [These data were included in the work on the complete 12-subunit yeast RNA polymerase II (Craighead et al., 2002).] (a) Particles of delta4/7 RNAPII were classified into 165 groups using reference projections generated from the crystal structure of delta4/7 RNAPII filtered to ~20 Å resolution. To assay the homogeneity of the groups, 10 rounds of reference-free averaging were performed on all 165 classes. For each of the classes, a final round of reference-free alignment on the 10 averages was used to obtain final reference-free class averages. (b) From the 165 reference-free class averages, 71 well-aligned images with high signal-to-noise ratio were chosen for common-lines determination of orientations ("angular reconstitution"). The 71 images were subjected to 200 cycles of common-lines search using the Penczek et al. (1996) algorithm, starting with a random set of Eulerian angles. (c) Back-projection reconstructions were performed using either the angles resulting from 3D projection matching, or those determined using common lines. The differences between the two sets of angles for the 71 averages ranged from 1 to 12°, except for one outlier at 30°. The average difference angle was 6.8°. From Craighead et al. (2002), reproduced with permission of Elsevier.

reconstitution). In each of these cases, the angles determined are inaccurate, and the resulting first reconstruction is far from optimal. There are different reasons for these inaccuracies:

(i)   In the case of the random-conical data collection, the selection of a particle subset with fixed 0° orientation is based on the classification of particles as they appear in the 0° micrograph. Since the discrimination of shapes and internal features by classification is limited, there exists an uncertainty in the actual size of the θ-angle that defines the inclination of the central section associated with the projection. This angular uncertainty can easily reach ±10° depending on particle shape and reduce the resolution of the reconstruction substantially (see the estimates given in section 7.5). Additionally, reconstructions from random-conical data are handicapped by the missing cone, the cone-shaped gap in Fourier space, which can only be filled by combining two or more data sets of particles exhibiting different views in the 0° micrograph (section 12).

(ii) The accuracy of angle determinations by multireference alignment is limited by the resolution of the reconstruction used as 3D reference, and by the size of the angular step. The two limitations work in tandem: for a given low-resolution reference, it makes no sense to go below a certain minimum angular step size, as the angular definition will no longer improve.

(iii) In the case of orientation determination by common lines, the angular accuracy is limited by the fact that each participating class average comprises a finite angular width. In addition, the formation of those averages eliminates high-resolution information, which is no longer available when the reconstruction is formed.

Iterative methods that have been developed to exhaust the information in the raw data are known under the summary term *angular refinement*. Successively, by matching the experimental projections with those computed from the reconstruction with smaller and smaller angular step sizes, a reconstruction is obtained that is optimally consistent with the data set (provided that no false minimum is reached). This principle of 3D projection matching is summarized in figure 5.23. By giving each experimental projection a chance "to find a better home" in terms of its orientation and shift, the resolution of the reconstruction is substantially improved. In the first applications of this technique to the 70S *E. coli* ribosome, an improvement from $1/47.5$ to $1/40 \, \text{Å}^{-1}$ was reported (Penczek et al., 1994). An angular refinement technique based on essentially the same principle, of using an existing lower-resolution reconstruction as a template, was formulated for the processing of virus particles [Cheng et al. (1994)—note that a later implementation of 3D projection matching, by Baker and Cheng (1996), used autocorrelation functions instead]. The two schemes resemble earlier schemes proposed by van Heel (1984) and Harauz and Ottensmeyer (1984a). In fact, all four schemes, although differing in the choice of computational schemes and the degree of formalization, can be understood as variants of the same approach (see section 7.2).

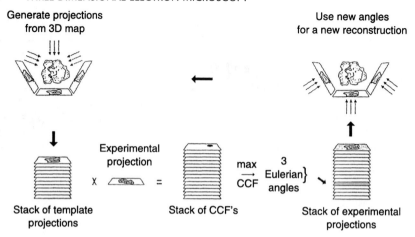

Figure 5.23 Principle of 3D angular alignment or refinement. From an existing reconstruction (top left), a large number of projections are obtained, covering the 3D orientation space as evenly as possible. These are used to find the Eulerian angles of experimental projections. This scheme is used both in the incorporation of new data, and in refinement of existing data: (i) 3D angular alignment of new data: a given experimental projection that is not part of the projection set from which the reconstruction was obtained is now cross-correlated with all trial projections. The direction (in terms of a triple of Eulerian angles) for which maximum correlation is obtained is then assigned to the new projection. In this way, an entire new data set can be merged with the existing data so that a new reconstruction is obtained. (ii) Angular refinement of existing data: each projection that is part of the experimental projection set is matched in the same way with the computed projections to find a better angle than originally assigned. However, in this case, the search range does not have to extend over the full space of orientations, because large deviations are unlikely. The best choice of search range can be gauged by histograms of angular "movements"; see section 7.5 and particularly figure 5.29. Adapted from Penczek et al. (1994).

7.2. Three-Dimensional Projection Matching

7.2.1. Principle

In the 3D projection matching scheme, reference projections are computed from the (low-resolution) 3D reference structure such that they cover the entire angular space evenly (figure 5.23). As reference, an existing 3D volume is used, which might have been obtained by merging several random-conical projection sets or from class averages aligned in 3D by a common-lines technique. In the application that Penczek et al. (1994) used to demonstrate the technique—the 70S ribosome from *E. coli*—5266 such reference projections were obtained, corresponding to an angular step size of 2°. Each of these reference projections is "indexed" by a triple of Eulerian angles (ψ, θ, and φ). (In practice, ψ is set to zero throughout, for reasons that become obvious in the next paragraph).

A given experimental projection is compared with all reference projections. This comparison amounts to a five-dimensional (5D) search, as it is performed for each pair of images (experimental versus reference projection indexed by the Eulerian angles ψ and θ) by a 3D search (Eulerian angle φ and x, y) using cross-correlation. In other words, the Eulerian angles (ψ, θ) giving the largest cross-correlation function (CCF) peak denote the desired projection direction, while the CCF between reference and current projection gives, at the same time, the shift (x, y) and the azimuthal orientation (φ) of the particle under consideration. Using this new set of parameters for each experimental projection, a new reconstruction is then computed, which normally has improved resolution. This refined reconstruction can now be used as new 3D reference, etc. Usually, the angles no longer change by significant amounts after 5–10 iterations of this scheme (see section 7.5 for a discussion of this point in the light of experimental results). In the demonstration by Penczek et al. (1994), additional 0° or low-tilt data could be used, following this approach, to improve the resolution from 1/40 to $1/29\,\text{Å}^{-1}$. The end results of the refinement, and the evenness of the distribution, can be checked by plotting a chart that shows the 3D angular distribution of projections (see figure 5.25). Before refinement, each random-conical set of projections will be mapped into a circle of data points on such a chart. It is seen that the angular corrections in the refinement have wiped out all traces of these circular paths.

7.2.2. Historical Note

The 3D projection matching technique as described above (Cheng et al., 1994; Penczek et al., 1994) is closely related to Radermacher's 3D Radon transform method (Radermacher, 1994; see section 7.4). The technique also has close similarities with projection matching techniques that were formulated some time ago, when exhaustive search techniques with the computer were quite time consuming and therefore still impractical to use. These relationships will be briefly summarized in the following [for a more detailed list, see Frank (1996)].

(i) van Heel (1984b) proposed a method designed to obtain the unknown orientations of a projection set using the following sequence: step 0: assign random orientations to the projections; step 1: compute 3D reconstruction; step 2: project 3D reconstruction in all directions on a finite grid in angular space to match experimental projections. The parameters for which best matches are obtained yield new orientations for the projections; compute differences between current model projections and experimental projections; if summed squared differences are larger than a predefined value then GO TO step 1, otherwise STOP.

This procedure thus contains the main ingredients of the algorithm of Penczek et al. (1994), except for the choice of starting point (step 0), which makes it difficult for the algorithm to find a satisfactory solution except in quite fortuitous cases where the random assignment of angles happens to come close to the actual values. However, the method appears to work when the structure has sufficiently high symmetry (P. Penczek, personal communication).

(ii) The approach used by Harauz and Ottensmeyer (1984a,b) makes use of a computer-generated model of the predicted structure, rather than an experimental reconstruction, as an initial 3D reference and a combination of visual and computational analyses to obtain an optimum fit for each projection. Because of the small size of the object (nucleosome cores whose phosphorus signal was obtained by energy filtering) and the type of specimen preparation used (air-drying), the result was greeted with considerable skepticism. There is also a principal question whether the use of an artificial 3D reference (as opposed to an experimental 3D reference) might not strongly bias the result. We recall the reports about reference-induced averages in the 2D case (Boekema et al., 1986; Radermacher et al., 1986b), and our previous discussion of this question in the introduction to reference-free alignment schemes (see chapter 3, section 3.5).

As van Heel (1984b), Harauz and Ottensmeyer (1984a,b) also make use of the summed squared difference (or generalized Euclidean distance), rather than the cross-correlation, to compare an experimental projection with the reprojections. It was earlier pointed out (chapter 3, section 3.3) that there is no practical difference between the summed squared difference and the cross-correlation as measures of "goodness of fit" when comparing 2D images that are rotated and translated with respect to one another. This is so because the variance terms in the expression of the generalized Euclidean distance are translation and rotation invariant. In contrast, the two measures *do* behave differently when used to compare projections of two structures, since the variance of a projection may strongly depend on the projection direction.

7.2.3. Applications

7.2.3.1. The progress of refinement   As an example for the progress of the refinement by 3D projection alignment, figure 5.24 shows the density map of the *E. coli* 70S ribosome at increasing resolutions, as documented by the FSC (see section 8 and definition of the 2D equivalent, FRC, in chapter 3, section 5.2.4). The convergence of the refinement around 5–10 iterations is typical for these kinds of data.

7.2.3.2. Distribution of angles   Three-dimensional projection matching assigns a triple of Eulerian angles to each particle, and answers the question of how uniformly the projections are distributed. The angular distribution can be plotted in various ways. In figure 5.25, the result of an initial classification into 83 quasi-equispaced directions is shown. The number of projections falling into each direction is represented by the area of a circle centered on that direction on a θ versus φ "starry sky" plot. Only two angles are shown, since the third angle, representing the azimuth of the particle around an axis normal to the plane of the support, can be regarded as trivial—although it needs to be tracked, for input into the 3D reconstruction program.

In figure 5.26, each projection is represented as a dot. This type of representation is appropriate after refinement, when angular increments have become quite small. Another version of an angular distribution plot was shown in figure 3.5 (chapter 3, section 1.6), at a place where orientational preferences of molecules in a layer of ice were discussed.

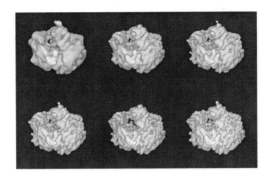

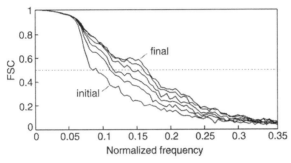

Figure 5.24 Progress in eleven cycles of refinement for a reconstruction of the 80S ribosome from *Thermomyces lanuginosus*. (This particular complex has ADP-ribosylated EF2 from yeast bound to the ribosome in the presence of the fungicide Sordarin.) Only every second reconstruction with corresponding resolution curve is shown. Top panel: from top left to bottom right, starting with the initial reconstruction, the ribosome (shown in side view, from the side accessed by EF2) exhibits greater and greater detail. Bottom panel: Fourier shell correlation curves corresponding to reconstructions in top panel. Resolutions according to the FSC = 0.5 criterion correspond to 18.9, 16.3, 15.6, 14.1, 13.1, 12.7 A, respectively. (From D. Taylor, J. Nilsson, R. Merrill, G.R. Anderson, and J. Frank, unpublished data.)

Figure 5.25 Number of projections of the eukaryotic multisynthetase complex falling into the 83 initial orientations. Each of the discrete directions is represented by a circle, whose size reflects the number of particles classified, by 3D projection matching, as falling into that direction. From Norcum and Boisset (2002), reproduced with permission of Elsevier.

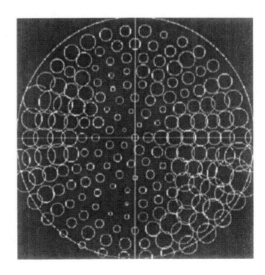

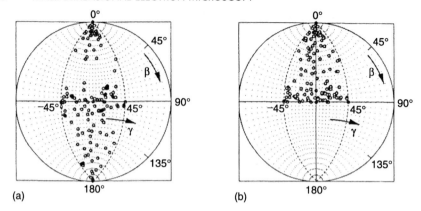

Figure 5.26 Distribution of orientations found by 3D projection matching for a human plasmin α₂-macroglobulin complex either stained with methylamine tungstate (a) or unstained, embedded in ice (b). In the case of this molecule, ice embedment apparently results in a much less uniform angular distribution than obtained with negative staining with methylamine tungstate. From Kolodziej et al. (1998), reproduced with permission of Elsevier.

A special application of the 3D projection matching, by Davis et al. (2002), illustrates the power of the method. RNA polymerase II (RNAPII), whose 3D structure is known, forms a complex with a large molecule called the *Mediator*. A 2D average of the RNAPII–Mediator complex was obtained by cryo-EM. Using 3D projection matching, the authors determined the orientation of RNAPII inside the complex (figure 5.27).

7.3. Numerical Aspects

The way the numerical solution to the 5D minimization implied in the 3D projection matching is found determines the accuracy and speed of the refinement. As we have seen, the 5D search is broken down into: (i) a comparison between the experimental projection and a set of precomputed reference projections that span the range of $\{\psi, \theta\}$ in quasi-equal increments, and (ii), for each such comparison, an exhaustive search in the 3D space spanned by the angle φ and the 2D shift **r**. Both aspects of the search prove to be extraordinarily demanding in terms of computational requirements: in the $\{\psi, \theta\}$ search, it is desirable to keep all precomputed reference projections (on the ψ, θ grid) in computer memory for instant access, but the number of projections increases very rapidly as the angular increment is decreased. However, since the numerical problem is of an "embarrassingly parallel" nature, the task can be broken down and distributed among a large number of processors, lending itself to manageable solutions even as the number of particles goes into the hundreds of thousands.

In the combined 2D rotational and translational search, the origins of reference and target projections are shifted with respect to each other

(a)

(b)

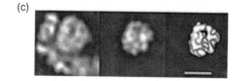

Figure 5.27 Determination of the orientation of RNA polymerase II within the RNAPII–Mediator complex. (a) 1600 projections, equispaced on a (θ, φ) grid, were calculated from the reconstruction of RNAPII (left). Four of these projections are shown as examples (right). (b) Cross-correlation coefficients between the RNAPII–Mediator projection and each of the RNAPII projections. The coefficients show a pronounced maximum (top left) on the angular grid. (c) 2D average of the RNAPII–Mediator complex (left), best-matching RNAPII projection (center), and corresponding 3D view of RNAPII (right). From Davis et al. (2002), reproduced with permission of Cell Press.

(c)

systematically by sampling units in x- and y-directions. For each shift, the target projection is resampled on a polar coordinate grid (the reference having been sampled once), and a rotational cross-correlation is performed for each ring by fast 1D Fourier transformations of reference and target, formation of conjugate cross-products (see convolution and correlation theorems; appendix 1), and inverse 1D Fourier transformation (Penczek et al., 1994; Joyeux and Penczek, 2002). The resulting 1D CCFs are then weighted and added up. In the end, the combination of rotation and shift giving a maximum is assigned to the target projection.

Refinement works most effectively if the size of the angular increment is gradually decreased with iteration count. However, below a certain size of angular increment it is no longer possible to keep all reference projections in memory.

On the other hand, particles toward the end of refinement are less likely to change their angles drastically. The refinement program is therefore usually laid out to restrict the highest-resolution search to a small angular range around the previous angle assigned to each particle.

For example, the finest overall grid might have a mesh of $1.5°$ at iteration N. Instead of decreasing the mesh size of the overall grid, a small conical range with a cone angle of $4.5°$ (or three grid points) around the previous grid point assigned to a given particle is now further divided into much smaller units, which could be as small as $0.2°$.

In closing, it should be mentioned that an integrative approach to the solution to the problem of orientation refinement, concurrent alignment of projections, and 3D reconstruction has been recently described (Yang et al., 2004). The strategy outlined before, widely practised now, consists of an iterative process in which the step of orientation determination and alignment is separated from the 3D reconstruction in which the newly refined parameters are used. This is not necessarily the most efficient way to obtain a stable, convergent solution. Yang et al. (2004) developed a unified procedure in which both the orientation parameters and the reconstruction are obtained by a simultaneous optimization using a quasi-Newton algorithm.

7.4. Three-Dimensional Radon Transform Method*

Although its principle is quite similar to the projection matching approaches described above, the Radon transform method (Radermacher, 1994) is distinguished from the former by its formal elegance and by the fact that it brings out certain relationships that may not be readily evident in the other approaches. Radermacher's method takes advantage of the relationship between the 3D Radon transform and the 2D Radon transform. The 3D Radon transform is defined for a 3D function $f(\mathbf{r})$ as

$$\hat{f}(p, \xi) = \int f(\mathbf{r})\delta(p - \xi^T\mathbf{r})d\mathbf{r} \qquad (5.38)$$

where $\mathbf{r} = (x, y, z)^T$ and $\delta(p - \xi^T\mathbf{r})$ represents a *plane* defined by the direction of the (normal) unit vector ξ. The 2D Radon transform is defined for a 2D function $g(\mathbf{r})$, in analogy with equation (5.38), as

$$\hat{g}(p, \xi) = \int g(\mathbf{r})\delta(p - \xi^T\mathbf{r})d\mathbf{r} \qquad (5.39)$$

where $\mathbf{r} = (x, y)^T$ and $\delta(p - \xi^T\mathbf{r})$ now represents a *line* defined by the direction of the (normal) unit vector ξ.

The discrete 2D Radon transform is also known under the name *sinogram* (e.g., van Heel, 1987b). Radermacher (1994) showed that the determination of the unknown orientation of a projection is solved by cross-correlating its discrete

2D Radon transform with the discrete 3D Radon transform of an existing 3D reference map. In figure 5.28, such a determination is demonstrated for the phosphofructokinase molecule from *Saccharromyces cerevisiae* (Ruiz et al., 2003). Translational alignment can be done simultaneously.

As we recall from section 3.4, the use of the CCF between 2D sinograms was proposed by van Heel (1987b) as part of a method termed *angular reconstitution*, as a means of determining the relative orientation between two or more raw, experimental projections. Applied to such data, that method generally fails because of the low SNR values normally encountered in electron micrographs of stained or frozen-hydrated specimens, unless high-point symmetries are present. However, the method has proved viable when applied to the matching of an experimental to an averaged projection, or the matching of two averaged projections to each other (see Orlova and van Heel, 1994; van Heel et al., 1994, 2000; Serysheva et al., 1995).

7.5. The Size of Angular Deviations*

It is difficult to estimate the size of the residual angular deviations and the frequency of misclassifications in the multireference classification, short of performing extensive computations with phantoms, but inferences can be drawn from the progress of the angular refinements with typical experimental data. As described in section 7.2, the refinement by 3D projection matching is done in several passes. After each pass, the change in angle of each particle is usually recorded in a document file. One can plot the average change in angles of all particles as a function of iteration count. (For this purpose, angular changes are calculated as distances on the sphere.) Such a plot, done for recent ribosome reconstructions (figure 5.29), shows a surprisingly fast convergence, from an initial change of several degrees to a fraction of a degree over the course of eight passes. The plot also reveals that at the end of the eighth pass, the majority of the particles are locked in their angular grid point, and further improvements might occur if the angular increment were further reduced.

Another type of statistical evaluation of these angular changes consists of a histogram displaying the numbers of particles (on the ordinate) undergoing particular changes in angle (the abscissa). For a well-behaved data set, such histograms show a rapid contraction from an initially wide field to a sharp spike representing little change for the large majority of all particles, which indicates a stabilization of most angles. For such a data set, the largest change happens at the very beginning of the refinement. In the first pass (figure 5.30a), one half of the particles "moved" by less than 10°, one-third moved by angles between 10° and 35°, and the rest moved by larger angles. In the second pass (figure 5.30b), a full 80% of the particles moved by less than 5° (50% are even within 1%; not shown in this figure), indicating approximate stabilization of the solution. One can, therefore, take the angular adjustments in the first pass as estimates for the actual angular deviations.

The size of these deviations—a full half of the particles are tilted by more than 10° away from the orientation of their class—is at first glance surprising.

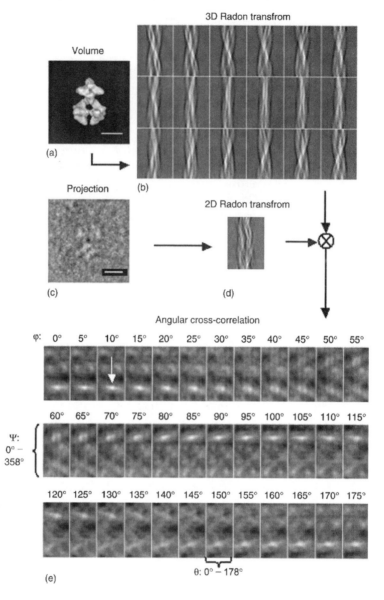

Figure 5.28 Demonstration of angle search with the 3D Radon transform, using the cryo-EM reconstruction of phosphofructokinase (Ruiz et al., 2003) as reference. (a) Reconstruction used as reference. (b) Slices through the 3D Radon transform of the reference volume, cut along the θ-coordinate in $10°$ intervals (horizontal: p-coordinate of the line in the Radon transform; vertical: φ-coordinate from $-180°$ to $179°$ in $1°$ intervals). (c) A projection of the phosphofructokinase molecule. (d) Radon transform of the projection. (e) Angular portion of the cross-correlation function between the Radon transforms of the projection and the reference volume. Shown are sections through the CCF in $10°$ intervals. The CCF was calculated over the full range of Eulerian angles in increments of $1°$. The maximum (indicated by arrow) occurs at $\varphi = 10°$, $\theta = 96°$, and $\psi = 271°$. (Data kindly made available by M. Radermacher.)

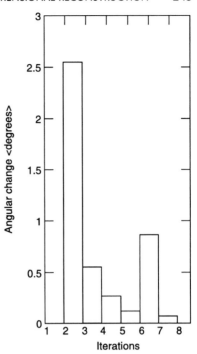

Figure 5.29 Diagram showing the average change in angle (computed as a distance on the sphere) as a function of iteration count during angular refinement for a ribosome data set of 50,000 particles. [The first iteration goes from the very coarse grid of initial classification (15°) to a 2° grid; the corresponding average change is not shown.] After the fifth iteration, the angular grid was changed from the initial 2° to 1.5° increment. The "bump" in angular change in the sixth iteration is due to the readjustment of angles to the new grid. The fact that the average angular change after the eighth iteration is very small compared to the mesh size indicates that with a finer mesh, further improvements in angular definition (and hence resolution) are possible. (K. Mitra, unpublished results.)

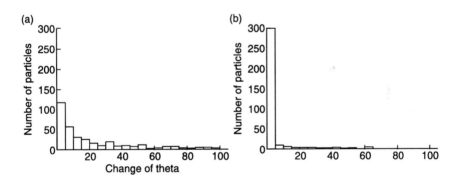

Figure 5.30 Histograms showing the change in theta angle during angular refinement of a ribosome reconstruction (Frank et al., 1995a) from 4300 projections. (a) First refinement pass; (b) second refinement pass.

It certainly explains (along with ψ and φ deviations not depicted in figure 5.30) the great gain in resolution that was achieved by correcting the angles. The deviations are the result of both misalignment and misclassification of the extremely noisy data. What it means is that the unrefined reconstruction is essentially a superimposition of a high-quality reconstruction (where the resolution limitation is due to factors unrelated to angular deviations, namely electron-optical limitations, conformational changes, etc.) and a blurred reconstruction, with the former based on projections whose θ angles are closely matching and the latter based on projections whose θ angles fall into a wide range. With this concept, it is now possible to understand the power of the angular refinement method: basically, a high-resolution reconstruction is already "hidden" in the unrefined reconstruction and it furnishes weak but essentially accurate reference information in the course of the angular refinement passes.

Histograms similar to those shown in figure 5.30 were already obtained, in model computations, by Harauz and Ottensmeyer (1984a), who gave the projections intentionally incorrect angular assignments at the start. Since these authors used a model structure, with all projection angles known, they could study the actual angular improvements as a function of the number of iterations. What is interesting in the results of Harauz and Ottensmeyer is that angular error limits of $\pm 10°$ were rapidly compensated, to a residual error below $2°$, while limits of $\pm 20°$ led to residual errors in the range of $8°$. Since the algorithm driving the correction of angles in this work differs somewhat from that employed by Penczek et al. (1994), the behavior of angular correction is not strictly comparable between the two methods. However, it is likely that there is again a threshold of root mean square angular deviation below which the angular refinement is well behaved and very efficient, but above which only small improvements might be achievable.

7.6. Model Dependence of the Reconstruction

The use of a 3D reference, either in the initial determination of orientations, or during refinement, raises the important issue of model dependence. If orientations are assigned to projections on the basis of their similarity to an existing 3D density map, then it is easy to see how features of a model can be preferentially selected and enhanced in a reconstruction. In the worst case, unless significance tests are used, the 3D reference can be recreated from pure noise (Grigorieff, 2000). X-ray crystallography has had to deal with a similar problem in validating the results of refinements. Brünger (1992) introduced the free R-value to address this problem. The idea is that a fraction of the data that is set aside for cross-validation will not be used in the phasing, but is in the end tested using the R-factor. Model dependence in cryo-EM refinement can be tested by leaving out data from the 3D reference either in real space or in Fourier space.

Leaving out data in real space is a common occurrence in studies of ligand binding to a large molecule, where a reconstruction of the large molecule is used as reference. The emergence of a significant density mass that has the correct volume of the ligand and is connected to the density of the large molecule is irrefutable evidence that the ligand has been localized, because that additional density built up even though it had no counterpart in the 3D reference (see, e.g., Gao et al., 2004).

Leaving out data in Fourier space has an advantage over the real-space method in that it provides the means of studying the spatial frequency behavior of the model dependence while the other does not. Such an approach will be described in section 8.3, as it is intimately linked with the measurement of resolution, which will be introduced first.

(Related to the question of model dependence is the problem of resolution assessment: if the data set is randomly split in half, and each half is subjected to alignment to the same 3D reference, then the reconstructions obtained are no longer statistically independent, and a comparison between them will entail false correlations. This means that resolution measured by FSC will be overestimated by an unknown amount.)

7.7. Consistency Check by Reprojection

It must be emphasized that cross-resolution comparison between the $0°$ average of a molecule set in a certain view and the $0°$ projection of the reconstruction, valuable as it is as a check for internal consistency of the numerical procedures (section 5), nevertheless fails to provide an adequate estimate of over-all resolution. It has been previously pointed out (Penczek et al., 1994) that the $0°$ projection of a random-conical reconstruction is insensitive to incorrect assignments of θ, since its associated central section in Fourier space is built up from 1D lines, each of which is an intersection between the $0°$-central section and a central section associated with the tilted projection.

Along similar lines, it must be pointed out that visual or CCF comparison between projections and reprojections in the same directions, often used to validate a reconstruction, is inconclusive and of questionable relevance. It is a test that follows circular reasoning, and cannot be failed by any data, since it amounts to the comparison of data placed into a central plane of the 3D Fourier transform, by virtue of the reconstruction procedure, with the very same data placed there. An analogous situation would be presented by a customer who, when asked in a bank to provide identification, produces a photograph of himself.

8. Resolution Assessment

8.1. Theoretical Resolution of the 3D Reconstruction

We recall that resolution is a boundary in reciprocal space defining the 3D domain within which Fourier components contribute significantly to the density map. Since a number of different effects (such as electron-microscopic contrast transfer function, recording medium, reconstruction algorithm) produce independent limitations, we can visualize the resolution limitation of the whole system as the effect of passing the Fourier transform of the structure through a series of 3D masks centered on the origin. It is always the most restrictive one that determines the final resolution. (Since each of the boundaries may be nonspherical, a more complicated situation can be imagined where the final boundary is a vignette, with restrictions contributed by different masks in different directions.)

In that context, "theoretical resolution" is only the resolution limitation imposed by the choice of reconstruction algorithm and data collection geometry. It will determine the outcome only if all other resolution limits are better (i.e., represented by a wider mask in 3D Fourier space).

The theoretical resolution of a reconstruction is determined by the formula given by Crowther et al. (1970) (section 2.1). It represents the maximum radius of a spherical Fourier domain that contains no information gaps. Corresponding formulas have been given by Radermacher (1991) for the case of a conical projection geometry. Here, we mention the result for an even number of projections:

$$d = 2\pi(D/N)\sin\theta_0 \qquad (5.40)$$

where $d = 1/R$ is the *resolution distance*, i.e., the inverse of the resolution, D is the object diameter, and θ_0 is the tilt of the specimen grid. Furthermore, for such data collection geometry, the resolution is direction dependent, and the above formula gives only the resolution in the directions perpendicular to the direction of the electron beam. In directions that form an angle oblique to those planar directions, the resolution is degraded. In the beam direction, the effect of the missing cone is strongest, and the resolution falls off by a factor of 1.58 (for $\theta_0 = 45°$) or 1.23 (for $\theta_0 = 60°$).

8.2. Practically Achieved Resolution

The account given thus far relates to the *theoretical* resolution, or the maximum resolution expected for a particular data collection geometry. There are, however, many effects that prevent this resolution from being realized. Some of these have already been mentioned in passing in the relevant sections on instrumentation and 2D data analysis. It is useful to have a list of these effects in one place:

(i) *Instrumental limitation*: the projections themselves may be limited in resolution, due to (i) partial spatial and energy coherence (Wade and Frank, 1977; see also chapter 2, section 3.3), and (ii) effects due to charging and mechanical instability (Henderson and Glaeser, 1985; Henderson, 1992; Gao et al., 2002). While concerns about partial coherence have been put to rest in recent years by the ready availability of instruments with field emission guns and high stability of voltage and lens currents, the effects in the other category (ii) are determined by the exact set-up and operating conditions in the laboratory.

(ii) *Heterogeneity*: the coexistence of particles with different conformations invalidates the entire single-particle reconstruction approach, as it leads to density maps in which features of different structures are intermixed. One of the noticeable effects is resolution loss, either in the entire density map or in parts of it. In fortuitous situations, this problem may be addressed by classification (section 9).

(iii) *Numerical errors*: rounding errors are possible in the course of the data analysis, and they may accumulate unless precautions are taken (chapter 3, section 3.4.3). A critical factor is the ratio between the resolution of the structural information and the resolution of the digital representation—scanning should be with a sampling increment that is half or less of the critical distance required by the *sampling theorem* (see chapter 3, section 2.2).

(iv) *Alignment errors*: due to the low SNR of the raw data, the parameters describing the geometric position of a particle in the molecule's coordinate frame may be incorrect. Even magnification errors that occur when data are merged from different instruments can have a substantial

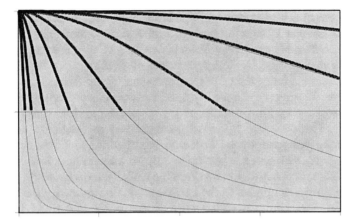

Figure 5.31 Predicted effect of angular errors, in terms of an envelope function affecting the Fourier transform of the 3D density map. Abscissa: sR, the product of spatial frequency and the radius of the particle. The curves are labeled by the standard deviation of the rotational error imposed. From Jensen (2001), reproduced with permission of Elsevier.

influence. These effects have been analyzed by Jensen (2001) in a series of tests with modeled data. A result of particular interest is the dependence of resolution on the error of angle assignment (figure 5.31). Similar errors of angular assignments affect the resolution if, following classification, particles within a finite angular neighborhood are assigned the same angles for the 3D reconstruction (Stewart et al., 1999), as in some reconstruction schemes following the angular reconstitution route (sections 3.4 and 6).

Discussions of the relative importance of these effects are found in many reviews of cryo-EM of single particles (see references in appendix 3). For articles specifically addressing the topic of resolution and resolution loss, see Glaeser and Downing (1992), Stewart and coworkers (1999, 2000), and Jensen (2001).

For all these reasons, the significant resolution of a reconstruction (i.e., the resolution up to which true, object-related features are actually represented in the 3D image) can differ substantially from the theoretical resolution deduced from the geometry, and needs to be independently assessed in each project.

As in the practical assessment of 2D resolution (section 5.2 in chapter 3), there are two different approaches; one goes through the evaluation of reconstructions done from randomly drawn halfsets of the projection data, the other a multiple comparison first devised by Unser and coworkers (1987, 1989), the spectral signal-to-noise ratio (SSNR).

8.2.1. Comparison of Halfset Reconstructions

Two reconstructions are calculated from two randomly drawn subsets of the projection set, and these reconstructions are compared in Fourier space using differential phase residual (DPR) or Fourier ring correlation (FRC) criteria. Obviously, in the 3D case, the summation spelled out in the defining formulas [chapter 3, equations (3.64) and (3.65)] now has to go over shells of

$k = |\mathbf{k}| = $ constant. This extension of the differential resolution criterion from two to three dimensions is straightforward, and was first implemented (under the name of *Fourier shell correlation*, abbreviated to FSC) by Harauz and van Heel (1986a) for the FRC and by Radermacher et al. (1987a, b) for the DPR.

One of the consistency tests for a random-conical reconstruction is the ability to predict the 0° projection from the reconstruction, a projection that does not enter the reconstruction procedure yet is available for comparison from the analysis of the 0° data. Mismatch of these two projections is an indication that something has gone wrong in the reconstruction, while the reverse line of reasoning is incorrect: an excellent match, up to a resolution R, is no guarantee that this resolution is realized in all directions. This is easy to see by invoking the projection theorem as applied to the conical projection geometry (see figure 5.13): provided that the azimuths φ_i are correct, the individual tilted planes representing tilted projections furnish correct data for the 0° plane, irrespective of their angle of tilt. It is, therefore, possible to have excellent resolution in directions defined by the equatorial plane, as evidenced by the comparison between the 0° projections, while the resolution might be severely restricted in all other directions. The reason that this might happen is that the actual tilt angles of the particles could differ from the nominal tilt angle assumed (see Penczek et al., 1994).

Most resolution tests are now based on the FSC curve (typical curves are shown in figure 5.24, where it is used to monitor the progress of refinement), although the value quoted can differ widely depending on the criterion used. In the following, the different criteria are listed with a brief discussion of their merits.

(i) *3-σ criterion* (Orlova et al., 1997): In this and the following criteria, the signal FSC on a given shell is compared with the correlation of pure noise, which is obtained by $1/\sqrt{N}$, where N is the number of independent points in the shell (Saxton and Baumeister, 1982). Initially, 2σ had been proposed (Saxton and Baumeister, 1982; van Heel, 1987a). Experience shows that the 3-σ criterion is still too lenient—reconstructions limited to this resolution show a substantial amount of noise. Indeed, the number of points on the Fourier grid is always substantially larger than the number of independent points, because of the mutual dependence of Fourier coefficients within the range of the shape transform. The problem is also evident when one pays attention to the behavior of the signal FSC in the vicinity of the 3-σ cutoff: it usually fluctuates significantly in this region, which indicates that the cutoff resolution itself becomes a matter of chance. Objections against the use of the noise curve have been voiced by Penczek (appendix in Malhotra et al., 1998) on theoretical grounds. A brief account of the controversy is given in Frank (2002). An insightful critique is also found in the appendix of the article by Rosenthal et al. (2003).

(ii) *5-σ criterion* (e.g., Radermacher, 1988; Radermacher et al., 2001; Ruiz et al., 2003): This criterion is tied to the same noise curve as the previous one, but attempts to address the problems by the use of a larger multiple of the noise.

(iii) *0.5 cutoff*: This cutoff was introduced by Crowther's group (Böttcher et al., 1997) as it became clear in their reconstruction of the hepatitis B virus that the 3-σ cutoff, even when corrected for the effect of the icosahedral symmetry, reported a resolution that proved untenable. Later, Penczek pointed out (appendix in Malhotra et al., 1998) that this choice of cutoff will limit the SNR in the Fourier shells to values equal to, or larger than, one. On the other hand, van Heel (presentation at the ICEM in Durban, 2002) has cautioned against the blind application of this criterion since at very low resolutions it can yield a value that is too optimistic. This problem is quite real for low-resolution maps of viruses but has no bearing on resolution assessments of single-particle reconstructions with resolutions better than 30 Å.

(iv) *0.143 cutoff*: Rosenthal et al. (2003) pointed out that, judging from experience, the 0.5 cutoff is too conservative. These authors suggest a rationale for the use of an FSC cutoff at 0.143 on the basis of a relationship (condition for improvement of the density map by including additional shells of data) that is known to be valid for data in X-ray crystallography. As yet, however, it is unclear whether this relationship holds for EM data, as well.

In trying to choose among the different measures, we can go by comparisons between X-ray and cryo-EM maps of the same structure (see chapter 6 for two examples). Such comparisons are now available for an increasing number of structures. Saibil's group (Roseman et al., 2001) argued, from a comparison of the appearance of the cryo-EM map with that obtained by filtering the X-ray structure of GroEL, that the cutoff should be placed somewhere between the 0.5 and 3-σ values. In its tendency, this would support Rosenthal et al. contention [point (iv) above]. On the other hand, a recent reconstruction of the ribosome from 130,000 particles (Spahn et al., 2005) shows the best agreement with the X-ray map when the latter is presented at the FSC = 0.5 resolution, namely at 7.8 Å (chapter 6, section 3.3). It seems prudent, therefore, to hold on to this criterion until a strong case can be built for an alternative one.

8.2.2. Truly Independent Versus Partially Dependent Halfset Reconstructions

The method of assessing resolution based on halfset reconstructions is not without problems: in the usual refinement procedure, the alignment of both halfsets is derived with the help of the same 3D reference. To the extent that each of the final halfset reconstruction is influenced by the choice of reference ("model dependence"), it also has similarity to its counterpart which is reflected by the FSC. True independence of the two halfset reconstructions, which would be a desirable feature of the test, cannot be achived under these circumstances. The statistical dependence of the halfset reconstructions results in a systematic increase of the FSC curve, which leads to an overestimation of resolution when any of the criteria (i)–(iv) are applied. (Fortuitously, the resolution overestimation is to some extent counteracted by its underestimation, resulting from the fact that the resolution measured pertains to the reconstruction that is obtained from one-half of the data only).

The problem of statistical dependence of halfset reconstructions originating from the same reference and possible remedies were extensively discussed by Grigorieff (2000), Penczek (2002a), and Yang et al. (2003). To circumvent the influence of model dependence, Grigorieff (2000) proposed to use, in a departure from the usual 3D projection matching procedure (Penczek et al., 1994), two distinctly different references, and to carry out the refinement of both halfsets totally independently. Evidently, convergence to a highly similar structure in such independent processing routes will attest to the validity and self-consistency of the result. What is unclear, however, is the nature of the competing references—how much should they differ to make the results acceptable?

The recently developed cross-validation method ("free FSC", Shaikh et al., 2003) provides a way to establish the degree of the model dependence and obtain a model-independent resolution estimate (see section 8.3).

8.2.3. Three-Dimensional Spectral Signal-to-Noise Ratio

Considerations along the lines of a definition of a 3D SSNR were initially developed by Liu (1993) and Grigorieff (1998). We recall that the 2D SSNR as originally defined by Unser and coworkers (1987, 1989; see also Grigorieff, 2000) is based on a comparison between the signal and the noise components of an image set in Fourier space that contribute to the average at a given spatial frequency, and is expressed as a ratio. Summation over subdomains (such as rings in two dimensions) of the spatial frequency range will give specific SSNR dependencies on spatial frequency radius, or angles specifying a direction. Penczek (2002a) considered how this concept can be generalized in three dimensions. The necessity to trace contributions from the 2D Fourier transforms of the projections to the 3D Fourier transform of the reconstruction singles out reconstruction algorithms based on Fourier interpolation, and practically excludes other algorithms such as weighted back-projection or iterative algebraic reconstruction in the generalization of the SSNR concept.

In Fourier interpolation (see section 4.3 in this chapter), for any given point of the 3D Fourier grid, contributions are summed that originate from all 2D Fourier transforms in its vicinity. These contributions are derived by centering the 3D shape transform—a convolution kernel reflecting the approximate shape of the 3D object or the volume it is contained in (see section 2.2)—on each 2D grid point of a central section. The fact that the 3D shape transform has, for practical purposes, a limited extent limits the number of projections and the number of 2D grid points that need to be considered in the summation.

Penczek (2002a), in derivations that cannot be properly paraphrased in this limited space, obtained approximate expressions for the 3D SSNR in two situations: in one, he considered the interpolation scheme with multiple overlapping contributions described above, and in the other, a poor nearest-neighbor interpolation scheme. It is noteworthy that the numerical application of the 3D SSNR (computed according to the more sophisticated interpolation scheme) yielded results that agreed very well with the FSC-based estimations in the range between $SSNR = 0$ and 100, or between $FSC = 0$ and 0.99. It is also important to note that the 3D SSNR provides a way to measure resolution for tomographic reconstructions (Penczek, 2002a; see also application in the study by Hsieh et al., 2002). Furthermore, it provides a very useful way for mapping out the direction dependence of resolution, to be discussed in section 8.2.4.

8.2.4. Direction Dependence of Resolution*

If the data coverage in angular space is isotropic, then the resolution boundary in Fourier space roughly follows the surface of a sphere (of course this is irrespective of the criterion used). Often, however, there exists a situation where the resolution is direction dependent. In that case, the resolution boundary in Fourier space is better approximated by an ellipsoid or a more complicated shape, having "good" and "bad" directions. For example, a random-conical reconstruction has data missing in a cone defined by the experimental tilt angle, so that the resolution boundary resembles the surface of a yo-yo.

There is as yet no convention on how to describe the direction dependence of the experimental resolution parametrically. Thus, the resolution sometimes relates to an average over the part of the shell within the measured region of 3D Fourier space; while in other instances (e.g., Boisset et al., 1993a, 1995; Penczek et al., 1994) it might relate to the entire Fourier space, without exclusion of the missing cone. It is clear that, as a rule, the latter figure must give a more pessimistic estimate than the former.

A tool for measuring the direction dependence of resolution in three dimensions is now available in the 3D SSNR (Grigorieff, 1998; Penczek, 2002a). In particular, Penczek (2002a) used this quantity to characterize and visualize the anisotropy of resolution in a single-particle reconstruction, of a molecule from some 3700 particle images, as the surface of the resolution domain in Fourier space. The resolution domain (whose boundary was defined by a threshold value for the 3D SSNR) had a shape vaguely resembling a potato, in its irregularity and unequal axes.

8.3. Cross-Validation Using Excision of Fourier Data from the 3D Reference

In section 7.6, we have already mentioned the problem of reference dependence and the idea of cross-validation. Particularly, we now proceed with an outline of the method introduced by Shaikh et al. (2003), where data are left out from the 3D reference in Fourier space.

The choice of shells as loci from where data are removed is motivated twofold: (i) each projection is affected equally, namely, according to the projection theorem, by excision of a corresponding ring in 2D Fourier space; and (ii) this choice also lends itself naturally to an evaluation by the FSC.

When such a modified 3D reference is used, the alignment of experimental data is not affected by the Fourier components lying on the selected shell. Hence, during refinement, Fourier components building up in the selected shell are free from model bias. The FSC curve of a reconstruction that is done using the modified 3D reference during refinement differs from the FSC obtained with the original 3D reference: we see a dip at the Fourier radius of the shell from which data have been excised (figure 5.32a). The size of the dip is an indicator for the degree of model dependence. Moreover, when the data consist of pure noise, the dip in the FSC extends all the way to the curve that indicates the FSC of pure noise (figure 5.32b).

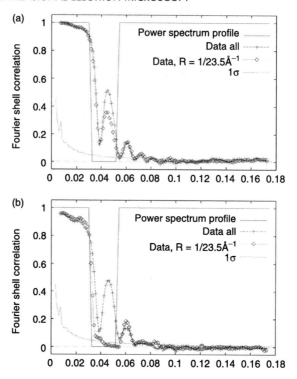

Figure 5.32 Cross-validation of refinement. A shell is removed from the 3D reference at the indicated place in the Fourier transform. In the refinement, the Fourier shell correlation (FSC) behaves differently inside from outside the removed shell, and the difference is most dramatic for pure noise. (a) Real data (ribosome particles from cryo-EM micrographs). The FSC within the shell decreases, reflecting the existence of a certain amount of model dependence. (b) Pure noise. In this case, the FSC falls to a value that is indeed expected for pure noise. From Shaikh et al. (2003), reproduced with permission of Elsevier Science.

As the SNR of the data decreases, the reference increasingly "takes over," up to the point, reported by Grigorieff (2000), where data representing pure noise give rise to a reconstruction that is a perfect copy of the reference. The cross-validation of Shaikh et al. (2003) makes it possible to spot those cases by the large drop in the FSC at the radius of the selected shell.

It is clear from the properties of the Fourier representation of a bounded object (see section 2.2) that the region in Fourier space from which data is to be excised must have a minimum width, which is given by the lateral extent of the object's shape transform. (We recall that each Fourier coefficient is surrounded by a "region of influence" within which neighboring Fourier coefficients are correlated with it. For the cross-validation to be true, we can only look at the behavior of Fourier coefficients that lie outside that correlation range from its data-carrying neighbors.) This extent roughly corresponds to the inverse of the object's diameter. For example, the ribosome has a globular shape with $\sim$250 Å diameter, so its shape transform has a width of $1/250\,\text{Å}^{-1}$. If the ribosome is contained in a box of size

500 Å, then the width of the shell should be at least 4 Fourier pixels (or $4/500 \, Å^{-1}$) so that the center of the shell is safely away from influence on both sides.

Excision of Fourier information over a shell may seem risky, as it could conceivably produce artifacts that affect the outcome of the 3D projection matching and the course of the refinement. A careful check (T. Shaikh, unpublished results) revealed that these effects are insignificant if the shell is thin enough, as chosen in Shaikh et al. (2003).

9. Contrast Transfer Function and Fourier Amplitude Correction

9.1. Introduction

We have seen (chapter 2, section 3) that the electron micrographs are degraded by the effects of the CTF. As a consequence, if uncorrected, the reconstruction will have exaggerated features in the size range passed by the CTF spatial frequency band. The theory of this degradation is well understood, and can be cast into the framework of linear transfer theory. The CTF leaves a "signature" in the power spectrum of the image if the object has a sufficiently broad spectrum to begin with. From this signature, the parameters of the CTF (defocus, axial astigmatism) can be recovered. Once the CTF is known, the reconstructed volume can be corrected, so that the original object is restored. Strictly speaking, the falloff in amplitude that is due to a multitude of effects such as partial coherence, mechanical instability, and charging is part of the CTF. Some of these effects are well understood, and are expressed analytically by envelope functions. Others can be heuristically approximated by analytical functions. But while these expressions are helpful in predicting the cumulative falloff for a model object, they are not very useful in the quantitative recovery of the object, since important parameters are usually unknown. It is more accurate and straightforward to use experimental data from other experiments, such as low-angle X-ray scattering, as a guide for the true radial dependence of Fourier amplitudes. Therefore, the following presentation is divided into two parts, one devoted to CTF correction without consideration of the amplitudes, the other to Fourier amplitude correction under the assumption that the CTF correction has already been done.

9.2. Contrast Transfer Function Correction

9.2.1. Introduction

Procedures for CTF correction in two dimensions have already been discussed in chapter 2. CTF correction can either be applied to the raw data in two dimensions (i.e., the individual projections, or the entire micrographs) or, as an isotropic correction depending on spatial frequency radius only, to the 3D volume. Effective CTF correction requires combining data sets obtained with two or more different defocus settings (each such data set is called *defocus group*), so that the gaps near the zeros of the CTF are covered.

In deciding whether to use a strategy in which the correction is applied before or after the computation of the 3D reconstruction, one has to consider the following pro's and con's: when applied to the 3D reconstructions, all procedures mentioned in section 3.9 of chapter 2 are very well behaved numerically because of the high SNR of the volume. The opposite is true when these procedures are applied to raw data. The low SNR in this case means that CTF correction is essentially limited to the flipping of phases in alternate zones of the CTF. For this reason, CTF correction after reconstruction is normally preferred. However, two caveats are in order:

For a random-conical reconstruction, correction *after* reconstruction runs into the difficulty that the projections from tilted-specimen micrographs have different defocus values (see Frank and Penczek, 1995). In order to proceed in this way, one would have to sort the raw data according to the distance from the tilt axis and perform separate reconstructions for each defocus strip. However, this difficulty is of minor importance since the reconstruction from a random-conical data set is rarely the final result—it is normally followed and superseded by 3D projection matching refinement using an exhaustive data set.

A more serious problem is encountered with every reconstruction method when going toward higher resolution. Within the same micrograph field, the particles may have different heights and thus, defocus values. By ignoring these variations, we would effectively introduce a defocus variation term, whose effect is equivalent to the attenuating effect of energy spread (see section 3.3.2 in chapter 2). This term becomes important at higher resolution, below $\sim 7\,\text{Å}$. To avoid this problem, we either have to make the defocus groups narrow enough (in terms of the defocus span), and proceed with the CTF correction after reconstruction, or, in a radical departure from the strategy used thus far, apply the CTF correction to each projection individually, *prior to the reconstruction*, according to its own defocus (see van Heel et al., 2000; Sander et al., 2003).

9.2.2. Wiener-Filtered Merging of Defocus Group Reconstructions

Given are N reconstructions from N data sets belonging to different defocus groups. Having originated from data with different average defocus, each of these reconstructions is affected by the CTF in a different way, characterized by different positions of the zeros and of the lobes with opposite signs. We now seek to optimally correct the CTFs and merge the reconstructions such that the resulting reconstruction comes close to the original object. The CTF correction and merging of such reconstructions by Wiener filtering was described by Penczek et al. (1997). The form of the Wiener filter follows from a generalization of the 3D Wiener filter introduced for $N = 2$ reconstructions (Frank and Penczek, 1995) to $N > 2$ [see the 2D version; equation (2.32a) in chapter 2]:

$$W_n(\mathbf{k}) = \frac{\text{SNR}_n(\mathbf{k}) H_n^*(\mathbf{k})}{\sum_{n=1}^{N} \text{SNR}_n(\mathbf{k}) |H_n(\mathbf{k})|^2 + 1} \tag{5.41}$$

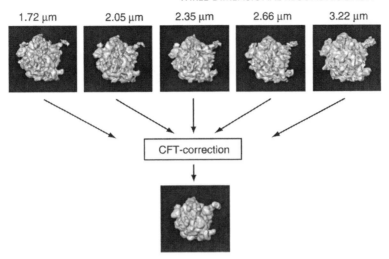

Figure 5.33 Principle of CTF correction by Wiener filtering/merging defocus group reconstructions.

where $H_n^*(\mathbf{k})$ is the CTF of the nth data set and $SNR_n(\mathbf{k}) = P_F(\mathbf{k})/P_N(\mathbf{k})$, and $P_F(\mathbf{k})$, $P_N(\mathbf{k})$ are the power spectra of the noise and signal of the nth image, respectively. [Often the spatial frequency-dependent ratio $SNR_n(\mathbf{k})$ is simply replaced by a single constant, SNR.] Assuming that the CTFs are rotationally symmetric (i.e., zero axial astigmatism), H_n and thus W_n are functions of spatial frequency radius $k = |\mathbf{k}|$ only, and the merged, CTF-corrected reconstruction becomes

$$F(\mathbf{k}) = \sum_{n=1}^{N} W_n(\mathbf{k}) F_n(\mathbf{k}) \tag{5.42}$$

Figure 5.33 shows how reconstructions from individual defocus group, which show characteristic distortions due to CTF flipping, are merged by Wiener filtering into the final, CTF-corrected density map.

9.2.3. Simultaneous CTF Correction and Reconstruction

For processing data from untilted specimens taken at different defocus settings, Zhu et al. (1997) developed a method of reconstruction that implicitly corrects for the CTF. This is done by including the CTF into the mathematical model describing the relationship between the 3D model and the observed projections.

In algebraic form, this relationship can be formulated as

$$\mathbf{p}_k = \mathbf{H}_k \mathbf{P} \mathbf{o} \tag{5.43}$$

where $\mathbf{p}_k$ is a matrix containing the projection data for the kth defocus setting, $\mathbf{o}$ is a vector representing the elements of the 3D object in lexicographic order,

P is a nonsquare matrix describing the projection operations, and $\mathbf{H}_k$ is the CTF belonging to the kth defocus.

In reality, however, the data are subject to noise, and the equation system underlying equation (5.43) is ill-conditioned. Zhu et al. (1997) made use of an approach of regularized least squares to find a solution to the expression:

$$\sum_k \frac{1}{N_k} |\mathbf{p}_k - \mathbf{H}_k \mathbf{Po}|^2 \to \min \tag{5.44}$$

where N_k is the number of projections in the kth set. A least-squares solution is found iteratively, by the use of Richardson's method:

$$\mathbf{o}^{(n+1)} = \mathbf{o}^{(n)} + \lambda \sum_k \frac{1}{N_k} \mathbf{p}_k^T \mathbf{H}_k^T \{\mathbf{p}_k - \mathbf{H}_k \mathbf{p}_k \mathbf{o}^{(n)}\} \tag{5.45}$$

where λ is a small constant controlling the speed of convergence. This method of "reconstruction-cum-CTF correction" was successfully used in the reconstruction of the ribosome from energy-filtered data that were taken with 2.0 and 2.5 μm defocus (Frank et al., 1995a,b).

9.2.4. Simultaneous CTF Correction and Fourier-Based Reconstruction*

Reconstruction by Fourier interpolation offers a convenient, tractable way to apply CTF correction in the same pass, as pointed out by Grigorieff (1998). This approach will be briefly sketched out in the following.

Fourier interpolation methods of reconstruction (see overview in section 4.3) infer the values on the grid points of the 3D Fourier transform from the contributions on the 2D grid points of the properly placed (i.e., according to the projection theorem) Fourier transforms of the projections. Sums are accumulated in each grid point of the 3D Fourier transform as more and more projections are added. Grigorieff formulated the recovery of the value of a 3D grid point R_i from its contributing 2D neighbors with their respective CTFs implicitly as a least-squares problem, which is solved by an expression that has the familiar Wiener filter form (see chapter 2, section 3.9):

$$R_i = \frac{\sum\limits_{n,s} w_n^2 b^2 c_{n,s} P_{in,s}}{f + \sum\limits_{n,s} (w_n b c_{n,s})^2} \tag{5.45a}$$

Here, $P_{in,s}$ represents a sample of the 2D Fourier transform of image n that contributes to R_i, $c_{n,s}$ is the CTF of image n at the sample point s, w_n is a weight describing a quality factor, and b is the *box transform*, an interpolation function which is calculated as the Fourier transform of the box that has the dimensions of the reconstructed volume. Since the 2D grid points of the projections rarely coincide with the 3D grid, the contributions of each 2D grid point $P_{in,s}$ to a 3D point in its vicinity, indexed i, must be calculated by centering the box transform on this grid point and evaluating its value at i. The symbol f represents a constant, as in the normal Wiener filter, that prevents undue noise amplification wherever $c_{n,s}$ approaches zero. The meaning of this constant in the Wiener filter context is 1/SNR. Grigorieff used this approach, incorporated in a program called *Frealign*, to reconstruct bovine NADH:Ubiquinone oxireductase (Complex I) at 22 Å from cryo-EM images. A similar approach was earlier used by Crowther's group to merge images of individual virus particles (Böttcher and Crowther, 1996; Böttcher et al., 1997).

9.3. Fourier Amplitude Correction*

9.3.1. B-Factor Formalism

In X-ray crystallography, amplitude falloff in reciprocal space due to thermal motion is described by a Gaussian term, the *temperature factor*, with a characteristic parameter B, a decay parameter, which has dimension $Å^2$. In analogy with this treatment, all resolution-limiting "envelope" terms, arising in practical EM (charging, specimen instability) as well as in the following image-processing procedures (alignment and interpolation errors), that affect the Fourier transform of the image in a multiplicative way, can be approximately lumped together into a single Gaussian term (Glaeser and Downing, 1992; Thurman-Commike et al., 1999):

$$E(k) = \exp[-B_{EM}k^2] \tag{5.46}$$

The parameter B_{EM} in this formulation is related to the parameter B in X-ray crystallography by

$$B_{EM} = B/4$$

(see Conway and Steven, 1999). If B_{EM} is known, then the true Fourier amplitudes can be restored by multiplication of the CTF-corrected Fourier transform of the cryo-EM density map, $F_{corr}(\mathbf{k})$, by a high-pass filtration term that is the inverse of equation (5.46):

$$F_{orig}(\mathbf{k}) = F_{corr}(\mathbf{k})\exp(B_{EM}k^2) \tag{5.46a}$$

Formally, equation (5.46a) can be seen as the application of a temperature factor with negative B_{EM}.

We note that, strictly speaking, the formulation in equation (5.46) is in conflict with the theory of partial coherence (chapter 2, section 3.3.2), according to which the damping term due to finite source size is defocus dependent. One can "rescue" the approximate validity of equation (5.46) by using a defocus-dependent $B_{EM} = B_{EM}(\Delta z)$, but in that case the amplitude correction would have to be applied separately to each defocus group. Such a revised formulation was given by Saad et al. (2001), who found that B_{EM} increases roughly linearly with defocus in the tested range 0.5–3 μm. Still, the crudeness of the Gaussian approximation of equation (5.46a) can be seen from the fact that among the envelopes contributing, the defocus spread term (chapter 2, section 3.3.2) goes with $\exp(-ak^4)$ while the spatial partial coherence term goes with $\exp(-bk^6-ck^4-dk^2)$, where a, b, c, and d are constants. The inadequacy of the B-factor formalism was pointed out in another context by Penczek et al. (2004a).

Since the amplitude decay is dependent on instrument quality and performance, it is not surprising that the values quoted for B_{EM} in the cryo-EM literature cover a wide range, from around $100\,\text{Å}^2$ [Böttcher et al., 1997 ($500\,\text{Å}^2$ quoted there corresponds to $B_{EM} = 125\,\text{Å}^2$; R.A. Crowther, personal communication); Conway and Steven, 1999; Miyazawa et al., 2003], over $250\,\text{Å}^2$ (Gabashvili et al., 2000), to $\sim500\,\text{Å}^2$ (Thurman-Commike et al., 1999). These values are usually derived by comparing the radial profiles of power spectra from electron micrographs with those obtained by low-angle X-ray solution scattering. However, Thurman-Commike et al. (1999) discovered that the subtraction of noise in the power spectrum results in substantially higher B values than without such a correction—since at higher spatial frequencies, the Fourier amplitude originating from the structure is usually overestimated due to the presence of the noise term. To avoid this complication, it would seem advantageous to use the power spectrum of the density map (in which the noise term is absent), rather than the power spectrum of the electron micrograph, for comparison with the X-ray scattering profile.

9.3.2. Empirical Amplitude Correction

The large number of effects that all work together in limiting resolution make it very difficult to determine the parameters of the associated analytical expressions, for example, the individual envelopes due to energy spread, partial coherence, and specimen instability. The temperature factor formalism described in the previous section was introduced to deal with these effects in a summary way. However, since the only reliable way to determine the parameter B_{EM} is by measuring low-angle scattering data anyway, it is much more straightforward to use, instead of an approximate Gaussian, the empirical function based on a comparison between the radial profile of the actual Fourier amplitude of the cryo-EM map (after the usual CTF correction) and the radial profile of the low-angle X-ray solution scattering amplitude, provided such data are available. An adjustment of this type is shown in figure 5.34 (Gabashvili et al., 2000).

Let $F(\mathbf{k})$ be the Fourier transform of the reconstructed volume, and $S(k)$ the measured low-angle X-ray scattering curve for the same specimen, then we first calculate the averaged radial profile $A(k)$ of $|F(\mathbf{k})|$, by averaging the latter over all angles. Next, we form the ratio between the target radial profile, $S(k)$ and observed profile, $A(k)$:

$$r(k) = S(k)/A(k) \tag{5.47}$$

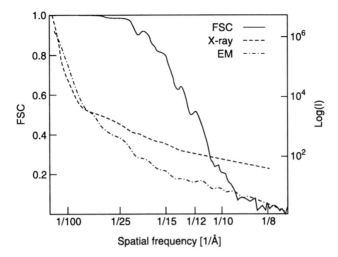

Figure 5.34 Fourier shell correlation for a data set of 73,000 particles, and empirical amplitude correction based on low-angle X-ray data. With increasing spatial frequency, the rotationally averaged cryo-EM Fourier amplitude falls off much more rapidly than the low-angle X-ray scattering amplitude (log scale on the right). In this method of correction, the spatial frequency-dependent ratio between the two values described by the curves is used to boost the Fourier amplitudes of the final reconstruction. However, it should be noted that this procedure is only valid up to the empirically determined resolution limit; beyond this limit, the amplitude adjustment would result in meaningless detail produced by the enhancement of noise. From Gabashvili et al. (2000), reproduced with permission of Cell Press.

The function $r(k)$ is the appropriate filter function (with high-pass properties, since the largest boost must occur at high spatial frequencies) that must be applied to $F(\mathbf{k})$ to restore the original values:

$$F_{\text{orig}}(\mathbf{k}) = r(k)F(\mathbf{k}) \tag{5.47a}$$

In the work by Gabashvili and coworkers, the ratio $r(k)$ rises from 1.0 at $\sim 1/50\,\text{Å}^{-1}$ to ~ 2.5 at $1/11.5\,\text{Å}^{-1}$. In another study, Penczek et al. (1999) obtained such a curve by comparing the density maps of the same structure (*T. thermophilus* ribosome) obtained by X-ray crystallography and cryo-EM in the same resolution range.

10. Three-Dimensional Restoration

10.1. Introduction

Three-dimensional restoration (as distinct from CTF correction; see section 3.9 in chapter 2 and section 9 in this chapter) is the term we will use for techniques designed to overcome the angular limitation of a reconstruction that leads to resolution anisotropy and an elongation of the molecule in the direction of the missing data. *Maximum entropy* methods present one approach to restoration (Barth et al., 1989; Farrow and Ottensmeyer, 1989; Lawrence et al., 1989). These methods are known to perform well for objects composed of isolated peaks, for example, stars in astronomical applications, but less well for other objects; see Trussell (1980). Another approach, based on a set theoretical formulation of the restoration and the enforcement of mathematical constraints, is known under the name of *projection onto convex sets* (POCS). POCS was developed by Youla and Webb (1982) and Sezan and Stark (1982) and introduced into EM by Carazo and Carrascosa (1987a,b). An overview chapter by Carazo (1992) addresses the general question of fidelity of 3D reconstructions and covers POCS as well as related restoration methods.

As was earlier mentioned, iterative reconstruction techniques allow nonlinear constraints to be incorporated quite naturally. In this case they actually perform a reconstruction-cum-restoration, which can also be understood in terms of the theory of POCS (P. Penczek, personal communication, 1993). All these methods, incidentally, along with multivariate data analysis, are examples of a development that treats image sets in framework of a general algebra of images. Hawkes (1993) gave an early glimpse into the growing literature of this field.

10.2. Theory of Projection onto Convex Sets*

What follows is a brief introduction to the philosophy of the POCS method which can be seen as a generalization of a method introduced by Gerchberg and Saxton (1971) and Gerchberg (1974). A similar method of *iterative single isomorphous replacement* (Wang, 1985) in X-ray crystallography is also known under the name of *solvent flattening*. Still another, related method of *constrained thickness* in reconstructions of 1D membrane profiles was proposed earlier on by Stroud and Agard (1979). An excellent primer for the POCS method was given by Sezan (1992).

Similarly as in multivariate data analysis of images (chapter 4), which takes place in the space of all functions with finite 2D support, we now consider the space of all functions with finite 3D support. In the new space (Hilbert space), every conceivable bounded 3D structure is represented by a (vector) end point. Constraints can be represented by sets. For instance, one conceivable set might be the set of all structures that have zero density outside a given radius R. The idea behind restoration by POCS is that the enforcement of known constraints that were not used in the reconstruction method itself will yield an improved version of the structure. This version will lie in the intersection of all constraint sets and thus closer to the true solution than any version outside of it. In Fourier space, the angular gap will tend to be filled. The only problem to solve is how to find a pathway from the approximate solution, reconstructed, for instance, by back-projection or any other conventional technique, to one of the solutions lying in the intersection of the constraint sets.

Among all sets representing constraints, those that are both *closed* and *convex* have proved of particular interest. Youla and Webb (1982) showed that for such sets the intersection can be reached by an iterative method of consecutive *projections. A *projection from a function $f(\mathbf{r})$ onto a set C in Hilbert space is defined as an operation that determines a function $g(\mathbf{r})$ in C with the following property: "of all functions in C, $g(\mathbf{r})$ is the one closest to $f(\mathbf{r})$." Here, "closeness" is defined by the size of a distance; for instance, by the generalized Euclidean distance in Hilbert space:

$$E = \|g(\mathbf{r}) - \mathbf{f}(\mathbf{r})\| = \sum_{j=1}^{J} |g(\mathbf{r}_j) - f(\mathbf{r}_j)|^2 \qquad (5.48)$$

(Note that, by implication, repeated applications of *projection onto the same set lead to the same result.) In symbolic notation, if $\mathbf{P}_i$ $(i = 1, \ldots, n)$ denotes the operation of *projection onto set C_i so that $f' = \mathbf{P}_i f$ is the function obtained by *projecting f onto C_i, the iterative restoration proceeds as follows: we start from a "seed" function $\hat{f}_0$, and obtain successively

$$\hat{f}_1 = \mathbf{P}_1 \hat{f}_0;$$
$$\hat{f}_2 = \mathbf{P}_2 \hat{f}_1 = \mathbf{P}_2 \mathbf{P}_1 \hat{f}_0;$$
$$\cdot$$
$$\cdot \qquad (5.49)$$
$$\cdot$$
$$f_n^{(1)} = \mathbf{P}_n \mathbf{P}_{n-1} \ldots \mathbf{P}_1 \hat{f}_0$$

as the result of the first pass of the *projection scheme. Next, the procedure is repeated, this time with $\hat{f}_n^{(1)}$ as the starting function, yielding $\hat{f}_n^{(2)}$, etc. As the geometric analogy shows (figure 5.35), by virtue of the convex property of the

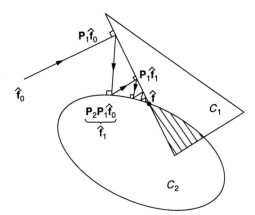

Figure 5.35 Principle of restoration using the method of *projection onto convex sets (POCS). C_1 and C_2 are convex sets in the space $\mathbf{R}^N$, representing constraints, and $\mathbf{P}_1$, $\mathbf{P}_2$ are associated *projected operators. Each element f_i is a 3D structure. We seek to find a pathway from a given blurred structure $\hat{f}_0$ (the seed in the iterative POCS) to the intersection set (shaded). Any structure in that set fulfills both constraints and is thus closer to the true solution than the initial structure $\hat{f}_0$. From Sezan (1992), reproduced with permission of Elsevier.

sets, each iteration brings the function (represented by a point in this diagram) closer to the intersection of all constraint sets. In this example, only $n = 2$ sets are used.

Carazo and Carrascosa (1987a,b) already discussed closed, convex constraint sets of potential interest in EM: (i) spatial boundedness (as defined by a binary mask); and (ii) agreement with the experimental measurements in the measured region of Fourier space, value boundedness, and energy boundedness. Thus far, in practical applications (see following), only the first two on this list have gained much importance, essentially comprising the two components of Gerchberg's (1974) method.[2] Numerous other constraints of potential importance [see, for instance, Sezan (1992)] still await exploration.

10.3. Projection onto Convex Sets in Practice*

The numerical computation in the various steps of POCS has to alternate between real space and Fourier space for each cycle. Both support (mask) and value constraints are implemented as operations in real space, while the enforcement of the "replace" constraint takes place in Fourier space. For a typical size of a 3D array representing a macromolecule (between $64 \times 64 \times 64$ and $128 \times 128 \times 128$), the 3D Fourier transformations in both directions constitute the largest fraction of the computational effort.

The support-associated *projector is of the following form:

$$\mathbf{P}_s f = \begin{cases} f(\mathbf{r}_i) & \mathbf{r}_i \in M \\ 0 & \text{elsewhere} \end{cases} \tag{5.50}$$

where M is the 3D mask defining the "pass" regions of the mask. In practice, a mask is represented by a binary-valued array with "1" representing "pass" and "0" representing "stop." The mask array is simply interrogated, as the discrete argument range of the function f is being scanned in the computer, and only those values of $f(\mathbf{r}_i)$ are retained for which the mask M indicates "pass." The support constraint is quite powerful if the mask is close to the actual boundary, and an important question is how to find a good estimate for the mask in the absence of information on the true boundary (which represents the normal situation). We will come back to this question later after the other constraints have been introduced.

The *value* constraint is effected by the *projector (Carazo, 1992):

$$\mathbf{P}_v f(\mathbf{r}_i) = \begin{cases} a & f(\mathbf{r}_i) < a \\ f(\mathbf{r}_i) & a \leq f(\mathbf{r}_i) \leq b \\ b & f(\mathbf{r}_i) > b \end{cases} \tag{5.51}$$

---

[2]Gerchberg's (1974) method, not that of Gerchberg and Saxton (1971), is a true precursor of POCS since it provides for replacement of both modulus and phases.

The *measurement* constraint is supposed to enforce the consistency of the solution with the known projections. This is rather difficult to achieve in practice because the projection data in Fourier space are distributed on a polar grid, while the numerical Fourier transform is sampled on a Cartesian grid. Each POCS *projection would entail a complicated Fourier sinc interpolation. Instead, the measurement constraint is normally used in a weaker form, as a "global replace" operation: within the range of the measurements (i.e., in the case of the random-conical data collection, within the cone complement that is covered with pro-jections; see figure 5.13), all Fourier coefficients are replaced by the coefficients of the solution found by weighted back-projection.

This kind of implementation is somewhat problematic, however, because it reinforces a solution that is ultimately not consistent with the true solution as it incorporates a weighting that is designed to make up for the lack of data in the missing region. Recall (section 4.2) that the weighting function is tailored to the distribution of projections in a given experiment, so it will reflect the existence and exact shape of the missing region.

A much better "replace" operation is implicit in the iterative schemes in which agreement is enforced only between projection data and reprojections. In Fourier space, these enforcements are tantamount to a "replace" that is restricted to the Fourier components for which data are actually supplied. The use of the global replace operation also fails to realize an intriguing potential of POCS: the possibility of achieving anisotropic super-resolution, beyond the limit given by Crowther et al. (1970). Intuitively, the enforcement of "local replace" (i.e., only along central sections covered with projection data) along with the other con-straints will fill the very small "missing wedges" between successive central sections on which measurements are available much more rapidly, and out to a much higher resolution, than the large missing wedge or cone associated with the data collection geometry. The gain in resolution might be quite significant.

What could be the use of anisotropic super-resolution? An example is the tomographic study of the mitochondrion (Mannella et al., 1994), so far hampered by the extremely large ratio between size (several micrometers) and the size of the smallest detail we wish to study (50 Å). The mitochondrion is a large structure that encompasses, and is partially formed by, a convoluted membrane. We wish to obtain the spatial resolution in any direction that allows us to describe the spatial arrangements of the different portions of the membrane, and answer the following questions: Do they touch? Are compartments formed? What is the geometry of the diffusion-limiting channels? The fact that the important regions where membranes touch or form channels occur in different angular directions makes it highly likely in this application that the relevant information can be picked up in certain regions of the object. The subject of tomography is outside the scope of this book, but similar problems where even anisotropic resolution improvement may be a bonus could well be envisioned in the case of macromolecules.

Examples for the application of POCS to experimental data are found in the work of Akey and Radermacher (1993) and Radermacher and coworkers (1992b, 1994b). In those cases, only the measurement and the finite support constraints were used. In the first case, the nuclear pore complex was initially reconstructed from data obtained with merely 34° and 42° tilt and thus had an unusually large missing-cone volume. Radermacher et al. (1994b) observed that POCS applied to a reconstruction from a negatively stained specimen led to a substantial contraction in the z-direction, while the ice reconstruction was relatively unaffected.

11. Reconstructions from Heterogeneous Data Sets

11.1. Introduction

Heterogeneous data sets obviously pose a challenge to the single-particle reconstruction approach, as each projection presents only a single aspect of a structure. Biochemical purifications generally result in preparations that have molecules of the same kind; heterogeneity is introduced by the coexistence of ligand-bound with ligand-free forms, or the coexistence of different conformational states that might equilibrate because they are separated by a small energy barrier. In terms of image-processing solutions, the problem posed by partial ligand occupancy is somewhat different than the problem of conformational heterogeneity, so that it is convenient to discuss them separately. We note, however, that there are many cases in which ligand binding induces conformational changes in the target molecule, making it possible to find the solution to one problem by solving the other.

11.2. Separating Ligand-Bound from Ligand-Free Complexes

We consider a situation where a ligand is nonstoichiometrically bound, whose mass is small compared to the target molecule. As an example, the mass of the tRNA ($\sim$70 kD) is just 3% of the total mass of the 70S ribosome (2.3 MD). With this assumption, the bias that is introduced by the ligand in the alignment and classification is negligible. This means that the orientation of the projection of a ligand-bound molecule is readily found by matching it to projections of a 3D reference. What follows is a description of a procedure that was developed and tested by Spahn and coworkers (C.M.T. Spahn, P.A. Penczek, and J. Frank, unpublished results).

For a ligand bound with partial occupancy, the reconstruction from the full data set contains the ligand density "diluted" according to the occupancy ratio; for instance, a reconstruction from data set of a protein complex with 60% occupancy will show the ligand with 60% of the average protein density. Features with such reduced density will not show up in the density map, but can be recovered by forming a difference map between the density map and a control map from a molecule without ligand binding.

It is then possible to design a 3D "ligand mask" that reflects the boundary of the ligand. As in the usual mask design via "Russian dolls" (chapter 4, section 3.2), the ligand density is first low-pass filtered, then a threshold is applied such that the resulting contour fully (and not too snugly) encloses the ligand. All values outside the 3D mask are set to "0," all values inside to "1." A strategy of separation is obtained as follows: project the 3D ligand mask into all directions on the angular grid used in the 3D projection matching. If the value of the mask in the interior is "1," and exterior "0," then the value range of the mask projection will be between "0" and the maximum diameter of the 3D mask, measured in pixels. From the mask projections, 2D masks are now obtained by applying a threshold, substituting "1" for any value >1.

In the next step, the 2D masks are used to define the precise area in each experimental projection ("the footprint") where a mass increase is expected if the ligand is bound. By focusing on the precise footprint, a much higher mass contrast is obtained than would be given by the ratio of total masses, since now we need to compare the mass of the ligand with the mass of the total ligand-shaped column of the target molecule above and below the ligand. In the above case of tRNA, the ratio is at least 30/250, or ~12%, the ratio between the width of the tRNA and the projection length through the surrounding ribosome. (It is normally even larger, for two reasons: one is that tRNA is not tightly surrounded by the ribosome, so that part of the beam will go through aqueous material, the other is that certain angles will yield oblique projections of the tRNA with much greater integrated mass.)

The ligand contrast may be high enough (as it is in the case used as our example) to allow classification based on the density. Each experimental projection is as usual classified by comparison with the 3D reference, which yields the likely orientation. Next the projection is integrated over the area of the 2D mask for that orientation. This integral is roughly proportional to the total mass of the ligand-shaped density column. For any angle, the histogram of this integrated mass is expected to have two distinguishable lobes, one for each subpopulation, provided that the ligand mass is large enough. A line dividing the two lobes can now be used for classifying the particles into the two groups.

Concern about possible bias introduced by the mask tailored to the ligand led to a modification of this procedure (Spahn et al., 2004a), where a general spherical mask (and resulting disk-shaped footprints) is used, which is centered on the ligand and sufficiently large to contain the ligand fully.

11.3. Separating Populations with Different Conformations

Separation on the basis of conformational differences is more difficult to achieve than on the basis of nonstoichiometric ligand binding, since the projections of the two conformers (to take the simplest case) are normally very similar. Nevertheless, the problem can be solved in particular situations where 3D references are available that represent the conformations in the different states: the experimental data are compared with projections from the different references, and the class assignment is made based on the reference that gives the highest correlation value.

Three-dimensional projection matching starts with a coarse grid. For ribosome data, the initial grid comprises 83 directions. In this initial step, more than one 3D reference can be used, giving each particle a further choice—for a given angle—which of the 3D references it "belongs to." A recent application will serve as an example: the reconstruction of the ribosome bound with the aminoacyl-tRNA•EF-Tu•GTP ternary complex in the presence of mRNA and the antibiotic kirromycin (Valle et al., 2002). This binding complex represents a stage in the decoding process just prior to the accommodation of the aminoacyl-tRNA into the A site. Apparently not all ribosomes were occupied in the specimen used for cryo-EM, leading to a fractured, scattered appearance of the ternary complex in the reconstruction from the full data set (figure 5.36a). To divide the particles into ribosomes that are occupied and those that are not, 3D projection matching was employed with two references, one presenting an empty ribosome (figure 5.36b), the other a ribosome bound with EF-G (figure 5.36a–c), a molecule known to have similar shape and similar binding position as the ternary complex. The reconstruction of particles with highest similarity to the EF-G-bound ribosome led to a density map in which the ternary complex was visible in its entirety (figure 5.36d). Comparison with a later reconstruction from a highly occupied specimen that did not require classification (Valle et al., 2003c) confirms that the application of the algorithm led to the correct decisions.

Valle et al. (2002), in their application of supervised classification with two 3D references made use of a histogram representation that reflected the preference of the particle (figure 5.36e). With the angle and class assigned to a particle indexed i according to the highest correlation value $\rho_+^{(i)}$, the lower correlation $\rho_-^{(i)}$ at that same angle with respect to the alternate reference is also looked up. Thus for each particle, a pair of numbers $(\rho_1^{(i)}, \rho_2^{(i)})$ is defined. Now the difference $\Delta\rho = \rho_1^{(i)} - \rho_2^{(i)}$ is plotted as a histogram. The histogram has a roughly normal distribution, with particles that display no preference for one reference or the other occurring with highest frequency. A similar analysis using two 3D references was employed by Gao et al. (2004) to separate two populations of ribosomes that differed in several features: binding of E-G, positions of the tRNAs, and the ratchet-like rotation of the small versus the large subunit.

In a situation where the two conformations are the only alternatives, particles with indifferent behavior are not acceptable and should be excluded. However, there are other situations where a number of transition states exist, that are ordered by decreasing similarity with one and increasing similarity with the other reference. In that case any subpopulation defined by a slice of the histogram attains significance, as it most likely originates from one of the transition states.

Such a situation was encountered by Heymann and coworkers (2003, 2004). A series of postulated structural intermediates were modeled by combining known end-point structures with different weights. These references were then used to classify the data into as many groups. Each group was subsequently refined by computing its reconstruction, to be used again in a next pass, and so on. Importantly, those particles that lacked commitment to any of the groups were eliminated. This study led to a model of the dynamics underlying virus capsid maturation.

In the study by Yang et al. (2003), in which the issue of heterogeneity is discussed, particles are sorted according to the prevalence of threefold versus sixfold symmetries. In other approaches, represented by the study of ATPase by Mellwig and Böttcher (2001), of ATP-bound states of GroEL (Ranson et al., 2001), and of heat shock proteins with different sizes (White et al., 2004), a heterogeneous population of molecules is separated by various recursive schemes using 2D classification and 3D reconstruction. It is difficult to generalize these

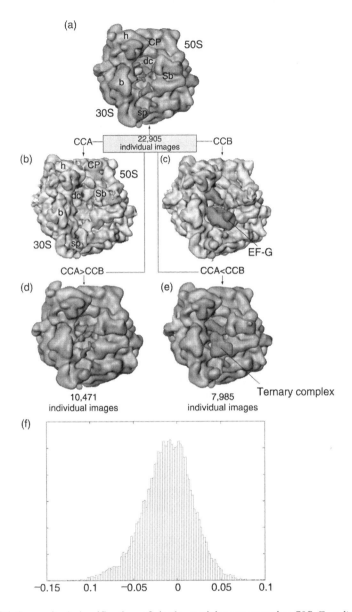

Figure 5.36 Supervised classification of single particles representing 70S *E. coli* ribosomes either occupied or unoccupied with the aminoacyl–tRNA•EF–Tu•GTP ternary complex. (a) Reconstruction from the whole data set without classification; (b) cryo-EM map of the empty ribosome, used as Reference I; (c) cryo-EM map of the ribosome occupied with EF-G, used as Reference II; (d) reconstruction from data with greatest similarity to Reference I; (e) reconstruction from data with greatest similarity to Reference II; (f) histogram of values obtained by forming the difference between CCF of each experimental projection with respect to References I and II. "Undecided" particles, equally dissimilar to Reference I as to Reference II, are grouped around 0. (a–e) from Valle et al. (2002), reproduced with permission of EMBO. (f) unpublished results.

kinds of approaches, however, as they rely on the soundness of numerous strategic judgments whose impact on the outcome is normally difficult to predict.

A particular case is presented by structural herterogeneity due to conformational variations of flexible molecules, which occur naturally in a thermal environment. It has been recently demonstrated (Tama et al., 2002, 2003; Wang et al., 2004) that large-domain motions of a molecule, often employed as part of the molecule's functional dynamics, can be predicted by normal-mode analysis (NMA) of a simplified representation of the atomic structure. Moreover, the same kind of analysis can also be performed on an elastic network representation of low-resolution density maps, again with remarkable predictive power (Chacón et al., 2003). Brink et al. (2004) went a step further, by combining NMA with the angular refinement procedure. NMA of a density map (fatty acid synthetase) used as initial reference yields multiple references, all slightly differing in conformation, which are all considered in the projection matching. The authors were able to show that some of the predicted conformations are experimentally realized.

12. Merging and Averaging of Reconstructions

12.1. The Rationale for Merging

Merging of different reconstructions is often necessary when any single one contains information gaps that can be complemented by one or more other reconstructions. CTF-related information gaps and phase reversals are in fact the rule, and are overcome, as we saw, by merging reconstructions obtained from different defocus groups by 3D Wiener filtering (see earlier section 9.1).

Another kind of information gap may occur due to a substantial gap in the angular coverage by projection data, which leads to anisotropic resolution. If the experiments can be designed such that the angular gap appears in different orientations of the coordinate system associated with the molecule, then it is possible to cover the gap by merging reconstructions from those experiments.

One of the important applications of this approach is in the random-conical reconstruction (section 5). Here, each reconstruction shows the molecule in an orientation that is determined by the orientation of the molecule on the specimen grid (and by a trivial "in-plane" rotation angle, whose choice is arbitrary but which also figures eventually in the determination of relative orientations). The goal of filling the angular gap requires that several reconstructions with different orientations be combined, to form a "merged" reconstruction in the final result. The procedure to obtain the merged reconstruction is easily summarized by three steps: (i) 3D orientation search, (ii) expression of the different projection sets in a common coordinate system, and (iii) reconstruction from the full projection set. Another application is tomographic reconstruction of single molecules (e.g., Walz et al., 1997b), a technique in which differently oriented molecules are reconstructed from a single-tilt series, so that for each reconstruction, the 3D Fourier transform is affected by the missing wedge in a different orientation.

The premise of the merging is that the reconstructions based on molecules showing different 0° views represent the same molecule, without deformation.

Only in that case will the different projection sets be consistent with a common object model. All evidence suggests that the flattening is normally avoided when ice embedment is used (e.g., Penczek et al., 1994). In addition, the high degree of preservation of 2D bacteriorhodopsin crystals in glucose-embedded preparations (Henderson and Unwin, 1975; Henderson et al., 1990) would suggest that single molecules embedded in glucose might also retain their shape, although to date this conjecture has been neither proved nor disproved by a 3D study. (The reason is that the low contrast between protein and glucose makes it difficult to find individual molecules in the micrograph field, normally preventing single-particle reconstruction from overcoming the very first hurdle.)

In contrast, negatively stained, air-dried molecules (section 12.2) often suffer a substantial deformation. For the latter type of specimens, the merging must be justified in every case, by applying a similarity measure: until such verification is achieved, the assumption that the particles reconstructed from different 0°-view sets have similar structure and conformation remains unproven.

12.2. Negatively Stained Specimens: Complications due to Preparation-Induced Deformations*

Molecules prepared by negative staining and air-drying often show evidence of flattening. Quantitative data on the degree of flattening are scarce; however, as an increasing number of comparisons between reconstructions of molecules negatively stained and embedded in ice become available, some good estimations can now be made. Another source of information is the comparison of molecules that have been reconstructed in two different orientations related to each other by a 90° rotation around an axis parallel to the grid. In interpreting such data, it is important to take incomplete staining into account. Unless the sandwiching technique is used, some portions of the particle that point away from the grid may "stick out" of the stain layer and thus be rendered invisible. On the other hand, the sandwiching may be responsible for increased flattening.

In all accounts, the behavior of the specimen is markedly different for conventional stains (such as uranyl acetate and uranyl formate) (section 12.2.1) when compared with methylamine tungstate (section 12.2.2).

12.2.1. Conventional Stains

Boisset et al. (1990b) obtained two reconstructions of negatively stained, sandwiched *Androctonus australis* hemocyanin, drawing either from particles lying in the top or in the side view, and reported a factor of 0.6 when comparing a particle dimension perpendicular to the specimen grid with the same dimension parallel to the grid. Since the flattening is accompanied by an increase in lateral dimensions, one can assume that the flattening is less severe than this factor might indicate. Another estimate, a factor of 0.65, comes from a comparison between two reconstructions of an *A. australis* hemocyanin–Fab complex, one obtained with negative staining (Boisset et al., 1993b) and the other using vitreous ice (Boisset et al., 1994b, 1995).

Cejka et al. (1992) reported a similar factor ($120\,\text{Å}/180\,\text{Å} = 0.67$) for the height (i.e., the dimension along the cylindric axis) of the negatively stained (unsandwiched) *Ophelia bicomis* hemoglobin. These authors were able to show that molecules embedded in aurothioglucose and freeze-plunged have essentially the same height (189 Å) when reconstructed in their top view as negatively stained molecules presenting the side view. The 50S ribosomal subunit, reconstructed in its crown view from negatively stained, sandwiched (Radermacher et al., 1987a,b), and ice-embedded preparations (Radermacher et al., 1992b), appears to be flattened according to the ratio 0.7:1. A factor of 0.6:1 holds for the calcium release channel, as can be inferred from a comparison (Radermacher et al., 1994b) of the side views of the cryo-reconstruction (Radermacher et al., 1994a,b) with the reconstruction from the negatively stained specimen (Wagenknecht et al., 1989a).

In summary, then, it is possible to say that as a rule, molecules prepared by negative staining with conventional stains (uranyl acetate, uranyl formate) are flattened to 60–70% of their original dimension. Consequently, the data sets stemming from two molecules lying in different orientations are difficult to merge since the corresponding reconstructions represent the molecule flattened in different directions. The problem is twofold: (i) the orientation between the two reconstructions cannot be accurately determined, and (ii) the data sets are inconsistent, so that their merger will result in a reconstruction that reflects the molecule in neither of the two shapes.

12.2.2. Methylamine Tungstate

Methylamine tungstate (MAT), in combination with Butvar support films, has been shown to avoid the specimen collapse to a great extent (Stoops et al., 1991b; Kolodziej et al., 1997). Most of these observations come from Stoops and coworkers, and were obtained for α_2-macroglobulin (Qazi et al., 1998, 1999), fatty acid synthetase (Kolodziej et al., 1997), pyruvate dehydrogenase complex (Zhou et al., 2001b), and calmodulin-dependent protein kinase II (Kolodziej et al., 2000). Comparison with structures obtained by X-ray crystallography provided the most powerful evidence for the high degree of preservation (Stoops et al., 1991b; Zhou et al., 2001b). The method was also used with success by Radermacher's group in the reconstruction of V_1-ATPase (Radermacher et al., 2001).

12.2.3. Cryo-Negative Stain

The behavior of molecules prepared with cryo-negative stain should be dominated by the properties of the ice matrix, and no orientation-related differences in particle dimension are expected. As yet, however, experimental confirmation of this expected behavior is missing.

12.3. Alignment of Volumes*

The reconstructed volumes have to be aligned both with respect to their translation and orientation. This is achieved by some type of computational search, either involving the 3D densities themselves or some derived function, in which the different parameters of shift and 3D orientation (six in all) are varied to maximize a functional embodying the concept of "similarity."

12.3.1. Cross-Correlation of 3D Densities

Here the cross-correlation coefficient is used as similarity measure (Knauer et al., 1983; Carazo and Frank, 1988; Carazo et al. 1989; Penczek et al., 1992; Koster et al., 1997; Öfverstedt et al., 1997; Walz et al., 1997b). The search range and the computational effort can be kept small if the approximate matching orientation can be estimated. Often it is clear from other evidence (e.g., interconversion experiments, see section 3.3; symmetries; or knowledge of architectural building principles, as in the case of an oligomeric molecule) how particular views are related to one another in angular space. Examples are the top and side views of *A. australis* hemocyanin, which are related by a 90° rotation of the molecule around a particular axis of pseudosymmetry (see, for instance, the appendix in Boisset et al., 1988).

A major difficulty in finding the relative orientation of different reconstructions is presented by the missing cone (or, in the case of tomographic reconstructions, the missing wedge). In Fourier space, the correlation between two volumes in any orientation is given by the sum over the terms $F_1(\mathbf{k})F_2^*(\mathbf{k})$ for all spatial frequencies $\mathbf{k}$ within the resolution domain. Here, $F_1(\mathbf{k})$ denotes the Fourier transform of the first volume in its original orientation, and $F_2^*(\mathbf{k})$ the complex conjugate of the Fourier transform of the second volume after it has been subjected to a "probing" rotation. Since the missing cones of data sets originating from different molecule views lie in different orientations, the orientation search is biased by a "vignetting effect." This effect results from the cones intersecting each other to different extents, depending on the angles. Real-space methods of orientation determination search generally fail to deal with this complication and may therefore be inaccurate. Fourier methods, in contrast, allow precise control over the terms included in the correlation sum and are, therefore, usually preferable.

In a case where three or more reconstructions are compared, the results of pairwise orientation searches can be checked for closure. Penczek et al. (1992) used this principle in the case of the 70S ribosome of *E. coli* where three reconstructions $\mathbf{S}_1, \mathbf{S}_2, \mathbf{S}_3$ were available: the combination of the rotations found in the orientation searches $(\mathbf{S}_1, \mathbf{S}_2)$, $(\mathbf{S}_2, \mathbf{S}_3)$ should give a result close to the rotation resulting from the search $(\mathbf{S}_3, \mathbf{S}_1)$. [Note that each of these rotations is expressed in terms of three Eulerian angles (see section 2.3), so that the check actually involves the multiplication of two matrices]. By adding up the angles, these authors found a closure error of 2°, which indicates almost perfect consistency. More generally, in the situation where N reconstructions $\mathbf{S}_i$ $(i = 1 \ldots, N)$ are compared, then any "closed" string of pairwise orientation determinations (i.e., a string that contains each volume twice, in any two successive comparisons):

$$(\mathbf{S}_{i_1}, \mathbf{S}_{i_2}), \ldots, (\mathbf{S}_{i_N}, \mathbf{S}_{i_1}) \tag{5.52}$$

should result in rotations that, when combined, amount to no rotation at all.

12.3.2. Orientation Search Using Sets of Projections*

Instead of the reconstructions, the projection sets themselves can be used for the orientation search (Frank et al., 1992; Penczek et al., 1994). In Fourier space, each random-conical projection set is represented by a set of central sections, tangential to a cone, whose mutual orientations are fixed (figure 5.37). It is obvious that the real-space search between volumes can be replaced by a Fourier space search involving the two sets of central sections. Instead of a single common line, the comparison involves N common lines (N being the number of projections in the two sets combined) simultaneously, with concomitant increase in SNR. The method of comparison uses a discrepancy measure $1 - \rho_{12}$, where ρ_{12} is the cross-correlation coefficient, computed over Fourier coefficients along the common line. Penczek et al. (1994) described the geometry underlying this search: the angle between the cones is allowed to vary over the full range. For any given angle, it is necessary to locate the "common line" intersections between all central sections of one cone with all central sections of the other.

While the location of the common line is constructed in Fourier space, the actual computation of the discrepancy measure is performed in real space, exploiting the fact that the 1D common line found for any particular pairing of central sections is the Fourier transform of a 1D projection. The method of orientation search using sets of projections has two advantages over the orientation search between reconstruction volumes: it makes the computation of reconstructions prior to merging unnecessary, and it offers a rational way of dealing with the angular gap.

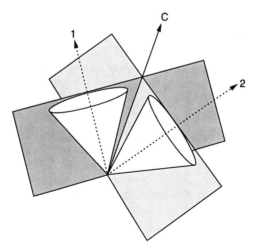

Figure 5.37 Principle of the method of orientation search using sets of projections (OSSP). The two cones around axes 1 and 2 belong to two random-conical data collection geometries with different orientations of the molecule. Two given projections in the two geometries are represented by central Fourier sections tangential to the cones. These central sections intersect each other along the common line C. The OSSP method simultaneously considers every common line generated by the intersection of every pair of central sections. From Penczek et al. (1994), reproduced with permission of Elsevier.

12.3.3. Fast Rotational Matching via Spherical Harmonics*

Unlike translational matching, for which fast Fourier methods are available, rotational matching as discussed before usually involves a "brute force" search, in which one of the two volumes is stepwise explicitly rotated and reinterpolated onto the grid of the other volume. By using an expansion of the density maps (provided they are centered) into spherical harmonics, it is possible to obtain the same kind of speedup as with translational search. Crowther (1972), the first to recognize this advantage, introduced an algorithm for a fast rotational match in two of the three angles. More recently, Pavelcik et al. (2002) and Kovacs and Wriggers (2002) showed that a match over all three angles can be achieved in this way. For the details, and the definition of spherical harmonics, the reader is referred to their article.

12.3.4. Rotational Matching Using Tensors of Inertia*

A method using the tensors of inertia of the two volumes to be matched was introduced by Lanzavecchia et al. (2001). In essence, orienting the volumes means to bring the axes of their associated inertia ellipsoids into mutual alignment. For a centered discrete density distribution d_{xyz}, the tensor of inertia is defined as

$$\begin{bmatrix} \sum_{x,y,z} d_{x,y,z}(y^2+z^2) & \sum_{x,y,z} d_{x,y,z}xy & \sum_{x,y,z} d_{x,y,z}xz \\ \sum_{x,y,z} d_{x,y,z}xy & \sum_{x,y,z} d_{x,y,z}(x^2+z^2) & \sum_{x,y,z} d_{x,y,z}yz \\ \sum_{x,y,z} d_{x,y,z}xz & \sum_{x,y,z} d_{x,y,z}yz & \sum_{x,y,z} d_{x,y,z}(x^2+y^2) \end{bmatrix} \quad (5.53)$$

It is a quadratic form representing the inertia ellipsoid of the volume d_{xyz}. Based on this representation, the steps required to bring two volumes into alignment are:

(i) Center each distribution
(ii) Calculate their tensors following the defining formula
(iii) Compute the eigenvalues and eigenvectors of each matrix, which give the three axes of the associated ellipsoids of inertia
(iv) Align the axes of the two ellipsoids

In their paper, Lanzavecchia et al. (2001) go into further details dealing with the conditions under which the solutions found are unambiguous.

12.4. Merging of Reconstructions through Merging of Projection Sets into a Common Coordinate Frame

Once the relative orientations of the projection sets are known, either from an orientation search of the reconstruction volumes, or from an orientation search of the projection sets (as outlined in the previous section), the Eulerian angles of all projections of the combined projection sets can be expressed in a common coordinate system. It is then straightforward to compute the final, merged reconstruction.

It is important to realize that because of the properties of the general weighting functions, it is not possible to merge reconstructions by simply averaging them, *unless all reconstructions are based on projections with the exact same tilt geometry*. (This would hold true only for tomographic reconstructions of molecules that are oriented the same way, so that the angular grids are exactly superimposable.) Instead, one must first go back to the projection data, apply the appropriate coordinate transformations so that all projection angles relate to a single coordinate system, and then perform the reconstruction from the entire projection set. The same is true when one uses iterative reconstruction techniques, where the solution is also tied to the specific geometry of a projection set.

12.5. Classification of 3D Volumes

Multivariate data analysis (PCA and CA) and classification, when applied *directly* to the discrete 3D density representations, faces problems when the reconstructions are obtained with different data collection geometries. In those cases, the structure-related variations may be overshadowed by other variations related to angular coverage. However, there are special cases where reconstructions are indeed strictly comparable in the way they are affected by the missing angular range. At one end of the spectrum, we have tomographic reconstructions of a molecule always oriented in the same way, and at the other end, single-particle reconstructions of a molecule in different functional states, each of which is obtained from thousands of randomly oriented particles—hence is not affected by an angular gap. In such cases, we can now ask—just as in the case of images that were to be combined—if there is any significant clustering or systematic variations among the 3D data that might warrant the application of classification prior to forming averages. The generalization of the techniques of multivariate data analysis and classification developed for two dimensions (chapter 4) to the 3D case is straightforward. Examples are the sorting of volumes reconstructed following common-lines procedures according to their handedness (Penczek et al., 1996), the analysis of variability exhibited by thermosome molecules reconstructed as tomograms (Walz et al., 1997), and the clustering of different repeats of rigor insect flight muscle in different stages of the functional cycle (Winkler and Taylor, 1999; Chen et al., 2002a; Pascual-Montano et al., 2002).

As an example, the case of the insect flight muscle is very informative. Tomography of a section of the flight muscle shows a regular array of structural motifs formed mainly by actin and myosin. Closer inspection shows that the repeats of the motif differ by the exact arrangement of the cross-bridges, so they could be called *pseudo-repeats*. Taylor and coworkers (Winkler and Taylor, 1999; Chen et al., 2002a; Pascual-Montano et al., 2002) attributed these differences to the power stroke cycle of insect flight muscle. According to this interpretation, each pseudo-repeat represents a different state of the power stroke. However, if the region is large enough, it is likely that several representatives of each state are encountered. Winkler and Taylor (1999) took advantage of this situation by classifying all 3D pseudo-repeats in a given tomogram. Averages of the resulting classes showed the important differences with much better SNR.

6

Interpretation of Three-Dimensional Images of Macromolecules

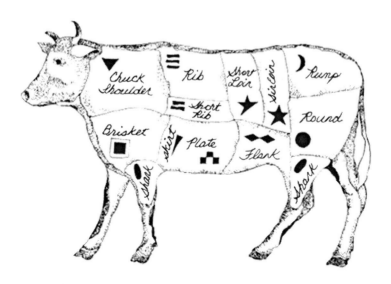

Figure 6.1 Segmentation and annotation of a domestic animal. From Rosso and Lukins (1989), reproduced with permission of Workman Publishing.

1. Introduction

Interpretation covers several important aspects of a study, once the density maps (normally a minimum of two, with one to serve as a control) have been obtained. The most obvious problem posed by a low-resolution density map is to

277

distinguish the different components of a structure, a problem known as segmentation (figure 6.1). Closely related are the issues of significance, experimental validity, and meaning.

Setting aside the problem of segmentation for the moment, it is necessary, at first, to ask questions regarding the relationships between the raw data, the three-dimensional (3D) reconstruction, and the molecule it is supposed to represent: (i) When, as a function of spatial separation and density separation, are differences in density between two points of a reconstruction significant? (ii) Next, if we obtain two reconstructions of a molecule in different states, what are the criteria for pronouncing the observed density differences statistically significant? (iii) What assurances do we have that the features we see in a reconstruction represents something real? And, if this is the case, what is the precise relationship between the density map and the input data?

The first two questions arise inevitably in any experiment where the result is derived by averaging over a number of measurements that are subject to noise. The question can be answered by application of statistical hypothesis testing, but this requires knowledge of another type of information not yet mentioned: the 3D variance. Both 3D variance estimation and significance assessment will be treated in section 2.

The third question arises because the single-particle approach to electron microscopic structure research involves a large battery of complex operations with a number of discretionary choices of parameters, for example, which particles to exclude, how many classes to seek in a classification, and which classes to choose for 3D reconstruction. The potential vagaries of contrast transfer function (CTF) determination and correction only add to the discretion. The question is, therefore, to what extent the end product is influenced by the processing methods chosen. This question is particularly important when a 3D density map is used as template to determine particle orientations.

Furthermore, there exists an inherent ambiguity in the interpretation of density maps that result from an averaging process, as low density in a given voxel may result either from averaging over contributions with consistently low density values at that spot, or from averaging over contributions that vary from one data item (single-particle projection) to the other. Most common is the case where a ligand is present in one subpopulation but not in the other ("partial occupancy"). However, the same situation can arise when a component of the structure is flexible, and therefore exists in different positions (resulting in "conformational averaging"). There also exists an ambiguity of a different kind, due to the degeneracy in the relationship between two-dimensional (2D) projections and the 3D object, which affects the 3D inferences drawn from 2D classification (section 4.11, chapter 4) and needs to be resolved by careful checks that might involve additional tilt experiments. Taken together, all these uncertainties call for some type of validation and testing of the results for experimental and mathematical consistency. Section 3 summarizes attempts that have been made to address some of those issues.

Once satisfied that the reconstructed 3D density distribution is statistically significant, reproducible in a strict sense (i.e., as deduced by independent experiments), and self-consistent, we proceed to the next question: What is the meaning of the 3D map? The search for the "meaning" of a 3D density distribution always involves expert interpretation, which relies on two tools that are jointly used: visualization (section 4) and segmentation (section 5). In low- and intermediate-resolution density maps, visualization is the crucial step in which expert knowledge comes to bear: characteristic shapes may be recognizable, leading to hypotheses as to the identity of corresponding components.

Segmentation is the process by which boundaries between 3D regions are determined that represent known components of the structure. Identification and designation of these components, or "annotation," happens in the context of other structural, biochemical, or bioinformatics knowledge. Despite the abundance of sophisticated tools and the wealth of knowledge now residing in public databases, the interpretation of a density map often occupies the largest share of the effort in a typical project.

The remaining two sections finally cover methods of docking and fitting (themselves tools of segmentation of a special kind, in which a portion of a low-resolution density map is identified with an atomic structure; section 6), and the classification of volumes (section 7).

2. Assessment of Statistical Significance

2.1. Introduction

There are two types of questions that involve the notion of statistical significance: (i) Is a density maximum observed in the reconstruction significant, given the set of observations, or can it be explained as the result of a random fluctuation? (ii) Let us assume that we need to localize a feature in three dimensions by forming the difference between two reconstructions, A–B, where A is the reconstruction obtained from the modified specimen (e.g., a molecule–ligand complex or the molecule in a modified conformational state) and B is the reconstruction obtained from the control. A and B are sufficiently similar in structure so that alignment of the common features is possible and meaningful, and the structural difference can be obtained by simple subtraction. A and B are derived from two different projection sets. We now wish to answer the question: is the observed density difference in the 3D difference map meaningful?

Both types of question can be solved through the use of the 3D variance map; in the first case, such a map must be computed for the single reconstruction volume considered, and in the second, such a map must be computed for both volumes to be compared. Once the variance map, or maps, have been obtained, we can proceed by applying the methods of statistical inference, just as we have done in two dimensions (section 4.2 in chapter 3). The only problem is that the 3D variance is normally not available in a similar way as it is in two dimensions, where it emerges naturally as a byproduct of averaging. We would have a similar situation only if we had an entire statistical ensemble of reconstructions, that is, N reconstructions for each of the experiments (i.e., molecular complex of interest and control). This situation is common in experiments where biological objects with helical symmetry are studied (e.g., Trachtenberg and DeRosier, 1987). In that case, we could derive an estimate of the 3D variance, in strict analogy with the 2D case, by computing the *average reconstruction*:

$$\bar{B}(\mathbf{R}) = \frac{1}{N} \sum_{i=1}^{N} B_i(\mathbf{R}) \tag{6.1}$$

and from that the variance estimate:

$$\bar{V}(\mathbf{R}) = \sum_{i=1}^{N} |B_i(\mathbf{R}) - \bar{B}(\mathbf{R})|^2 \tag{6.2}$$

Experiments on single macromolecules that came closest to this situation were first done in Walter Hoppe's group (see Öttl et al., 1983; Knauer et al., 1983), albeit with numbers N that were much too small to allow meaningful statistical testing. Automated electron tomography (Koster et al., 1992) is now producing larger sets of reconstructions where an estimation along the lines of equation (6.2) is becoming meaningful (Öfverstedt et al., 1997; Walz et al., 1997a,b). Comparable applications are found in radiological imaging of the brain, where abnormalities are determined relative to an "averaged" brain (see, e.g., Andreasen et al., 1994).

However, the more common situation, encountered in all single-particle reconstruction projects, is that only one projection is available for each projection direction; in other words, that each observed projection originates from a different particle. Thus, only one reconstruction results from each experiment, and the 3D variance information has to be inferred from an analysis of projections only.

Two approaches have been formulated to deal with this problem: an estimation method based on comparison of adjacent projections (Liu, 1991, 1993; Frank et al., 1992; Liu and Frank, 1995; Liu et al., 1995), and a bootstrapping method, according to which different projections are left out in repeated reconstructions (Penczek, 2002b). These methods will be introduced below.

2.2. Three-Dimensional Variance Estimation from Projections*

2.2.1. Comparison of Adjacent Projections

The estimation from projections starts with the concept of the projection noise (Liu and Frank, 1995). (The following derivation assumes uncorrelated voxels, an assumption that is unrealistic, but can be enforced by low-pass filtering the volume to its actual resolution and then resampling it.) Each experimental projection is seen, in a *gedankenexperiment* (an experiment in concept only), as a member of a (virtual) statistical ensemble of M projections in the same ith direction. Hence, the noise belonging to projection j in direction i is

$$n_j^{(i)}(\mathbf{r}) = p_j^{(i)}(\mathbf{r}) - \bar{p}_j^{(i)}(\mathbf{r}) \tag{6.3}$$

where $\bar{p}_j^{(i)}(\mathbf{r})$ denotes the ensemble average. The contribution of the noise to a reconstruction based on the weighted back-projection algorithm is obtained by application of the weighting function [see equations (5.27) and (5.28)]:

$$n_{jW}^{(i)}(\mathbf{r}) = n_{jW}^{(i)}(\mathbf{r}) * W_i(\mathbf{r}) \tag{6.4}$$

Now, drawing projections at random from the virtual bins (each containing the projection ensemble relating to a different direction), we can produce N different virtual reconstructions $\hat{B}_i$ and again define the variance according to

$$V(\mathbf{R}) = \sum_{i=1}^{N} [\hat{B}_i(\mathbf{R}) - \bar{\hat{B}}(\mathbf{R})]^2 \tag{6.5}$$

The 3D variance $V(\mathbf{R})$ can be estimated by back-projecting the weighted-projection 2D noise variances (now leaving out the drawing index j):

$$\tilde{V}(\mathbf{R}) = \sum BP\{[n_w^{(i)}(\mathbf{r})]^2\} \tag{6.6}$$

The only remaining task is to estimate the projection noise $n^{(i)}(\mathbf{r})$ for each projection direction. This can be done in two different ways. One way is by obtaining a reconstruction from the existing experimental projections by weighted back-projection, and then reprojecting this reconstruction in the original directions. The difference between the experimental projection and the reprojection can be taken as an estimate of the projection noise.

The other way (figure 6.2), closely examined by Liu and Frank (1995) and Liu et al. (1995), is by taking advantage of the fact that in normal data collections, the angular range is strongly oversampled, so that for each projection, close-neighbor projections that differ little in the signal are available. In other words, an ensemble of projections, of the kind assumed in the *gedanken-experiment*, actually exists in a close angular neighborhood of any given projection.

It is obviously much easier to go the first way, by using reprojections (see figure 6.2). The detailed analysis by Liu and coworkers, in the works cited above, needs to be consulted for an evaluation of systematic errors that might favor the second way in certain situations. It is also important to realize that there are certain systematic errors, such as errors due to interpolation, inherent to all reprojection methods, that cause discrepancies unrelated to the 3D variance (see Trussel et al., 1987). We have implicitly assumed in the foregoing that the weighted back-projection algorithm is being used for reconstruction.

It should be noted that the estimation of the 3D variance from the 2D projection variances works only for reconstruction schemes that, as the weighted back-projection, maintain linearity. Iterative reconstruction methods such as the algebraic reconstruction technique (ART) (Gordon et al., 1970) and the simultaneous iterative reconstruction technique (SIRT) (Gilbert, 1972) do maintain the linearity in the relationship between projections and reconstruction, unless

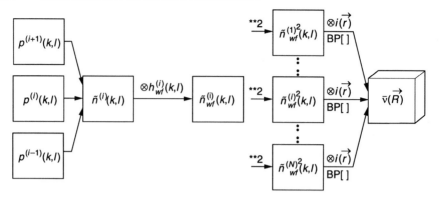

Figure 6.2 Flow diagram of 3D variance estimation algorithm. Provided the angular spacing is small, the difference between neighboring projections can be used as noise estimate $\tilde{n}^{(i)}(k, l)$. When such an estimate has been obtained for each projection, we can proceed to compute the 3D variance estimate by using the following steps: (i) convolution with the inverse of the back-projection weighting function (i.e., the real-space equivalent to the back-projection (BP) weighting in Fourier space); (ii) square the resulting weighted noise functions, which results in estimates of the noise variances; (iii) back-project (BP) the noise variance estimates which gives the desired 3D variance estimate. From Frank et al. (1992), reproduced with permission of Scanning Microscopy International.

they are modified for the purpose of incorporating nonlinear constraints (e.g., Penczek et al., 1992). Weighted back-projection, as we have seen earlier (chapter 5, section 4), has the property of linearity for evenly spaced projections, for which $|k|$-weighting holds, while general weighting functions often incorporate (nonlinear) application of thresholds.

An application of the variance estimation method by adjacent projections is shown in figure 6.3 in section 2.3.2.

2.2.2. Bootstrap Method*

Penczek (2002b) developed a method of variance estimation in which multiple reconstructions are created by repeatedly drawing a set of projections at random from the existing pool. In addition to an accurate variance estimation, the method allows estimation of covariances (or correlations) between regions of interest in 3D volume. An application of this method is described in Spahn et al. (2004a).

Specifically, the method is based on the assumption that the number of available projections is large; thus, if we consider the weighted back-projection algorithm we can assume that the weighting function (see chapter 5, section 4.2) will not change significantly with a small change in the number or distribution of projections. Thus, a 3D reconstruction can be considered a weighted sum of projections, with weights being approximately constant. In this case, the variance/covariance can be calculated using a statistical bootstrap technique: given a set of k projections, n different reconstructions are obtained by repeatedly

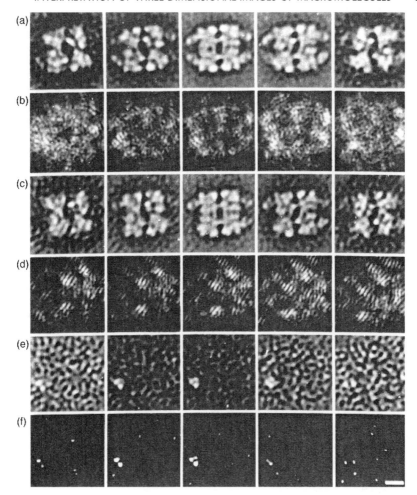

Figure 6.3 Example of variance estimation, applied to the assessment of significance of a feature in a 3D difference map. Two reconstructions are compared: one (selected slices in row a) derived from hemocyanin molecules that are labeled at all four sites, and another (selected slices in row c) at three sites only. Rows b and d: 3D variance estimation computed for reconstructions a and c, respectively, following the procedure outlined in the flow diagram of figure 6.2. Row e: difference volume a−c. Row f: 3D t-test map, computed for a 99% confidence level, showing the 3D outline of the Fab molecule. From Boisset et al. (1993b), reproduced with permission of Elsevier.

selecting k projections at random with replacement. This means that in each such permutation of the original projection set, some of the projections are missing and some occur repeatedly. From the n reconstructions, a voxel-by-voxel variance σ_n^2 is calculated. The sought-for "correct" variance is then estimated as

$$\sigma_t^2(\mathbf{R}) = k\sigma_n^2(\mathbf{R}) \tag{6.7}$$

It is important to know how large n must be made for the variance estimate to be reliable. Penczek (2002b) found that in a typical case, about 30 reconstructions will suffice. Despite the high speed of today's computers, the computation of this number of reconstructions is still substantial. Penczek (2002b) therefore suggested that the Fourier interpolation method of reconstruction be used instead of the weighted back-projection, for increased speed. Fourier interpolation, even though it results in a much cruder reconstruction quality, has the advantage that it allows the different reconstructions from permutations of the projection set to be obtained in a very simple way, by subtracting or adding Fourier transforms of projections on the appropriate central planes according to the realization of the random draw. [Note, though, that the recent addition of a fast gridding method (Penczek et al., 2004b) promises a way to obtain reconstructions both fast and accurately, which means that the accuracy of the variance computation does not need to be compromised.]

It must be noted that the variance obtained by the bootstrapping method is different from, and normally larger than, the variance of the original volumes. The reason is that in addition to the desired structural component, there are two terms, one stemming from the reconstruction method itself and its numerical errors, the other from the unevenness of the angular distribution of projections.

An application of the bootstrapping method to a realistic problem is found in Spahn et al. (2004a): given was a heterogeneous data set of ribosome images. The source of the heterogeneity was the partial occupancy of a ligand, EF-G: a certain percentage of the ribosomes were bound with EF-G, the rest were empty. As expected, the variance estimate showed high variance in the region of EF-G binding.

2.3. Use of the 3D Variance Estimate to Ascertain the Statistical Significance

2.3.1. Significance of Features in a 3D Map

To ascertain the significance of a feature in a 3D map, we proceed in a way similar to that outlined in the 2D case (chapter 3, section 4.2). We need to know the statistical distribution of the deviation of the 3D image element $p(\mathbf{R}_j)$ from the averaged element $\bar{p}(\mathbf{R}_j)$, or the distribution of the random variable

$$t(\mathbf{R}_j) = \frac{p(\mathbf{R}_j) - \bar{p}(\mathbf{R}_j)}{s(\mathbf{R}_j)} \tag{6.8}$$

where $s(\mathbf{R}_j) = \sqrt{V(\mathbf{R}_j)/N}$ is again the standard error of the mean. $s(\mathbf{R}_j)$ can be estimated either from the variance estimate $\bar{V}(\mathbf{R}_j)$ [equation (6.2)]—which is normally not available, unless we can draw from many independent 3D reconstructions (see section 2.2 for the detailed argument)—or from the estimate $\tilde{V}(\mathbf{R}_j)$ [equation (6.6)], which is always available when weighted back-projection or one of the linear iterative reconstruction algorithms is used as the method of reconstruction.

Having constructed the random variable $t(\mathbf{R}_j)$, we can then infer the confidence interval for each element of the 3D map, following the considerations outlined in the 2D case earlier on (section 4.2, chapter 3). On this basis, the local features within the 3D reconstruction of a macromolecule can now be accepted or rejected [see, for instance, the use of t-maps by Trachtenberg and DeRosier (1987) to study the significance of features in reconstructions of frozen-hydrated filaments].

2.3.2. Significance of Features in a Difference Map

The most important use of the variance map is in the assessment of the statistical significance of features in a 3D difference map. Due to experimental errors, such a difference map contains many features that are unrelated to the physical phenomenon studied (e.g., addition of antibodies, deletion of a protein, or conformational change). Following the 2D formulation, equation (3.52) in chapter 3, we can write down the standard error of the difference between two corresponding elements of the two maps, $\bar{p}_1(\mathbf{R}_j)$ and $\bar{p}_2(\mathbf{R}_j)$ as follows:

$$s_d[\bar{p}_1(\mathbf{R}_j), \bar{p}_2(\mathbf{R}_j)] = [V_1(\mathbf{R}_j)/N_1 + V_2(\mathbf{R}_j)/N_2]^{1/2} \tag{6.9}$$

where $V_1(\mathbf{R}_j)$ and $V_2(\mathbf{R}_j)$ are the 3D variances. In studies that allow the estimation of the variance from repeated reconstructions according to equation (6.2) (see, e.g., Milligan and Flicker, 1987; Trachtenberg and DeRosier, 1987), N_1 and N_2 are the numbers of reconstructions that go into the computation of the averages $\bar{p}_1(\mathbf{R}_j)$ and $\bar{p}_2(\mathbf{R}_j)$. In the other methods of estimation, from projections, $N_1 = N_2 = 1$ must be used.

On the basis of the standard error of the difference, equation (6.9), it is now possible to reject or accept the hypothesis that the two elements are different. As a rule of thumb, *differences between the reconstructions are deemed significant in those regions where they exceed the standard error by a factor of three or more* (see section 4.2 in chapter 3).

Analyses of this kind have been used by Milligan and Flicker (1987) to pinpoint the position of tropomyosin in difference maps of decorated actin filaments. In the area of single-particle reconstruction, they are contained in the work by Liu (1993), Frank et al. (1992), and Boisset et al. (1994a). In the former two studies, the 50S-L7/L12 depletion study of Carazo et al. (1988) was re-evaluated with the tools of the variance estimation. The other application concerned the binding of an Fab fragment to the *Androctonus australis* hemocyanin molecule in three dimensions (Liu, 1993; Boisset et al., 1994a; Liu et al., 1995). In the case of the hemocyanin study (figure 6.3), the appearance of the Fab mass on the corners of the complex is plain enough, and hardly requires statistical confirmation. However, interestingly, the t-map (figure 6.3f) also makes it possible to delineate the epitope sites on the surface of the molecule with high precision.

Among the groups that have most consistently used formal significance tests in assessing the validity of structural differences are DeRosier's (Trachtenberg and DeRosier, 1987; Thomas et al., 2001), Wagenknecht's (Sharma et al., 1998, 2000; Samsó et al., 1999), and Saibil's (Roseman et al., 1996).

3. Validation and Consistency

Three questions are raised with cryo-electron microscopy (cryo-EM), as with any new method of structural research: (i) Are 3D reconstructions that are done from the same data set but with different algorithms and numerical procedures reproducible? (ii) To what extent is the entire experiment, from the preparation of the specimen to the reconstruction, reproducible? (iii) Does cryo-EM come to the same answer as other methods of structure research, such as X-ray crystallography and nuclear magnetic resonance (NMR), when applied to the same specimen? In addition, there is the more mundane question of whether the programs used in the data processing and reconstruction are correct and consistent implementations of the algorithms underlying the reconstruction.

Although there are not many systematic studies addressing these questions, there are enough results in the literature, if only anecdotal, to give an assuring picture. Nonetheless, in one instance, different groups reported markedly different results: the IP3 receptor (Jiang et al., 2002; da Fouseca et al., 2003; Serysheva et al., 2003; Sato et al., 2004; reviewed by Taylor et al., 2004), presumably due to the large flexibility and conformational heterogeneity of this molecule.

3.1. Internal Consistency

Internal consistency of the reconstruction procedure has to do with "housekeeping" matters such as the consistency in the algorithmic implementation of the projection and back-projection operations, the consistency in the definition of angles throughout the programs, etc. In that sense, a comparison between a set of experimental projections (e.g., obtained by averaging images falling into different classes) and corresponding reprojections of the reconstructed density map is helpful and reassuring, since the absence of such agreement indicates flaws in the program. However, as we have noted earlier (chapter 5, section 7.7), the proven agreement between input projections and reprojections cannot be taken as an indication that the structure is correct, since they amount to little more than a comparison between data placed into the 3D Fourier transform and data extracted at the same place. By contrast, a comparison between experimental tilted-specimen projections (not used in the reconstruction) with reprojections of a single-particle reconstruction obtained for untilted projections, such as the one done by Hegerl (1991) represents a very informative, legitimate validation (figure 6.4).

3.2. Reconstructions from the Same Data Set with Different Algorithms

The most comprehensive study of reconstruction algorithms, suitable for single-particle reconstruction, using a number of fidelity criteria, was by done Sorzano et al. (2001). The study was initially motivated by reports of artifacts due to directional overabundance of projections in single-particle reconstructions (Boisset et al., 1998), discounted and explained in the new study, but was expanded beyond this narrow scope. The algorithms compared—using phantom data throughout—were ART with blobs, SIRT, and weighted back-projection.

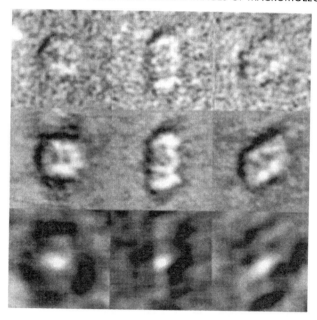

Figure 6.4 Comparison of experimental tilted-specimen projections of proteasomes (top row) with reprojections of the random-conical reconstruction (obtained by the simultaneous iterative reconstruction technique, middle row) and corresponding cross-correlation functions (bottom row). The comparison shows the extent of agreement obtained without angular refinement. The comparisons were done to check the accuracy of translational alignment, which was found to be within a pixel. From Hegerl et al. (1991), reproduced with permission of Elsevier.

Altogether, 21 different figures of merit were tabulated based on the comparison of the phantom model—from which the projection set was generated—with the reconstruction. These figures of merit encompass more common measures, such as the mean squared error, but also measures sensitive to anisotropic resolution and to foreground versus background separation.

Another comparison is found in the recent work by Penczek et al. (2004b), on the occasion of an assessment of the quality of the gridding-based direct Fourier reconstruction (GDFR) algorithm, a new reconstruction method based on gridding (see chapter 5, section 4.4). The techniques being compared were weighted back-projection using two different weighting functions, namely (i) general weighting (Radermacher et al., 1986a, 1987b), and (ii) the so-called exact weighting (Harauz and van Heel, 1986a), furthermore (iii) SIRT, and (iv) GRFR. In this comparison, GDFR came out superior to the other algorithms in terms of fidelity (measured by FSC against the phantom) and reproducibility (measured by FSC of randomly split halfsets).

3.3. Consistency with X-Ray Structures

The density maps obtained by cryo-EM and X-ray crystallography both originate from elementary scattering events on the atoms of the structure. In X-ray crystallography, with few exceptions, the experimentally obtained, phased high-resolution density map is interpreted in terms of the underlying structural motifs, and the resulting publicly deposited Protein Data Bank (pdb) coordinate files represent the end result of the structural analysis. If the same structure is studied by both methods, we should obtain equivalent results when the X-ray coordinates are reinterpreted as electron density maps, and filtered to the same resolution as the cryo-EM map. Residual differences are extremely interesting as they shine light on deficiencies in either method, help in improving the methodology, and carry the potential of supplying additional information. Such differences arise from the following circumstances:

(i) Differences between electrons and X-rays in their interaction with the structure. Indeed, the physical density distributions retrieved with X-ray (electron density) and cryo-EM (Coulomb potential) are not exactly equivalent. Unlike scattering by X-rays, scattering by electrons is affected by the distribution of electric charges and the local ionization state. The effects may show up at sufficiently high resolution, as demonstrated in the molecular structure of the proton pump (Fujiyoshi, 1999).

(ii) Local disorder, which may have its root in conformational heterogeneity resulting from the coexistence of different states (cryo-EM) or flexibility of a peripheral component not fixed by crystal contacts (X-ray). It must be realized that such blurred regions in the density map are treated completely differently in cryo-EM and X-ray crystallography. In cryo-EM, they are kept at face value, as part of the density map, while in X-ray crystallography, they are suppressed in the course of the refinement procedure, so that they appear as entirely "empty" or nonexistent in the pdb coordinates.

In cryo-EM, any motion of a peripheral component, when comparing all molecules of the ensemble, will result in a vignetting effect in the 3D density map: a given voxel will have 100% protein or RNA density only if the corresponding point in the molecule is occupied with density throughout the ensemble. Any point in the molecule that is occupied only in a subset of the ensemble, on account of the flexing of the peripheral component, will have a proportionately reduced density in the reconstruction. Characteristic for such regions is the lack of a sharp delineation of the boundary, and a pronounced dependency of features on the choice of threshold for contouring. An example is provided by the L1 and L7/L12 stalks of the 70S and 80S ribosomes, which are mobile, and often appear truncated or even doubled (Malhotra et al., 1998; Gomez-Lorenzo et al., 2000; Valle et al., 2003).

As an aside in this context, the high redundancy of X-ray diffraction information is exploited to obtain local information, in terms of a temperature factor for each atom, from which conformational dynamics or disorder can be deduced. In the cryo-EM case, information on the distribution of local inconsistencies is obtained by computing the 3D variance map (see section 2.2).

(iii) Differences in the amplitude falloff. Here, the X-ray structure serves as the standard, which is initially not met by the cryo-EM map, because of a number of experimental factors including charge build-up and mechanical instabilities. The ratio curve obtained by comparison of amplitude falloff curves between X-ray and CTF-corrected cryo-EM maps of the same structure (see Penczek et al., 1999) can, therefore, be used to assess and monitor the quality of cryo-EM imaging.

(iv) A point addressed below, in the context of segmentation based on molecular volume (section 5), is that the visual appearance of boundaries, especially internal boundaries, may differ between an X-ray and a cryo-EM map since the electron scattering density of protein regions with side chains is not much higher than that of ice. Hence, depending on the choice of threshold, regions free of the main chain but packed with side chains may appear empty in cryo-EM, leading to their interpretation as channels or tunnels.

After these preliminaries, two case studies will be presented here, of large widely studied molecules known both from X-ray crystallography and cryo-EM: GroEL and the prokaryotic 70S ribosome. Another well-documented comparison is provided by Harris et al. (2001) for the decameric structure of peroxiredoxin-II. The reader should also note that many such comparisons are now available in the virus structure field; see, for instance, Baker et al. (1999) and San Martin et al. (2001).

3.3.1. Example of GroEL

GroEL, the bacterial heat shock protein, forms a "molecular bottle," a protected space in which incorrectly folded proteins can refold without interference from proteolytic agents in the cell. GroEL in complex with various ligands including ATP has been studied extensively by cryo-EM, and the X-ray structure of the ATP-bound form has been solved in Paul Sigler's group (Boisvert et al., 1996). Roseman et al. (2001) juxtaposed the cryo-EM map with the X-ray structure at varying resolutions, and found a detailed agreement in molecular features (as visually assessed) for a choice of low-pass filter for a filter radius that is below the choice Fourier shell correction (FSC) = 0.5 advocated by Böttcher et al. (1997) and Malhotra et al. (1998), but above the 3-σ level of Orlova et al. (1997). A more recent reconstruction by Ludtke et al. (2004), at 6 Å resolution, offers the most detailed comparison yet (figure 6.5).

3.3.2. Example of 70S Bacterial Ribosome

The ribosome, a large macromolecular complex composed of RNA and proteins, is responsible for protein synthesis in all organisms. The prokaryotic ribosome has been visualized by cryo-EM in many functional complexes (reviewed by Frank, 1998, 2000, 2001, 2002; Orlova, 2000; Stark, 2002). Most of these were obtained for *Escherichia coli*. The high stability and large size (diameter ~200 Å) of the complex are some of the reasons why resolutions in the range of 10 Å could be achieved (Gabashvili et al., 2000, Valle et al., 2003a) despite the absence of symmetry.

In an earlier study, Penczek et al. (1999) compared a 19-Å cryo-EM density map (and the data it was derived from) with a 9-Å X-ray density map of the 50S ribosomal subunit from *Haloarcula marismortui* using FSC. In one test reconstruction (map #1), the X-ray map was used as reference for the cryo-EM data. In another (map #2), the cryo-EM reconstruction was used as initial reference in a

(a)

(b)

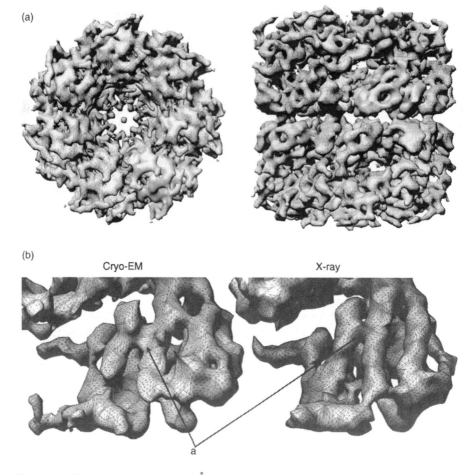

Figure 6.5 Comparison between a 6-Å density map of GroEL, derived from cryo-EM and X-ray crystallography. (a) Cryo-EM map at 6 Å resolution [Fourier shell correlation (FSC) = 0.5]. (b) Detail of the map: comparison between cryo-EM map and density map derived from pdb coordinates of X-ray map, filtered to 6 Å resolution. From Ludtke et al. (2004), reproduced with permission of Cell Press.

five-cycle refinement by 3D projection matching with a final $2°$ angular interval. Both reconstructions have virtually identical FSC curves, indicating matching resolution. However, interestingly, the FSC of map #1 versus the X-ray map shows a high degree of similarity extending to $1/12\,Å^{-1}$, while the FSC of map #2 versus the X-ray map shows much diminished similarity. Thus, the self-consistency of the cryo-EM data extends to a higher resolution than the consistency between X-ray and cryo-EM data. This result indicates that, despite a good overall resemblance between the X-ray and cryo-EM maps, the detailed phase information differs significantly.

The basis for a higher-resolution comparison was a density map of the 70S ribosome with the highest resolution achieved to date, 7.8 Å (FSC = 0.5; Spahn et al., 2005). X-ray structures are now available for the individual 30S and

50S ribosomal subunits from a number of extremophilic species. For the 70S ribosome, a single X-ray structure has been obtained (Cate et al., 1999; Yusupov et al., 2001), but the low resolution of the X-ray map (5.5 Å in the latter study) meant that its atomic interpretation had to rely on the existing published atomic structures of the subunits.

The comparison is therefore handicapped by the fact that no data exist for identical complexes: there are differences between species, states of association (single versus associated into the 70S), and exact ligand-binding states. Species differences exist in the precise sequences—not visible at the current cryo-EM resolution—and in the presence of insertions/deletions mostly in peripheral regions, which do show up. The state of association is important, since cryo-EM has revealed that the structures of the subunits change on association. Similarly, the binding of ligands such as tRNAs and elongation factors result in conformational changes of the ribosome, and there is even a difference in the behavior of the ribosome depending on whether a complete tRNA is bound or, as in several X-ray studies, only an anticodon stem loop is used.

Nevertheless, the known high degree of structural homology in the intersubunit space (where most ligand binding occurs) allows us to focus on some of the features to ascertain reproducibility at the resolution of the cryo-EM map. In view of the considerations given above, the X-ray structure of the 70S ribosome from *Thermus thermophilus* is the structure best suited for comparison with the 7.8-Å cryo-EM map from *E. coli*.

For comparison, the pdb coordinates of the X-ray structure are converted into electron densities and cut off with a Butterworth filter (see appendix 2) at the resolution of the cryo-EM map. [Following the strict theory, the conversion should take into account anomalous scattering, which is stronger for RNA than for protein (see section 3.4, chapter 2), but this effect was not considered, so that certain discrepancies can be expected in addition to those caused by the different species.] As in all low-pass filtrations, a smooth filter such as the Butterworth filter is required to avoid truncation artifacts in real space. Different resolution cut-offs were used to determine the best match (figure 6.6).

The resolution of the cryo-EM map appears to be different in different parts of the map, probably as a result of local conformational averaging. In some parts, the map appears more similar to the 10-Å X-ray map, in other parts closer to the 7.8-Å X-ray map.

3.4. Concluding Remarks

The purpose of this section was to stress that the application of image processing procedures must be accompanied by a careful validation. Such validation is all the more important as the molecule that is not constrained in a crystal lattice assumes a much larger variety of conformational states; this means that in a population of molecules on the grid, different conformations and binding states may coexist, and changes of protocol in all steps of the analysis may have an impact on the final result. Known tools of statistical hypothesis testing should be employed where possible, but the required 3D variance information is often difficult to obtain.

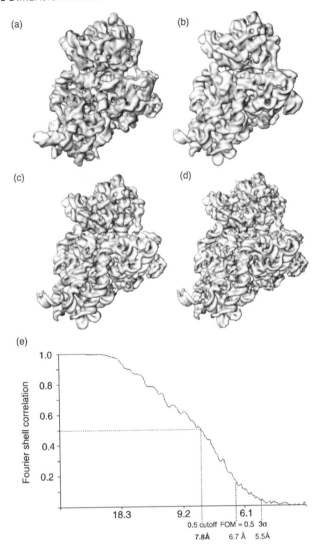

Figure 6.6 Comparison between density maps of the *E. coli* ribosome derived from cryo-EM and X-ray crystallography. (a) 30S Subunit portion of 7.8-Å cryo-EM map (Spahn et al., 2005). (b, c, d) Density maps created by limiting the resolution of the X-ray coordinates of the 30S subunit from *T. thermophilus* (Cate et al., 1999) to 10.0, 7.8, and 6.7 Å resolution, respectively. (e) FSC curve, indicating a resolution of 7.8 Å, using FSC = 0.5.

Self-consistency is a thorough test for validity. The most convincing case of self-consistency is made when a component of the macromolecular complex reconstructed separately exhibits the same structure, to the resolution to which this can be checked, as it has within the complex. Finally, the most convincing validation is provided when the structure or one of its components has been solved in an independent experiment by other methods of structure research, such as X-ray crystallography or NMR.

4. Visualization and Rendering

A presentation of the reconstruction as a series of slices (figure 6.7a) is the most comprehensive but least accessible display of the 3D information. By the use of visualization and rendering techniques, 3D models are usually built,

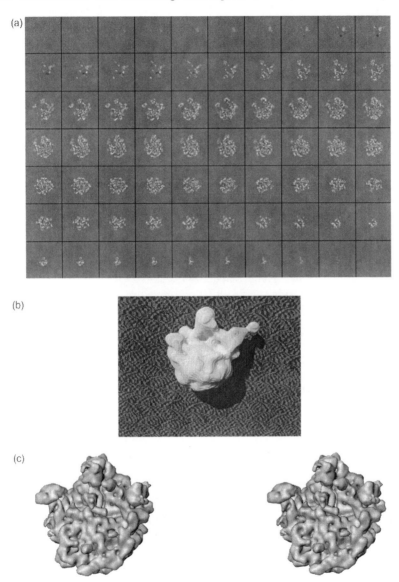

Figure 6.7 Methods for visualization of 3D density maps. (a) 10-Å cryo-EM reconstruction of the 50S ribosomal subunit from *E. coli*, represented as a series of slices. (b) Physical model of a 3D reconstruction of the 50S ribosomal subunit from negatively stained specimens, at 20 Å resolution (Radermacher et al., 1987a, b); (c) stereo-view of 50S ribosomal subunit, surface-rendered (Gabashvili et al., 2000).

enabling us to comprehend the structure as an object with depth and perspective, as objects in our physical world.

Only 20 years ago, the visualization of electron microscopic objects reconstructed by the computer still relied largely on physical model building. Computer output was primarily in the form of crude printouts and contour maps, from which slices could be cut. Meanwhile we have witnessed an explosion of graphics capabilities that have reached the personal computer. A juxtaposition of the hand-built model (figure 6.7b) with a current computer-rendered model, visualized in stereo (figure 6.7c), marks the tremendous progress in visualization. This section cannot possibly cover the many options of representation currently available in both commercial and public-domain software, but can only describe the underlying principles briefly. (For an early overview of the different techniques with potential importance in EM, see Leith, 1992.) Many visualization techniques in biomedical imaging, whose progress is featured in annual conferences (e.g., *Visualization in Biomedical Computing*; see Robb 1994), can obviously be adopted to EM. Visualization methods important in 3DEM have been the subject of special issues of 3DEM journals (see appendix 5).

4.1. Surface Rendering

The objective of surface rendering is to create the 3D appearance of a solid object by the use of visual cues such as intensity, cosine-shading, reflectance, perspective, and shadows. Radermacher and Frank (1984) developed a combination of intensity and cosine-shading first employed to obtain a surface representation of ribosomes (Verschoor et al., 1983, 1984). Similar algorithms were developed by other groups (van Heel, 1983; Vigers et al., 1986a). The intensity is used to convey a sense of distance, with closer parts of the structure made to appear brighter than more distant parts. Cosine-shading mimics the appearance of matte surfaces that are inclined relative to the viewing direction, with the highest brightness attached to surface elements whose normal coincides with the viewing direction, and zero brightness to surface elements whose normal is perpendicular to the viewing direction.

The schematic diagram in figure 6.8 explains how in general the distance of the surface from the viewer is derived from the 3D density distribution stored in the volume. First, a threshold density d_0 must be defined, to create the distinction between "inside" ($d > d_0$) and "outside" ($d < d_0$). The choice of this threshold is critical in the visualization of the structure, as different choices may bring out different aspects of it (see below). With the given threshold, the algorithm scans the volume starting from a reference plane (which can be inside or outside the volume), counting the (normal) distance between that plane and the first encounter of a voxel that is equal to, or larger than, d_0. The precise distance must then be evaluated by interpolation in the close vicinity of that boundary. The resulting distances are stored in the so-called distance buffer. On completion of the scan, the distance buffer contains a faithful representation of the topography of the molecule as seen from the direction perpendicular to the reference plane. The desired surface representation is now achieved by using the distance buffer in the computation of the two representational components,

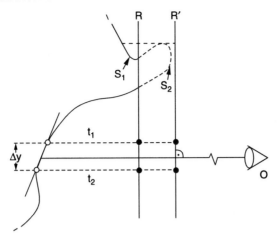

Figure 6.8 Principle of surface representation combining shading and depth cue. The representation of a surface element is based on its distance from a reference plane measured normal to this plane. R and R', reference planes chosen inside and outside the structure, respectively, S_1 and S_2, two portions of the surface. S_1 is hidden by S_2 unless the interior reference plane R is used. O denotes the position of the observer; t_1 and t_2, the distances from two adjacent surface elements separated in the vertical direction by Δy. The slope of the plane seen from the viewing point, $|t_1 - t_2|/\Delta y$, can be used to control the local shading, while the distance itself can be used to make elements close to the observer appear bright, and those far from the observer, dark. From Frank and Radermacher (1986), reproduced with permission of Springer Verlag.

intensity and shading, and mixing the two resulting signals. A 50:50 mixture of the two components is usually quite satisfactory.

Much more elaborate methods of representation incorporate one or two sources of light, a choice of surface texture, and the use of colors. In addition, ray tracing makes it possible to create effects associated with reflections of the surroundings on the surface of the structure. However, one must bear in mind that beyond conveying shape information, surface representations reflect no physical property of the molecule and that any extra effect has merely esthetic appeal. Examples of spectacular representations employing color to indicate segmentation are especially found in the virus field (e.g., Yeager et al., 1994; Böttcher et al., 1997; Che et al., 1998; Zhou et al., 2001a), but color is also indispensable in representing macromolecular assemblies of high complexity (e.g., Gabashvili et al., 2000; Stark et al., 2001; Davis et al., 2002). Particularly challenging is the need for the simultaneous representation of a density map and an X-ray structure fitted into it; in this case, one resorts to semitransparent surface renderings, often with reflections that reinforce the impression of three dimensions (e.g., Spahn et al., 2001a).

Surface representations are especially effective when shown in the form of stereo pairs. To achieve this, the representation is calculated for two viewing directions separated by 6° around the vertical axis, and the two images are mounted side by side with a distance equal to the average horizontal distance between the two eyes, which is in the region of 6.5 cm (see figure 6.7c). Stereo viewing at a workstation is achieved by presenting the two frames alternately to the left and the right eye at a suitable rate, making use of synchronized switching of polarized glasses.

4.2. Definition of Boundaries

The definition of molecular boundaries is a special case of segmentation, and some of the considerations presented in section 4.1 apply here as well. Here, the large difference in density between protein and ice (or the even larger one between RNA and ice) leaves no doubt about the density threshold range, although the precise value is difficult to ascertain. Several approaches to the determination of the threshold have been tried and discussed (see below).

4.2.1. Density Histogram

The density histogram is a discrete function that shows the number of voxels falling into a density value range, $d_i \leq d \leq d_{i+1}$, as a function of the density d_i (see figure 6.9). Molecular boundaries are characterized as regions where the density falls from the protein or RNA value to the value of water. Voxels with densities intermediate between the two values are relatively rare. In the density histogram, we therefore expect to find two peaks, one for water, the other for protein/RNA, separated by a minimum. For density maps with sufficiently high resolution, the minimum may be easily determined, and can be used for

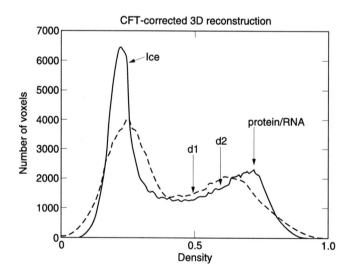

Figure 6.9 Density histogram of a 25-Å reconstruction of the ribosome from cryo-EM data. Dashed line shows the histogram before CTF correction; solid line, after correction. The densities marked as d_1 and d_2 represent the volumes 3.4×10^6 and 2.4×10^6 Å^3, respectively. From Zhu et al. (1997), reproduced with permission of Elsevier.

an objective determination of the density threshold representing the molecular boundary. Lower resolution maps present a problem since the ranking of densities is scrambled. For example, RNA bordering ice exhibits densities that range all the way from one extreme to the other, since sharp boundaries are "washed out" into a margin of a width corresponding to the resolution distance. The histogram of such maps may, therefore, not contain clues about the placement of the threshold.

4.2.2. Volume Constraint

For cryo-EM maps with resolutions of 15 Å or worse, molecular volume is a rather unreliable guide to the selection of density threshold, since the limited resolution leads both to an infilling of interior channels and crevices, and a truncation of thin components sticking out into the solvent. This becomes quite evident when comparing a structure displayed at different resolutions (see figure 6.6, where X-ray structure coordinates are used to generate density maps of the ribosome at different resolutions). The volume is a quantity with fractal properties (see section 5.3.2), and hence very sensitive to resolution.

The easiest way to obtain the density associated with a desired molecular volume is by computing a *cumulative histogram*, which expresses the number of voxels above a given density d_{th} as a function of d_{th}. It is related to the density histogram described above by integration. The threshold d_{th} can be immediately looked up as that density for which the associated contour encloses a given molecular volume.

An example for the use of volume constraint is contained in the work of de Haas and coworkers (1996c, 1997), who used it to determine the thresholds for the display of, respectively, *Riftia pachyptila* hemoglobin and *Lumbricus terrestris* extracellular hemoglobin, as obtained by single-particle reconstruction. In the first of these papers, the authors go into a comparison of the volume obtained from an examination of the density histogram with the volume estimated from molecular mass, and find a large discrepancy (a factor of 1.7 in favor of the volume estimated from the histogram). It is quite likely that at resolutions as low as 36 Å, used in the study, volume deduced from the density map simply ceases to be useful as a guide.

4.2.3. Multiple-Sigma Criterion

As in X-ray crystallography, a multiple of the standard deviation above the mean density is frequently used to determine the molecular boundary. Use of an objective standard for thresholds is particularly important in comparisons between density maps obtained in different experiments (e.g., de Haas et al., 1999; Gilbert et al., 2004).

4.3. Volume Rendering

Volume rendering is the term used for techniques that give a view of the entire volume, from a given viewing direction, to which all voxels are able to contribute.

Thus, the structure appears transparent or semitransparent, and all interior features appear overlapped. The simplest rendering of this kind is obtained by parallel projection (summation along rays in the viewing direction). However, more complicated schemes have been devised where interior densities are reassigned according to a classification scheme and gradient information is allowed to contribute.

The value of volume rendering is still controversial in many applications of image processing. The problem is that such rendering corresponds to no known experience in the physical world around us. Perhaps the closest experience is watching a jellyfish from a close distance and seeing its organs overlap as the animal moves. However, this example shows at the same time the value of watching a transparent structure in motion in order to appreciate the spatial relationships of its components. Stereo representation also helps in conveying these relationships, but experience shows that a presentation as a "movie" is by far the most powerful tool.

Somewhere in between surface and volume rendering falls the use of semi-transparent surfaces in conjunction with solidly rendered interior density objects or atomic models. Surfaces represented as a contour mesh have an advantage over other semitransparent surface renderings in that they present sharply defined clues to the eyes in stereo presentations (see figure 6.7 in section 6.4). Seemingly more realistic and esthetic effects including reflections are, however, obtained by ray tracing (see for instance, Spahn et al., 2001a).

5. Segmentation of Volumes

Segmentation is the general term we will use for the separation and identification of known components in the reconstruction volume. Since the most straightforward visualization of a structure already involves a decision about the placement of the boundary against the background (i.e., the solvent, support, or embedding medium), segmentation and visualization (section 4) will always go hand in hand.

The concept of segmentation can be illustrated with the well-known charting of beef for culinary purposes (figure 6.1): different regions are demarcated by boundaries drawn on the surface of the animal, and corresponding labels are attached. There is an important difference, though: the regions we have in mind are 3D, and they may be located anywhere within a volume, not just at the surface. The boundaries defining them are not closed lines (with an implied depth) but closed surfaces.

At atomic resolution ($\sim 3 \text{Å}$ and better), molecular motifs are distinct and recognizable, forming the basis of a "natural segmentation." Below atomic resolution, 3D density distributions show internal divisions to different degrees, making meaningful segmentations a challenge. Down to resolutions of 1/7 to $1/10 \text{Å}^{-1}$, the outlines of helical regions can still be recognized, but below that, the connection between secondary structure and the shape of domains of proteins is lost. Still, domain separations are still discernable by density troughs at the domain boundaries. RNA double-helical regions are exceptional in that they are recognizable down to a resolution of $\sim 15\,\text{Å}$ (Müller and Brimacombe, 1997; Malhotra et al., 1998).

Segmentation can address two different goals: one is the separation of 3D regions on the basis of the density alone, the other is the identification of these regions with known components of the structure. It must be noted that the second goal can be achieved without the first, since surfaces can be labeled with antibodies or other chemical tags. Conversely, however, separation of regions on the basis of density is of limited use without a way of identifying these regions in terms of the underlying structure. The importance of density-based segmentation is evident from the fact that some molecular fitting methods require separated densities (e.g., SITUS; Wriggers et al., 1999, 2000).

5.1. Manual (Interactive) Segmentation

Expert knowledge is the easiest and most efficient guide in segmentation. The problem is that few tools are available that allow an interaction with a 3D volume. Perhaps best known are packages used in medicine allowing the surgeon to operate by guiding a "virtual scalpel" in a 3D representation of the patient's brain. A similar tool, "Tinkerbell" (Li et al., 1997), has been devised for use in the context of 3DEM, specifically, electron tomography. Here, the idea is that two volumes are visualized on the workstation in stereo; one (volume I) is the volume containing the density map to be analyzed, the other (volume II) is a "target volume." Volume rendering (see section 4.3) is used throughout. A small manually controlled 3D cursor (which can have the shape of a sphere) can be moved interactively within volume I. Material that the cursor overlaps with can be copied, at any given time, and pasted into volume II, where a substructure of the density accumulates. Thus, the program makes it possible to isolate or demarcate any region, a prerequisite of segmentation. One of the options of the program allows the 3D cursor to leave a colored trace at the selected sites in volume I, or to erase the material at the moment it is copied. In this simultaneous copy/erase and paste mode, material appears to be shuffled from one volume to the other.

5.2. Segmentation Based on Density Alone

Segmentation can be achieved by the detection of troughs of density separating the usually compact domains, or by taking advantage of differences in the electron scattering density of components (e.g., RNA versus protein), or differences in texture. The first kind of separation is achieved by some version of *density thresholding* (section 5.2.1); the other kind can be achieved by employing either of two image-processing techniques that have been recently introduced into the field of EM: the *watershed transform* and *level set* related techniques (Sethian, 1996). The third property of the density map that can be exploited for segmentation is texture. Although its applications may be mainly confined to low-resolution density maps obtained by electron tomography, a texture-sensitive method, employing an eigenvalue analysis of the so-called affinity matrix, is described below as well.

Low-resolution density maps are by nature fuzzy—exact definitions of boundaries are lacking, and it is only "after the fact," when an atomic structure becomes available, that the actual physical boundaries can be ascertained. Some newer approaches to segmentation, so far only tried in computerized axial tomography-scan or magnetic resonance imaging, start from a representation of a fuzzy 3D image, and the extent to which its elements hang together are expressed by a *fuzzy connectedness* (see Udupa and Samarasekera, 1996). It may be worth trying these newer methods in 3DEM, as well.

5.2.1. Density Thresholding

Domains of proteins are often compact, but divided from other domains by a trough of density. Depending on resolution, this dip in density often helps in the delineation of the domains, by the application of a density threshold that is above the density of the trough.

In principle, molecules consisting of materials with different electron-scattering densities can be divided into their components by application of an appropriate density threshold. A typical example is RNA and protein, a mixture in which RNA stands out on account of its higher scattering density due to the high content of phosphorus atoms. For example, the ribosome embedded in ice is a three-component system, in which three kinds of density transitions are realized: those at the boundaries ice–protein, ice–RNA, and protein–RNA (Kühlbrandt, 1982; Frank et al., 1991). Low-resolution density maps pose a problem, since the sharp density steps are washed out, making the determination of the precise boundary uncertain (Frank et al., 1991). The same situation arises in all visualizations of RNA by cryo-EM, as in spliceosomal components (Stark et al., 2001) and RNA polymerase II (Craighead et al., 2002).

This problem can be discussed in terms of the density histogram (figure 6.9), which is a display of the number of voxels in which a certain density is realized, versus the correponding density value. For a high-resolution density map, the density histogram of the ribosome in the above example would consist of three well-separated peaks: one for water (very high because of the large volume within the box around the particle), one for protein, and one for RNA (Zhu et al., 1997).[1] However, low resolution will cause each of the peaks to spread and overlap with its neighbor. It is easy to see in the example shown that there are only few small density ranges where the identity of the density mass can be determined with certainty. For example, densities above the density identified as d_{RNA} can only originate from RNA, meaning that the mass obtained by eliminating all density below that density value (i.e., by application of a density threshold) is solely RNA. However, this process has not only eliminated protein and water, but a large part of the RNA, as well. Similarly, there is no density range allowing pure protein to be isolated, since there exists a low-density tail of RNA, mostly originating from boundaries where RNA directly borders water and causes all

---

[1]Note that in Zhu et al. (1997), the legends of two key figures showing histograms were erroneously switched: the legend of figure 11 belongs to figure 12 and vice versa.

density values intermediate between water and RNA to be seen in the low-resolution map.

5.2.2. Density Thresholding Combined with Region Growing

The limitation of the simple thresholding method can be overcome by use of knowledge from other experiments. Spahn et al. (2000) introduced a method of RNA/protein separation based on thresholding, availability of generic histogram information, and region growing.

While a cryo-EM density map, at resolutions below $1/10\,\text{Å}^{-1}$ (i.e. worse than $10\,\text{Å}$), of a structure composed of RNA and protein has a broad density histogram without trace of a separation of the underlying components, the histograms of regions solely containing RNA or protein (*signature histograms*) look distinctly different. Even though there is substantial overlap, there exists a lower-density range that must definitely originate from protein, and a higher-density range that must originate from RNA. Thus, knowledge of those signature histograms makes it possible to identify, by thresholding, those "core regions" in the map that are protein or RNA with absolute certainty. Starting from the RNA core, Spahn et al. (2000) took advantage of the fact that, in each ribosomal subunit, RNA forms a single contiguous structure: 16S within the small subunit; 23S and—tightly associated with it—5S RNA. They used an image-processing method called *region growing*. First, the voxel with highest density is used as initial seed. The density threshold is then stepwise reduced. In each step, only those voxels are added in the growth process that are (i) above the threshold and (ii) directly connected with the already existing group of voxels. This process is stopped when the known volume of the RNA is reached.

As we have seen, the method uses signature histograms, derived from an analysis of any two regions of the density map exemplary for the two components. In the case shown, the knowledge of where to look for densities stemming from RNA and protein came from a published 5.5-Å X-ray map in which the RNA and some of the published proteins were delineated in the small ribosomal subunit.

5.2.3. Watershed Transform

This approach to segmentation is most easily understood in the application to 2D images: the image is considered as a landscape whose inverse height profile represents the image intensities. Mountains represent low intensities, valleys high intensities. Initially, consider the entire landscape dry. Water is then allowed to seep in from the bottom. By gradually raising the water level, one finds that the water body is initially divided into separate water basins. These merge along certain lines, called watersheds, once a critical level is reached. Now in the metaphoric description, the contents of the water basins stand for separate objects. They are allowed to expand until they touch along the watershed line but they are not allowed to merge. The same process, not as easily conceptualized, can be applied to 3D density maps as well. Here, of course, a 4D landscape needs to be invoked to explain the emergence of 3D density blobs—the water seeping in—that expand and merge into one another as the total water volume expands. Volkmann (2002) shows some applications of this technique widely known in the general image-processing field.

5.2.4. Eigenvalue Methods Using the Affinity Matrix

Frangakis and Hegerl (2002a,b) introduced a very effective segmentation tool into 3DEM. It originated with an approach originally formulated by Shi and Malik (1997). For simplicity, the approach will first be described for 2D images, followed by a generalization to 3D data.

Segmentation of an image may be considered as a division of the pixels into subsets, such that pixels within subsets are "closer to one another" by a suitable measure than pixels belonging to different subsets. This relationship of "closeness" is expressed by the term "affinity," which is meant to capture both spatial proximity and similarity in density values. That is, *two pixels have high affinity if they are lying close to each other and if their densities are closely matched.*

The concept of affinity can be easily understood by picturing the image as a landscape, whose height profile represents the densities. "Affinity" would here correspond to the 3D distance between two points on the surface of this landscape. This distance depends on two components: one is the distance in the x–y plane, the other the vertical distance, which combine according to the Pythagorean rule.

To obtain a quantitative expression of affinity, we introduce weights of the following form:

$$w_{ik} = \exp[-|\mathbf{r}_i - \mathbf{r}_k|^2/2\sigma_d^2]\exp[-|p_i - p_k|^2/2\sigma_p] \tag{6.10}$$

where $\mathbf{r}_i, \mathbf{r}_k$ $(i, k = 1 \ldots, N)$ are the lexicographically ordered coordinates of the pixels and p_i, p_k are their density values. The parameters σ_d and σ_p define the "range of influence" within which pixels are seriously considered in the affinity calculation. Thus, the pairwise affinities of all pixels in the image are expressed by the affinity matrix $\mathbf{W}$ with elements $\{w_{ik}\}$.

Now the task of finding two subsets A and B with least affinity is equivalent to minimizing the following expression:

$$S = \sum_{i \in A, k \in B} w_{ik} \tag{6.11}$$

It can be shown that this minimization is equivalent to solving the eigenvector equation:

$$(\mathbf{D} - \mathbf{W})\mathbf{y} = \lambda \mathbf{D}\mathbf{y} \tag{6.12}$$

where $\mathbf{D}$ is a diagonal $N \times N$ matrix formed by the elements $d = \sum_{k=1}^{N} w_{ik}$. The first nontrivial solution corresponds to the second eigenvalue. The associated eigenvector ("segmentation eigenvector") $\mathbf{y}$ has the meaning of an index that indicates the partition: it has high value for those pixels that belong to set *A*, and low value for those of set *B*. Appropriate thresholding separates the two domains. To obtain more than two sets, one repeats the process by applying eigenvector analysis to set *B* obtained in the first analysis, splitting off a subset *C*, etc.

For the implementation, Frangakis and Hegerl (2002a,b) noted that the large computational problem of diagonalizing an N^2 matrix (for a volume of 256^3 voxels, this matrix would have $2^{24} \approx 10^8$ elements) can be reduced to a manageable scale in two ways: (i) by making use of the fact that the relations between pixels are effective over a small distance $R \ll N$ only; (ii) by exploiting the fact that in each step, only a single eigenvector—namely the second one—is required, which makes it possible to use the fast Lanczos algorithm (Sorensen, 1992).

An application of this segmentation to the slice of a tomographic volume (figure 6.10) demonstrates the power of the method: in a stepwise, hierarchical fashion, the image is first divided into "interior" and "exterior" of the cell, then these regions are further divided, leading to a precise delineation of the exterior structure attached to the cell. While application of the algorithm to the cell's interior produces only an uninteresting gradation of density, its further application to subregions (as defined by thresholding of the segmentation eigenvector) would lead to the segmentation of the vesicles, as well. Other applications, especially useful in the reconstruction of macromolecular assemblies, are in the automated separation of ice and carbon portions of a micrograph, or in automated particle picking (see Frangakis and Hegerl, 2002a,b; also section 2 in chapter 3).

5.3. Knowledge-Based Segmentation, and Identification of Regions

The following is a list of possible sources of information for identification of regions, roughly in the order of increasing certainty: (i) recognition of morphological features, (ii) known volume of a component with higher density, (iii) 3D antibody labeling, (iv) cysteine and oligonucleotide labeling, (v) difference mapping, (vi) use of element-specific signals in the imaging (e.g., electron energy-loss spectroscopy, EELS), and (vii) finally, beyond a certain resolution, it becomes possible to identify secondary structural motifs.

5.3.1. Recognition of Morphological Features

Expert recognition of morphological features is by its nature subjective, but it may provide a basic orientation. Often this recognition involves an inference from a labeling study in two dimensions, and in this case it is possible to fortify the evidence by projecting the structure in the orientation that matches the views in which the labels were identified. Examples for this type of inference are found in the earlier identification of morphological features such as "L1 stalk" and "L7/L12 stalk" in the reconstruction of the 50S ribosomal subunit (Radermacher et al., 1987b). Morphological features with obvious functional implications are the channels immediately recognized in the 3D image of the

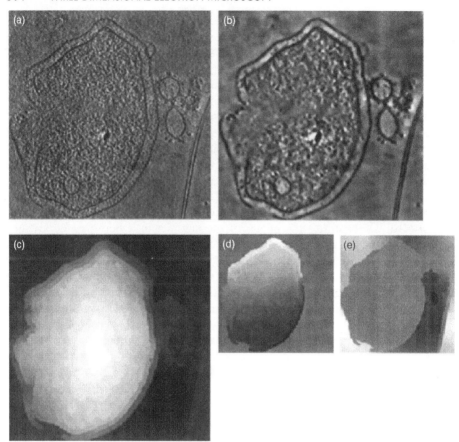

Figure 6.10 Method of hierarchical segmentation by the eigenvector method, here applied to the slice of an electron tomogram of a cell. The reconstruction contains diverse elements inside and outside of the cell. (a) Original reconstruction; (b) denoised version of (a); (c) first segmentation eigenvector obtained from (b) separates the cell from its surroundings; (d) second segmentation eigenvector, calculated from the interior of the cell, as defined by thresholding of the first segmentation eigenvector (c). This eigenvector merely shows a graduation of density in the cell's interior; (e) third segmentation eigenvector, calculated from the exterior of the cell, as defined by thresholding the first segmentation eigenvector (c). This eigenvector differentiates between the exterior structures attached to the cell and the surrounding matrix of ice. From Frangakis and Hegerl (2002b), reproduced with permission of Elsevier.

calcium release channel (Wagenknecht et al., 1989a; Radermacher et al., 1994b), the mRNA channel in the ribosome (Frank et al., 1995a), and the compartments seen in GroEL (Roseman et al., 2001).

5.3.2. Known Volume of Component with Higher Density

When a component possesses higher scattering density than the rest of the structure, due to its atomic composition, segmentation can be based on its known partial volume. Frank et al. (1991) and Stark et al. (1995) estimated the region occupied by ribosomal RNA within the 70S ribosome in this way. Obviously, the validity of this kind of segmentation relies on the accuracy of volume information and the assumption that the density ranking among the voxels in the 3D map reflects true ranking in the object. The validity of the second assumption must be examined separately in each case because the contrast transfer function produces some scrambling of the density ranking, due to its suppression of low spatial frequencies.

Volume has fractal properties and thus depends strongly on the resolution of the density map. At high resolution, it would be best defined by the volume enclosed in the van der Waals' surface. At lower resolution, tunnels and crevices fill in, while thin protuberances sticking out into the solvent disappear. The first effect increases the volume, the second decreases it, and the net effect on the volume is determined by the number and geometry of one kind of feature compared to the other. A brief discussion of the vagaries arising when going by the molecular volume is found in Frank (1997).

Interestingly, internal boundaries may be considerably different when seen by cryo-EM compared to the X-ray structure (P. Nissen, personal communication). Extended regions filled with side chains but bare of the main chain have an average scattering density not much higher than water, so that application of the usual threshold may result in the appearance of bubbles or channels where none are seen in the atomic structure. This is the probable reason for the discrepancy between the appearance of a second polypeptide exit tunnel in cryo-EM (Frank et al., 1995a,b; Gabashvili et al., 2001a) at a place where none is found in the X-ray structure (Moore and Steitz, 2003). Only when atomic resolution is achieved with cryo-EM will these particular discrepancies disappear.

5.3.3. Antibody Labeling

This method (also known as 3D immuno-EM) involves a separate experiment for each protein or epitope to be identified, each experiment yielding a 3D map. Comparison with the unlabeled control structure then allows the labeling site to the identified with a precision that is normally higher than the resolution (see Boisset et al., 1995). Studies of this kind are well known in the virus field (e.g., Smith et al., 1993; Che et al., 1998). For molecules with lower symmetry considered in this book, some of the first 3D labeling studies were done by Srivastava et al. (1992b) and Boisset and coworkers (1992b, 1993b) on negatively stained specimens (3D mapping of protein L18 on the 5S ribosomal subunit and of the four copies of protein Aa6 on the surface of *A. australis* hemocyanin, respectively.) The first such studies on ice-embedded specimens were done by Boisset et al. (1995), again on the *A. australis* hemocyanin–Fab complex.

Antibody labeling can only point to the location of a single accessible site on the molecule's surface. Double labeling is feasible by attaching heavy metal clusters with different scattering properties of two antibodies at different sites and employing energy filtering for discrimination (Koeck et al., 1996); see next section. Multiple labeling experiments with different monoclonal antibodies would be necessary to cover the exposed area of a protein and thereby determine its surface boundaries. Even then, one must use additional information (known volume, approximate shape) in order to infer the boundaries of the molecule in three dimensions.

Before the X-ray structure of the ribosome was solved, 2D immuno-EM of negatively stained ribosomes (Stöffler-Meilicke and Stöffler, 1988; Oakes et al., 1990) provided information on the surface location of most ribosomal proteins, and thereby a roadmap for the interpretation of biochemical data. Later, 3D cryo-EM maps could be interpreted in terms of the locations of proteins such as L1 and L11 by comparison with the 2D immuno-EM projection maps in various orientations.

5.3.4. Use of Gold Clusters

Specific sites of domains can be tagged, for positive identification and localization of the domains. Such tags comprise a site-specific part, a linker, and a heavy metal cluster. In proteins, cysteine can be used to attach such strongly scattering labels. Normally, a mutant form of the protein is obtained that has all naturally occurring cysteines replaced by inert amino acid residues, and new cysteine residues are introduced at the specific sites to be tagged. Gold and silver tags, in combination with energy loss spectroscopy, can be used to differentiate between the antibodies of a double-labeled protein (Koeck et al., 1996). In RNA, oligonucleotides provide the specific part of a tag, to which a gold cluster is covalently linked. Naturally occurring ligands too small to be visualized directly can also be linked to a heavy metal cluster and thereby rendered visible.

A popular marker is monomaleimido-$Au_{1.4nm}$, which forms a 14-Å cluster containing 55 gold atoms (*Nanogold*; Hainfeld and Furuya, 1992; see section 2.7 in chapter 2). An example for its use is provided in the localization, by Wagenknecht et al. (1994), of the sites of calmodulin binding on the calcium release channel (see also the commentary by Franzini-Armstrong, 1994). Although this localization was 2D only, additional experiments with a biotin–streptavidin–$Au_{1.4nm}$ compound narrowed down the sites to the cytoplasmic face of the receptor. Other examples are the localization of thiol–ester bonds in α_2-macroglobulin (Boisset et al., 1992a), the localization of the elusive, highly flexible protein L7/L12 within the 70S ribosome (Montesano-Roditis et al., 2001) and the localization of the PsbH subunit within Photosystem II using a His-tag and Ni^{2+}–NTA gold cluster (Brüchel et al., 2001). In 2D electron crystallography, the reader is referred to the work by Wenzel and Baumeister (1995).

5.3.5. Difference Mapping

Difference mapping, involving the removal of a component in a separate experiment (or partial reconstitution of the complex), produces accurate 3D segmentation information, provided that the component does not have a role in stabilizing the entire macromolecular complex. The literature is full of examples for applications of difference mapping in establishing the location of a structural component.

Early examples are provided in the long-term project of exploring the molecular motor of bacteria by single-particle reconstruction of flagellar basal bodies using mutants that lack certain proteins (Stallmeyer et al., 1989a,b; Sosinski et al., 1992; Francis et al., 1994; for newer literature, see Thomas et al., 2001).

Cryo-EM of mutant ribosomes depleted of the protein L11 of the large subunit (a protein that plays a crucial role in the binding of the elongation factors) was used in the positive localization of L11 (Agrawal et al., 2001) by subtraction of two cryo-EM maps, one with, the other without the protein. Another ribosomal protein, S1, was located and identified by subtracting an X-ray map of the *T. thermophilus* 30S subunit (which had been treated to remove all S1) from the 30S portion of the cryo-EM map of the 70S ribosome from *E. coli* (Sengupta et al., 2001). Similarly, RACK1 was located and identified on the 40S ribosomal subunit of yeast by difference mapping (Sengupta et al., 2004).

Molecular processes involved in transcription have also been explored by single-particle reconstruction and difference mapping (see tabulation in appendix 3). Näär et al. (2002) mapped the binding of the C-terminal domain of RNA polymerase II on the CRSP coactivator. In addition, the role of difference mapping in elucidation of the structure of transcription preinitiation complexes was discussed by Nogales (2000).

5.3.6. Use of Element-Specific Signals

Element mapping has not been realized except in a few applications by two groups (Harauz and Ottensmeyer, 1984a,b; and see references listed in the tabulation of methods, appendix 3), because it requires specialized electron spectroscopic devices that are not standard in today's instruments. It should be possible to make out the location of RNA components on account of their high amount of phosphorus. One serious limitation of the method is that only a very small fraction of the dose (less than 1%) is utilized, so that the build-up of a statistically significant image requires a total specimen dose that exceeds the low dose (as defined in section 4.1, chapter 2) by a large factor. In contrast, spectroscopic 3D imaging is becoming increasingly important in materials science applications where electron dose is not a major concern (e.g., Midgley and Weyland, 2003; Möbus et al., 2003).

5.3.7. Recognition (Template Matching) of Structural Components

When the structure of a component is known, 3D template matching can be employed to determine its position within a density map. This technique works in strict analogy with the correlation-based 2D alignment techniques introduced earlier (chapter 3, section 3), except that the number of parameters to be determined has increased from three to six (three components of a translational vector and three Eulerian angles). Again, the analysis of the Fourier relationships underlying cross-correlation helps in understanding the limitation of the approach. The value of the correlation peak is determined by the integral over $F_1(\mathbf{k})F_2^*(\mathbf{k})$ over the Fourier domain within which the two transforms are highly correlated, and the larger the volume of this domain (which goes with the third power of the radius in Fourier space), the higher the chance that the correlation peak stands out from the noisy background, and the narrower and better localized is the peak.

With this context it is clear that the ability to find a structural motif within a larger density map is always limited by the lower resolution of the two. The typical situation is that the motif is given by a set of coordinates from the Protein Data Bank, while the (larger) density map originates from cryo-EM. There exists a hierarchy of interpretable structural features as a function of resolution, which roughly runs from the overall shape of a molecule (20–25 Å), over shapes of individual domains or subunits, and RNA double helixes with identifiable major and minor grooves (15 Å), alpha helixes (7–8 Å), individual strands in beta sheets (4.5 Å), to recognizable and traceable chains (3.5–4 Å).

Following this hierarchy, a resolution of 20 Å might be sufficient to determine the location of an entire protein within a molecular assembly, but not to delineate its domains. A resolution of 15 Å might be sufficient to locate domains, but not any elements of secondary structure such as alpha helixes, etc. Different types of fully automated motif-searching programs are geared toward these different levels of the structural hierarchy. To be effective, programs for general 3D motif search by cross-correlation must use local normalization; otherwise, the fluctuations of the structure surrounding the motif will create contributions to the CCF that often dwarf the correct peak due to exact structural match (Volkmann and Hanein, 1999; Roseman, 2000; Rossmann, 2000). "Local" here refers to a region that is defined by the shape and position of a 3D mask. Local normalization entails the computation of the variance of the density map in each position of the mask within the volume.

Two avenues have been taken, appropriate at different resolutions. In one, a subunit, or fragment of a protein, or an entire protein is searched within a larger context provided by the density map of a macromolecular assembly (Volkmann and Hanein, 1999; Roseman, 2000; Rossmann, 2000; Rath et al., 2003; fold hunter: Jiang et al., 2001). In the other, a secondary-structure element such as an alpha-helix or beta-sheet is searched within the density map in efforts to compose an *ab initio* structural interpretation (helix hunter: Jiang et al., 2001; Zhou et al., 2001a; Chiu et al., 2002; sheet tracker: Kong and Ma, 2003; Kong et al., 2004). The second route is more limiting since it requires a resolution beyond 10 Å. (More details on both approaches are found in section 6.)

6. Methods for Docking and Fitting

It has been pointed out in an earlier section that CTF-corrected cryo-EM maps represent, in good approximation, the Coulomb potential distribution of the molecule, which can be quantitatively compared with coordinates or electron density maps from X-ray crystallography. Many applications of cryo-EM deal with complexes formed by macromolecules of known structure with some ligand whose atomic structure is also known. Examples are GroEL (Saibil, 2000c) and complexes formed by a virus and antibodies against one of its coat proteins (Smith et al., 1996), by a virus and adhesion molecules (Rossmann, 2000), actin and myosin (Rayment et al., 1993), and between the ribosome and its functional ligands (Agrawal et al., 1998; Valle et al., 2003).

In each of those cases, the determination of the precise atomic contacts is of eminent value in the understanding of functionally relevant macromolecular interactions. In addition, by using multiple docking of protein and RNA structures into a cryo-EM density map, combined with homology modeling, "quasi-atomic" models of very large structures such as the ribosome of yeast can be built (Spahn et al., 2001). Methods for computer-aided docking and fitting of atomic structures into low-resolution density maps are, therefore, an important part of the interpretative tools. A brief overview over some of the existing algorithms and programs is contained in Roseman (2000).

Baker and Johnson (1996) were among the first to recognize and herald the advent of a "resolution continuum from cells to atoms," which has been made possible by the combination of cryo-EM results with those from NMR and X-ray crystallography. The way in which data obtained by cryo-EM and X-ray crystallography complement one another has been discussed by Mancini and Fuller (2000). Integration of data from both fields as well as fluorescence spectroscopy and other techniques of structural probing ("hybrid methods of structural analysis") is increasingly seen as necessary for approaching an understanding of the workings of complex molecular machines. In this integration, the ability of fitting one modality of data to the other is of critical importance (see also Editorial, 2002).

6.1. Manual Fitting

Initially, many interpretations of density maps from EM are done subjectively, by manual fitting of X-ray coordinates into the density contours, using molecular fitting programs such as "O" (Jones et al., 1991). Such programs are able to present atomic models imported as pdb coordinates in the form of 3D "ball-and-stick" or "Ribbons" (Carson, 1997) cartoons, and at the same time present density maps as a mesh of contours. Stereo presentation and full dial control over relative movements between X-ray structures and cryo-EM maps facilitate the interactive fitting.

The extra density mass due to the ligand binding can be found by subtracting a control map from the cryo-EM map of the complex, or by using a segmentation tool. The objective of the manual fitting is to place as much of the atomic structure as possible fully into the ligand density, avoiding both *internal conflicts*, arising from overlaps with the atomic structure of the control, as well as *external conflicts*, arising from violations of the outer contour of the map where portions of the ligand (often side loops) structure stick out into the solvent. In balancing these requirements, external conflicts are normally given a lower weight, for two reasons: the resolution limitation leads to the blurring of small structural features, and peripheral loops are more likely to be flexible, so their

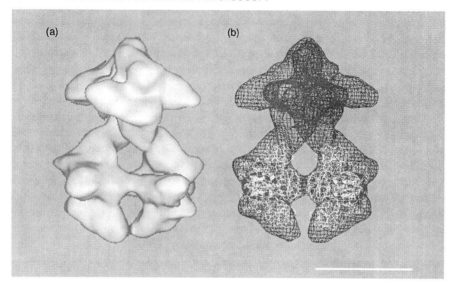

Figure 6.11 Fitting of cryo-EM map of the phosphofructokinase from *Saccharomyces cerevisiae* with the X-ray coordinates from *E. coli*. (a) Cryo-EM map of the tetrameric structure at 10.8 Å resolution; (b) cryo-EM map fitted with the X-ray structure of phosphofructokinase from *E. coli* (only bottom of the tetramer was fitted for clarity). The fitting was done manually with the molecular graphics program O. Lateral displacement by a single pixel (3.6 Å) led to clearly visible mismatches. Scale bar, 100A. Adapted from Ruiz et al. (2003).

contribution to the reconstructed density could be small. An example for manual fitting that led to a very convincing solution is the interpretation of the tetrameric phosphofructokinase molecule, shown in figure 6.11.

It is in the nature of such a subjective procedure as manual fitting that the result varies from one person to the next. Methods for computerized, automated fitting can overcome this limitation, but the answer still depends on the type of approach taken and the algorithm used. Manual fitting is often combined with a cross-correlation scoring of alternative placements (e.g., Agrawal et al., 1998; Gilbert et al., 1999).

A serious problem is posed by the fact that the binding of the ligand molecule may introduce conformational changes in either target or ligand structure, or both. Irresolvable misfits and steric incompatibility at the docking site are indications for such changes. In these situations, a procedure of quantitative flexible fitting must be applied, in which the target structure (or, as applicable, the ligand structure) is "molded" into the observed density.

Because of these complications, manual fitting is appropriate as the first step of the interpretation only. However, most quantitative fitting techniques to be discussed below are based on time-consuming multidimensional (three translations and three angles) searches, and run into the risk of finding false minima if the searching grid is too coarse. For this reason, manual, knowledge-based fitting has an important role in defining the starting position of a search that can be restricted to a much smaller local multidimensional range.

6.2. Quantitative Fitting

6.2.1. Preliminaries: Computation of Densities from X-Ray Coordinates

Quantitative fitting procedures are based on a voxel-wise comparison of densities between the cryo-EM map and the X-ray map. X-ray structures are available as sets of coordinates found in a public databank. The comparison can be done in two ways: (i) the X-ray structure is converted into a density map, which is then rotated and translated with respect to the cryo-EM density map (Stewart et al., 1993) or (ii) the X-ray structure is rotated and translated simply as a rigid body by transforming the coordinates of its atoms (Rossmann, 2000). In the latter case, the computation of the cross-correlation or other measures of similarity is done by a summation over the contributions from each of the atoms. Since atoms will generally fall into positions that are not on the 3D grid on which the cryo-EM density map is represented, an interpolation must be used to estimate the cryo-EM density at most points.

What is important here is that we wish to determine the density distribution due to electron, not X-ray scattering, since it is the former that is seen in the cryo-EM map. Stewart et al. (1993) used the following simple method: (i) impose a 3D grid onto the structure; (ii) sum the Z-values (atomic numbers) of all atoms falling within each voxel. Conveniently, the 3D grid should be chosen to match the 3D grid of the cryo-EM density, which is usually $\sim 1/3$ to $1/4$ of the distance corresponding to the resolution (following the general rule of thumb regarding sampling, in chapter 3, section 2.2). The initially high-resolution density distribution can then be low-pass filtered to the required search resolution.

A note of caution should be added here, related to the distinction between X-ray and electron scattering. One of the approximations made in the approach by Stewart et al. (1993) is that the density computed from the atomic numbers represents the phase contrast portion only, and that the amplitude contrast is ignored. Discrepancies between the simulated and actual density increase with atomic number, and, for example, lead to mismatches in the modeling of densities of structures containing both proteins and RNA. Another approximation concerns the atomic form factor: atoms have radial profiles characteristic for each species, but this distinction is entirely ignored in the simple density modeling. (In this approach, atoms are tacitly modeled as Gaussian functions, if a Gaussian filter is used in the computation of the density map at the resolution of the cryo-EM map.) The *atomic form factor*, a term known from X-ray crystallography, describes the X-ray scattering falloff for a single atom in reciprocal space. Similarly, atomic form factors for electron scattering have been calculated. Tables are available for two widely used potential distributions: the Hartree-Fock model (Haase, 1970a) and the Thomas–Fermi–Dirac model (Haase, 1970b). For more details on this issue, see chapter 2, section 3.5.

A problem that occurs in each fitting application is a mismatch in scale. Electron microscopy magnification is only known with $\sim 5\%$ accuracy, unless a calibration is used routinely. The correlation function, being based on a summation over the product of widely separated matching pairs of numbers, is extremely sensitive to variations of scale. It is, therefore, necessary to include a scale search in the procedure. This problem was first recognized in the comparisons of virus structures independently visualized by X-ray crystallography and cryo-EM (Aldroubi et al., 1992; Dryden et al., 1993). The appropriate mathematical tool for scale matching, the *Mellin transform*, is discussed in the article by Aldroubi et al. (1992). Rossmann (2000) gives an example for the sensitivity of the scale search, in his case performed as an exhaustive search in increments of 1%.

We will discuss several computerized approaches to docking and fitting in the following numbered sections: (6.2.2) "classical" correlation search looking for a single maximum; (6.2.3) correlation search defining a solution set, combined with bioinformatics constraints; (6.2.4) the method of fitting using vector quantization; (6.2.5) multidomain fitting and docking using real-space refinement; (6.2.6) fitting guided by Normal Mode Analysis, and (6.2.7) hunting for secondary structure motifs.

It should be noted that, in principle, all search algorithms allow a choice between an exhaustive grid search (which underlies all our formulations thus far) and a Monte Carlo search. In the latter method, known from applications where multidimensional conformational space is sampled (Allen and Tildesley, 1987), sample points are generated randomly in the space of Eulerian angles and 3D shifts. Some experience with the Monte Carlo search approach was reported by Wu et al. (2003).

6.2.2. Cross-Correlation Search

A global correlation search, to find the exact docking position, can be realized in a straightforward way by the use of a CCF that incorporates both translational and rotational search. The density representing the small ligand structure is placed in the origin of a 3D array that has the same size as the array in which the cryo-EM map is housed. For different rotations of the ligand structure, the translational CCF (chapter 3, section 3.3) has to be repeatedly computed via the fast Fourier transform algorithm.

Experience has shown that for the success of such a search, the method of normalization is of critical importance. Global normalization, using the global variance of the larger structure as in the definition of the cross-correlation coefficient (chapter 3, section 3.3), will often fail since a correlation peak arising from the correct match may be dwarfed by peaks caused by the occurrence of local density maxima in the cryo-EM map of the complex. The solution to this problem is provided by a locally variable normalization [Volkmann and Hanein, 1999; Rossmann, 2000 (incorporated in the program EMFIT); Roseman, 2000]. In Roseman's notation:

$$C_L(x) = \frac{1}{P}\sum_{i=1}^{N} \frac{(S_i - \bar{S})M_i(T_{i+x} - \bar{T})}{\sigma_S \sigma_{MT}(x)} \tag{6.13}$$

where S_i and T_i are the samples of the X-ray search object and the target map, respectively, P is the number of points in the search object, and M_i is a mask function that defines the boundary of the search object. The product $M_i T_{i+x}$

defines the 3D region of the target map as the search object assumes all possible positions symbolically indicated by x. Finally, $\bar{S}$, $\bar{T}$ are the respective means, and σ_S, $\sigma_{MT}(x)$ are the local variances of the two densities within the exact region of intersection in the current relative position.

Chacón and Wriggers (2002) introduced a modified cross-correlation search to deal with the problem of low accuracy encountered with low-resolution (20–30 Å) density maps. The modification lies in the application of a Laplacian filter to the target structure, with the purpose of enhancing the contributions from the molecular boundaries. A modification of a different kind was proposed by Wu et al. (2003) in the *core-weighted fitting method*. These authors introduce the concept of the *core of a structure* within a low-resolution density map, as a region whose density is unlikely to be affected by the presence of an adjacent structure. A "core index" is defined, which quantifies the depth of a grid point within the core. With the help of this index, a weight is defined, which, when applied to the cross-correlation function, has the property of exerting a "pull" with the tendency to make the core regions of search structure and target structure coincide. Test computations indicate that this fitting method allows correct fitting at resolutions where unweighted cross-correlation fails.

In the form in which it is stated in equation (6.13), the correlation search is computationally expensive since it does not lend itself to the use of the fast Fourier transformation in the calculation of the local normalization. Of particular interest is Roseman's (2003) modified version of the algorithm, which makes it possible to perform the computation of the local variances in Fourier space. To understand this approach, one has to realize that the local mean and local variance of the target density in equation (6.13) can be written as convolution products of the mask with the target density or its square, respectively (see also van Heel, 1982). This means, according to the convolution theorem (appendix 1), that they can be computed in Fourier space by performing a simple scalar multiplication. An implementation of Roseman's fast algorithm also exists in the form of a SPIDER script (Rath et al., 2003). Here, the method was demonstrated by a search of several proteins and an RNA motif within the cryo-EM map of the *E. coli* ribosome. Also, as mentioned earlier in the context of volume matching (chapter 5, section 12.3.3), a spherical harmonics expansion can be used to greatly accelerate the orientation part of the search.

6.2.3. Correlation Search with Constrained Solution Set

Volkmann and Hanein (1999) start with the premise that the maximum of the cross-correlation search rarely gives the best solution. Rather, considering the statistics of the cross-correlation term defined in equation (3.15), an entire neighborhood of the peak should be considered as a set of possible solutions, all having different rotations and shifts. Thus, the outcome of the search is not just a single position, but a *solution set*, from which the most plausible solution can be selected only by applying additional constraints. These constraints are provided by the consistency with data available from bioinformatics. For instance, those solutions are prohibited that would lead to direct steric clashes or clashes considering a van der Waals' molecular envelope. Consultation of such rules gives not only "permissible" or "nonpermissible" verdicts, but can also provide a scoring for each member of the solution set, resulting in a probability distribution. The maxima of this distribution can then be taken as belonging to the most plausible, energetically favored solutions.

This approach is promising but would seem to require more study to ascertain the appropriateness of the statistical models underlying the definition of the solution set. Applications of this method have been in the docking of various actin-binding molecules to actin: fimbrin (Hanein et al., 1998), myosin (Volkmann et al., 2000), and the Arp2/3 complex (Volkmann et al., 2001).

6.2.4. Method of Vector Quantization*

Wriggers and coworkers (1998, 1999) introduced a method termed *vector quantization* to tackle the problem of fitting a high-resolution into a low-resolution structure. Subsequently, this method was further developed to deal with conformational changes (flexible fitting: Wriggers et al., 2000) and is embodied in the software system SITUS (Wriggers et al., 1999). It is difficult on these pages to explain the approach in any degree of detail, since it involves a number of new concepts hitherto unknown in structure research. Our description below follows the outline given in Wriggers et al. (1999).

Vector quantization is a technique that makes it possible to approximate a 3D density distribution $m(r)$ with a finite number of vectors, the so-called codebook vectors $\{w_i; i = 1, \ldots, N\}$. We form the mean square deviation of the discrete set of codebook vectors from the encoded 3D data:

$$E = \sum_j m(\mathbf{r}_j) ||\mathbf{r}_j - \mathbf{w}_{i(j)}||^2 \tag{6.15}$$

where j is the grid index of the density map, $m(\mathbf{r}_j)$ is the density at that point, and $i(j)$ is the index of the codebook vector $\mathbf{w}_{i(j)}$ closest to $\mathbf{r}_j$ in the Euclidean distance sense. We would like to find a unique set of N codebook vectors that minimizes E. The trick to do this without getting trapped in local minima lies in the use of topology-representing self-organizing networks (Wriggers et al., 1998). These avoid local minima by a stochastic gradient descent on an initially smooth energy landscape that is gradually changed to the energy E in the above expression.

The only free parameter in the approximation is the number N of codebook vectors; N determines the level of detail represented, but it needs to be smaller than a certain critical number that corresponds to the structural complexity, to avoid overfitting. When a set of codebook vectors is found for two density maps that are low- and high-resolution versions of the same structure, it can be used to find the proper fit between the two maps. This is the basis for a rigid-body fitting and docking technique. Flexible fitting, first formulated in an application to the docking of elongation factor G onto the ribosome (Wriggers et al., 2000), goes one step beyond this *ansatz*. Conformational changes are now allowed between the high-resolution atomic structure and the low-resolution density map. Essentially this is done by deforming the atomic structure, using a molecular mechanics refinement approach, such that the two sets of codebook vectors are brought into optimum agreement.

6.2.5. Real-Space Refinement*

Real-space refinement methods were developed by Michael Chapman for application in X-ray crystallography (Chapman, 1995, and see recent review by Fabiola and Chapman, 2005). The module RsRef, designed to work with the package TNT (Tronrud et al., 1987) used in X-ray crystallography, performs a least-squares fit of a structure represented as a set of rigid components that are flexibly connected; the objective of the procedure is to place each component optimally into the local density, while minimizing the strain produced at the junctions and the number of atom–atom overlap conflicts.

In the first application to EM, myosin and actin coordinates were fitted to the actin–myosin matrix of insect flight muscle, visualized by electron tomography of negatively stained plastic sections (Chen et al., 2002b). However, staining and plastic embedment compromise the quality of the agreement achievable in the fit. The study was later repeated on samples prepared by slam-freezing and freeze-substitution (Liu et al., 2004). An application to cryo-EM maps of the ribosome (Gao et al., 2003; Gao and Frank, 2005) has brought out the full potential of the technique, as it yielded quasi-atomic models representing the ribosome in different states related by the "ratchet" movement, a relative rotation between the two subunits during tRNA translocation. (The term quasi-atomic model has often been used to refer to a model not based on density maps that are in the atomic resolution range. As cryo-EM approaches that range, the legitimacy of the model becomes comparable to that of an atomic model obtained by X-ray crystallographic techniques, and the "quasi-" prefix can then be dropped.)

6.2.6. Fitting Guided by Normal-Mode Analysis

Normal-mode analysis (NMA) is a computational exploration of the "natural" motion of a structure given as a set of coordinates (Brooks and Karplus, 1983; Tama et al., 2002) or as a low-resolution density map (Chacón et al., 2003). In both cases the structure is represented as an elastic network, and the dynamical properties of the resulting object are investigated near its equilibrium. The analysis results in a set of principal motions or flexing of entire domains. A few low-energy modes are often typically represent motions or flexing of entire domains. A few low-energy modes are often sufficient to arrive at a satisfactory description of the experimentally observed dynamic behavior of large molecules, as demonstrated for some functionally relevant conformational changes of large macromolecular assemblies, such as virus swelling (Tama and Brooks, 2002) and the ribosomal ratchet motion (Tama et al., 2003; Wang et al., 2004; see also Ma, 2005). The demonstrated ability of NMA to predict coarse (i.e., relation to motions of large domains) dynamic features of a molecule has been exploited by the development of a new fitting method (Tama et al., 2004a,b), termed normal-mode flexible fitting (NMFF). While fitting by real-space refinement, described in the previous section, starts out with a representation of the atomic structure as a set of flexibly connected rigid components, fitting by NMFF leaves the atomic structure intact and attempts to explain the observed density map by deforming the X-ray structure using a superimposition of suitably weighted normal modes. In a process of refinement, the cross-correlation coefficient between the experimental map and the deformed simulated map is maximized. Application to cryo-EM maps of RNA polymerase and elongation factor G as bound to the ribosome lead to excellent fits (Tama et al., 2004b).

6.2.7. Hunting for Secondary Structure Motifs

A significant fraction of most biological structures known from X-ray crystallography is organized as a combination of two secondary structure motifs, α-helices and β-sheets. Once the resolution extends past a critical limit, which lies in the range of 10 Å, it is possible to conduct a 5D search for these motifs and locate them in the cryo-EM map. Jiang et al. (2001) pointed out that knowledge of the size and location of these two motifs would be sufficient to classify a protein into one of four families distinguished in SCOP (structural classification of proteins). The authors developed a program, called *Helix-hunter*, that sifts through the density using locally normalized cross-correlation with cylinders of variable length. For applications, see, for instance, the articles by Zhou et al. (2001a) and Ludtke et al. (2004).

Similarly, Ma and his group (Kong and Ma, 2003; Kong et al., 2004) developed programs, *Sheet-miner* and *Sheet-tracker*, to address the even more challenging problem of finding β-sheets. The programs make use of successive rounds of exclusion, eliminating density identified by Helix-hunter right at the start, and using principal component analysis of voxel neighborhoods to find putative strands.

One can see the localization of α-helical and β-sheet elements in a density map as a start of a hierarchical interpretation process that might enable the researcher to identify subfamilies as the resolution improves, and narrow down the identification as additional, nonstructural information is consulted (Jiang et al., 2001; Chiu et al., 2002).

7. Classification of Volumes

Volumes, represented as a lexicographic string of density values arranged on a 3D grid, can be subjected to the same techniques of multivariate data analysis (MDA) and classification as earlier discussed for images in chapter 4. Again, the fundamental requirement is that they are perfectly aligned, so that in each case the lexicographic index of the array refers to the same point in space. Three-dimensional masking is used to focus the analysis on the relevant region of the molecule while discarding the surrounding densities.

These volumes can be multiple reconstructions of the same molecule from different single-particle data sets or from the same data set but with different starting conditions of a common-lines algorithm (Penczek et al., 1996). In the latter case, classification is used to find a core group of density maps most consistently reproduced. They can be single-particle reconstructions representing the same molecule in different experimental conditions or states, in which case it might be desirable to explore the similarity relationships between the density maps that could contain clues on the way the states evolve over time. They can also be separate tomographic reconstructions of several molecules, where the objective is to find those that are most consistently preserved (Walz et al., 1997b; Böhm et al., 2001). Finally, they can be pseudo-repeats of a 3D ordered structure, such as the insect flight muscle (Winkler and Taylor, 1999; Chen et al., 2002b; Pascual-Montano et al., 2002), representing the force-generating interaction between actin and myosin in different states. In this case, MDA and classification are able to group the different motifs, so that they can be separately averaged. The resulting density maps provide a view of the dynamically changing structure.

Appendix 1

Some Important Definitions and Theorems

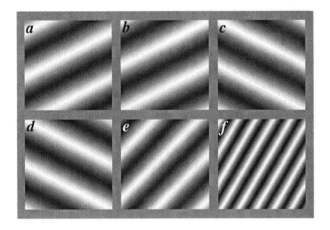

Figure A1.1 Some examples of the two-dimensional sine waves which form the basis of the Fourier representation of an image (white represents $+1$, black -1). (a) $\varphi=0°$; $l=1$, $m=2$; (b) $\varphi=90°$; $l=1$, $m=2$; (c) $\varphi=90°$; $l=-1$, $m=2$; (d) $\varphi=90°$; $l=1$, $m=-2$; (e) $\varphi=180°$; $l=2$, $m=2$; (f) $\varphi=45°$; $l=5$, $m=3$.

1. The Discrete Two-Dimensional Fourier Transform

An image (such as the projection of a molecule) is defined as a two-dimensional (2D) array representing samples of the optical density distribution on a regular grid:

$$p_{ik} = p(x_i, y_k);\ i = 1\ldots, I;\ k = 1\ldots, K \qquad (A1.1)$$

319

The coordinates are multiples of the sampling distance $\Delta x = \Delta y$; $x_i = i\Delta x$; $y_k = k\Delta x$. The Fourier transform is a mathematical representation of the image using sine waves or complex exponentials as basis functions. One can see the Fourier transform as an alternative representation of the information contained in the image. What singles out the Fourier transform, among many other expansions using sets of orthonormalized basis functions, is that it allows the influence of instrument aberrations and the presence of periodicities in the object to be readily analyzed. It also provides a key to an understanding of the relationship between the projections and the object they originate from, and to the principle underlying three-dimensional (3D) reconstruction.

The basis of the Fourier representation of an image is a set of "elementary images"—the basis functions. In the case of the *sine transform*, to be used as an introduction here, these elementary images have sinusoidal density distribution:

$$e_{lm|ik} = e_{lm}(x_i, y_k) = \sin[2\pi(u_l x_i + v_m y_k)] \tag{A1.2}$$

Each of these images (examples in figure A1.1), represented by a discrete set of pixels at positions (x_i, y_k) of a square lattice, is characterized by its discrete spatial frequency components (u_l, v_m)—describing how many full waves fit into the frame in the horizontal (l) and vertical (m) directions:

$$u_l = \frac{l}{I\Delta x}, l = 0, \ldots, I-1; v_m = \frac{m}{K\Delta y}, m = 0, \ldots, K-1 \tag{A1.3}$$

Any discrete image $\{p_{ik}; i=1,\ldots,I; k=1,\ldots,K\}$ can be represented by a sum of such elementary images, if one uses appropriate amplitudes a_{lm} and phases φ_{lm}. The amplitudes determine the weights of each elementary image, and the phases determine how much it has to be shifted relative to the normal position (first zero assumed at $x_i = y_k = 0$; note that in figure A1.1 this occurs in the upper left corner since a left-handed coordinate system is used):

$$p_{ik} = p(x_i, y_k) = \sum_{l=0}^{I-1} \sum_{m=0}^{K-1} a_{lm} \sin\left[2\pi\left(\frac{li}{I} + \frac{mk}{K}\right) + \phi_{lm}\right] \tag{A1.4}$$

[Here, we have rewritten the elementary image (A1.2) by substituting u_l by $1/(I\Delta x)$ and x by $i\Delta x$, etc.) To understand equation (A1.4), consider one of its elementary images, indexed ($l=1$, $m=0$): as i increases from 1 to I, the argument of the sine function goes from 0 to 2π exactly once: this elementary image is a horizontally running sine wave that has the largest wavelength (or smallest spatial frequency) fitting the image frame in this direction.

The 2D sine transform is a 2D scheme, indexed (l, m), that gives the amplitude a_{lm} and the phase shifts φ_{lm} of all elementary sine waves with spatial frequencies (u_l, v_m), as defined in equation (A1.3), that are required to represent the image.

The Fourier representation generally used differs from the definition (A1.4) in that it makes use of "circular" complex exponential waves:

$$\exp\left[-2\pi i\left(\frac{li}{I} + \frac{mk}{K}\right)\right] = \cos\left[2\pi\left(\frac{li}{I} + \frac{mk}{K}\right)\right] + i\sin\left[2\pi\left(\frac{li}{I} + \frac{mk}{K}\right)\right] \tag{A1.5}$$

(Note that "i" is the symbol for the imaginary unit, to be distinguished from the index i.) The reason is that this representation is more general and leads to a more tractable mathematical formulation. Hence, the Fourier representation of an image is

$$p_{ik} = \sum_{l=0}^{I-1} \sum_{m=0}^{K-1} F_{lm} \exp\left[-2\pi i\left(\frac{li}{I} + \frac{mk}{K}\right)\right] \tag{A1.6}$$

or, symbolically,

$$p(x, y) = \mathcal{F}\{F(u, v)\} \text{ ("Fourier synthesis")} \tag{A1.7}$$

Conversely, given an image, the Fourier coefficients F_{lm} are obtained by a very similar reciprocal expression:

$$F_{lm} = \sum_{l=0}^{I-1} \sum_{m=0}^{K-1} p_{ik} \exp\left[2\pi i\left(\frac{li}{I} + \frac{mk}{K}\right)\right] \tag{A1.8}$$

or, symbolically,

$$F(u, v) = \mathcal{F}^{-1}\{p(x, y)\} \text{ ("inverse Fourier transformation")} \tag{A1.9}$$

2. Properties of the Fourier Transform

Linearity. This property follows from the definition: if an image is a linear combination of two images,

$$f(\mathbf{r}) = c_1 f_1(\mathbf{r}) + c_2 f_2(\mathbf{r})$$

then its Fourier transform is

$$F(\mathbf{k}) = c_1 F_1(\mathbf{k}) + c_2 F_2(\mathbf{k}) \tag{A1.9a}$$

where $F_1(\mathbf{k})$, $F_2(\mathbf{k})$ are the respective Fourier transforms of the images $f_1(\mathbf{r})$ and $f_2(\mathbf{r})$.

Scaling of argument. Given two images of the same object with different magnifications, $f_2(\mathbf{r}) = f_1(s\mathbf{r})$, where s is a scaling factor, then their Fourier transforms are related by $F_2(\mathbf{k}) = F_1(1/s\ \mathbf{k})$.

Symmetry property. Images are normally real-valued. It is easy to see from equation (A1.8) that for real-valued functions, the normally complex-valued Fourier transform has the property:

$$F_{-l-m} = [F_{lm}]^* \tag{A1.10}$$

which is known as Friedel symmetry in X-ray crystallography. The asterisk stands for taking the complex conjugate:

$$[F_{lm}]^* = \{\Re e\{F_{lm}\}, -\Im m\{F_{lm}\}\} \tag{A1.11}$$

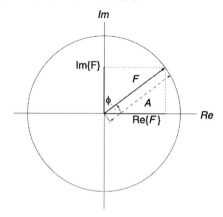

Figure A1.2 Representation of a complex number F as a vector in the complex plane. A (amplitude, or modulus) is the length of the vector, while φ (the phase) is the angle (counterclockwise, or mathematically positive) that the vector is rotated by with respect to the real axis.

Thus, since for real-valued images, half the Fourier transform determines the values of the other half, only one half is normally required in all computations and storage.

Amplitude and phase. The complex coefficient F_{lm} can be represented by a vector in the complex plane (figure A1.2). Its length,

$$|F_{lm}| = \left[|\Re e\{F_{lm}\}|^2 + |\Im m\{F_{lm}\}|^2\right]^{1/2} \tag{A1.12}$$

is the amplitude of the Fourier transform, while its phase is

$$\varphi_{lm} = \arctan[\Im m\{F_{lm}\}/\Re e\{F_{lm}\}] \tag{A1.13}$$

Shift Theorem. When the image is shifted by a vector $\Delta \mathbf{r} = (\Delta x, \Delta y)$, its Fourier transform changes according to equation (A1.8) by multiplication with an exponential factor:

$$\mathfrak{F}^{-1}\{p(x + \Delta x, y + \Delta y)\} = F(u, v)\exp[2\pi i(u\Delta x + v\Delta y)] \tag{A1.14}$$

3. Convolution Product and Convolution Theorem

Suppose we were able to form an image of a single point of the object, located at (x_i, y_k). Its image will appear as an extended disk, on account of resolution limitations, at the corresponding point (x_i, y_k) in the image plane. In bright-field electron microscopy (EM), applied to weak phase objects (which is a good approximation in cryo-electron microscopy of biological macromolecules), image formation can be described by linear superimposition: the image of two object points can be obtained by adding the images that *would be obtained* by imaging the two points independently and separately. An even more restrictive property holds in EM, the property of isoplanasy: the image of a point looks the same irrespective of its location in the image field. Under those conditions, the image is related to the object by

$$p(x_i, y_k) = \sum_{l=1}^{I} \sum_{m=1}^{K} o(x_i, y_k) h(x_l - x_i, y_m - y_k)$$ (A1.15)

where i, k, l, m relate to positions on a regular sampling grid and $h(x, y)$ is the point-spread function—it is the image of a single object point placed at $(0, 0)$. The relationship (A1.15) follows immediately from the properties of linearity and isoplanasy. The right-hand side of equation (A1.15) represents the convolution product of $o(x, y)$ with $h(x, y)$. We will make use of the symbolic notation $o(x, y) \circ h(x, y)$. The *delta-function* is a particular point spread fuction consisting of a single point, $h(x_l - x_i, y_m - y_k) = \delta(x_l - x_i, y_m - y_k)$, which has the value one at $x_l = x_i$, $y_m = y_k$ and zero elsewhere. Use of this function in (A1.15) gives the result $p(x_i, y_k) = o(x_i, y_k)$; i.e., *convolution with the delta-function reproduces the original function.*

The *convolution theorem* is of fundamental importance for the computation of convolution expressions. The theorem states that the *Fourier transform of a convolution product of two functions is equal to the product of their Fourier transforms.*

Hence, the computation of the convolution product of two functions $o(x, y)$ and $h(x, y)$ can be computed in the following way:

(i)   Compute $H(u, v) = \mathfrak{F}\{h(x, y)\}$
(ii) Compute $O(u, v) = \mathfrak{F}\{o(x, y)\}$
(iii) Form the scalar product of these two transforms

$$P(u, v) = O(u, v) \cdot H(u, v)$$ (A1.16)

(iv) compute the inverse Fourier transform of the result:

$$p(s, y) = \mathfrak{F}^{-1}\{P(u, v)\}$$ (A1.17)

This is one of the many instances where the seemingly complicated step of Fourier transformation proves much more efficient than the direct evaluation of a linear superimposition summation. The time saving is due to the very economic organization of the fast Fourier transform (FFT) algorithm in numerical implementations (see Cooley and Tukey, 1965). Another example will be found in the evaluation of the auto- and cross-correlation functions below.

4. Resolution in the Fourier Domain

There exists a practical limit to the number of sine waves to be included in the Fourier representation (A1.6) of an image. Obviously, sine waves (or complex exponentials) with wavelengths smaller than the size of the smallest features contained in the image carry no information.

This information limit can be expressed in terms of the spatial frequency radius $R = \sqrt{u^2 + v^2}$; beyond a certain radius $R = R_0$ (i.e., outside of a roughly circular domain in the Fourier plane) no meaningful Fourier components are encountered. This boundary is called *the bondlimit* or *resolution limit*. [Note that usage of the term resolution is inconsistent in the literature; it is used to denote either the smallest distance resolved (in real space units, dimension *length*) or the resolution limit defined above (in spatial frequency units, dimension 1/length.]

The resolution limitation invariably can be traced back to a physical limitation of the imaging process: either there is an aperture in the optical system that limits the spatial frequency radius of the object's Fourier components (e.g., the objective lens aperture in the electron microscope), or the recording of the image itself may give rise to some blurring (as, for instance, the lateral spread of electrons in the photographic emulsion).

By virtue of the convolution theorem, any such spread or blurring in the image is expressed by a multiplication of the object's Fourier transform with a function that has a finite radius in the spatial frequency domain.

5. Low-Pass and High-Pass Filtration

For a noncrystalline object, which we are exclusively dealing with, the signal and noise components of the Fourier transform are superimposed and normally inseparable. However, the two components frequently have different behaviors as a function of spatial frequency radius R: while the signal component falls off at the resolution limit (see section 4), the noise component has significant contributions beyond that limit. Hence, multiplication of the Fourier transform of such an image with a cutoff function:

$$L(u, v) = \begin{cases} 1 & R = \sqrt{u^2 + v^2} = R_0 \\ 0 & \text{elsewhere} \end{cases} \qquad (A1.18)$$

will eliminate part of the noise, thus enhancing the signal-to-noise ratio. Application of such a function, or variants of it having a smooth transition instead of a sharp cutoff, is termed *low-pass filtration*, as it *passes* only Fourier components with *low* spatial frequencies.

Functions with smooth radial transition are usually preferred, since application of equation (A1.18) would cause an artificial enhancement of image features whose size corresponds to the spatial frequencies at the cutoff (e.g., see Frank et al., 1985). The most "gentle" function used for Fourier filtration is one with a Gaussian profile (e.g., Frank et al., 1981b):

$$L(u, v) = \exp\left[-(u^2 + v^2)/R_0^2\right] \qquad (A1.19)$$

Its falloff behavior is controlled by the parameter R_0: at a spatial frequency radius of $R = \sqrt{u^2 + v^2} = R_0$, the filter function reduces the Fourier amplitude to $1/e$ of its original value.

6. Correlation Functions

Translational cross-correlation function. The cross-correlation function is defined in the following way:

$$\Phi_{12}(x_l, y_m) = \sum_i \sum_k p_1(x_i, y_k) p_2(x_i + x_l, y_k + y_m) \tag{A1.20}$$

It can be shown that this function can also be evaluated by using a Fourier theorem closely related to the convolution theorem. Thus, again the tedious calculation of the sum can be bypassed: *the Fourier transform of the cross-correlation function is equal to the conjugate product of the Fourier transforms of these images.* This suggests a fast way of computing equation (A1.20):

(i) Compute $P_1(u, v) = \mathfrak{F}\{p_1(x, y)\}$
(ii) Compute $P_2(u, v) = \mathfrak{F}\{p_2(x, y)\}$
(iii) Take the complex conjugate of $P_2(u, v)$

$$P_2^*(u, v) = \mathfrak{Re}\{P_2(u, v)\} - i\,\mathfrak{Im}\{P_2(u, v)\} \tag{A1.21}$$

(iv) form the conjugate product of the two Fourier transforms:

$$C(u, v) = P_1(u, v) P_2^*(u, v) \tag{A1.22}$$

(v) finally, take the inverse Fourier transform of the result, which gives

$$F_{12}(x, y) = \mathfrak{F}^{-1}\{C(u, v)\} \tag{A1.23}$$

Autocorrelation function. As a special case, the autocorrelation function is obtained by letting $p_2(x, y) = p_1(x, y)$. In this case, the computational sequence above is reduced to the following:

(i) Compute $P_1(u, v) = \mathfrak{F}\{p_1(x, y)\}$
(ii) Compute its absolute square, $A(u, v) = |P_1(u, v)|^2$
(iii) Take the inverse Fourier transform of the result

$$F_{11}(x, y) = \mathfrak{F}^{-1}\{A(u, v)\}$$

Appendix 2

Profiles, Point-Spread Functions, and Effects of Commonly Used Low-Pass Filters

Figure A2.1 Test image ("Lena") used for demonstration of the effects of various low-pass filters; see figures A2.2–A2.4.

The effect of three commonly used filter functions is demonstrated in the following, using the image of Lena (figure A2.1) as an example. In each case, the following information is provided for three choices of a characteristic parameter: (i) filter function, in the range of Nyquist frequencies $0 \ldots 0.5$; (ii) point-spread

327

function (PSF); and (iii) the resulting images. Among the artifacts produced by application of the filter, the *ringing* artifact is the most noticeable.

1. Gaussian Low-Pass Filter

Gaussian low-pass filters (LPFs) are two-dimensional (2D) Gaussian functions in the Fourier domain. One-dimensional profiles are shown below, at several different values for the filter radius. The radius specifies the 2-σ level of the Gaussian. The corresponding PSFs are in real (or image) space. The wider the PSF, the more blurring. It should be noted that, in contrast with the Fermi and Butterworth filters below, even the smallest radius leads to no ringing artifacts (figure A2.2).

2. Fermi Low-Pass Filter

The Fermi LPF is defined as $1/(1+\exp[(F-\text{radius})/T])$, which negotiates between "top hat" and Gaussian characteristics, depending on the value of the

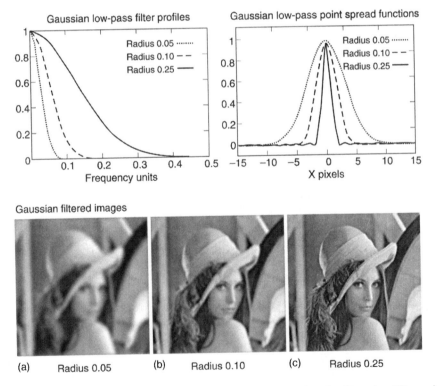

Gaussian filtered images

(a)   Radius 0.05   (b)   Radius 0.10   (c)   Radius 0.25

Figure A2.2 Top: filter profiles and point spread functions for the Gaussian filter, with three different filter radii: (a) 0.05, (b) 0.10, and (c) 0.25. Bottom: low-pass filtered versions of the test image.

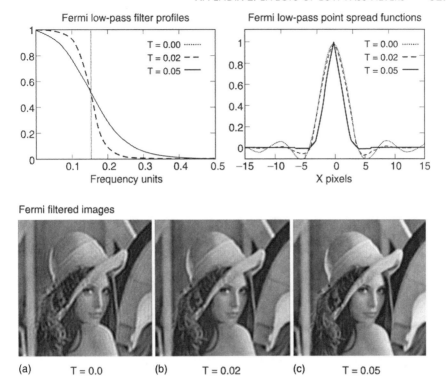

(a)   T = 0.0   (b)   T = 0.02   (c)   T = 0.05

Figure A2.3 Top: filter profiles and point-spread functions for the Fermi filter, with a radius of 0.15 and three different choices for the "temperature" parameter: (a) 0.0, (b) 0.02, (c) 0.05. Bottom: low-pass filtered versions of the test image. Note that $T=0$ produces a filter with a "top-hat" profile similar to the choice of **PB** (pass band) = 0.15 and **SB** (stop band) = 0.15 in the Butterworth filter (figure A2.4). Both filters result in the "ringing" artifact in the output image.

temperature factor, T. The temperature defines the reciprocal distance (in terms of frequency units), within which the filter falls off ($T=0$ gives a sharp cutoff). The radius in all three filters below has been set to 0.15. The sharp cutoff ($T=0$) causes "ringing" artifacts in the filtered image. The PSFs of all three filters are similar, but the filters that decay more slowly reduce the undulations in the PSF (figure A2.3).

3. Butterworth Low-Pass Filter

The Butterworth LPF can be roughly thought of as a smoothed trapezoid. The corners of the trapezoid are controlled by the passband and the stopband. Frequencies lower than the passband are unaltered, frequencies beyond the stopband are set to zero, while those in between these limits are attenuated.

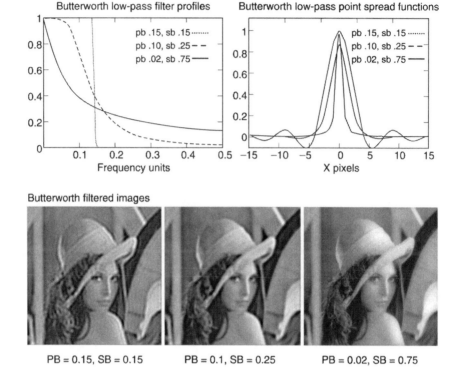

Butterworth filtered images

PB = 0.15, SB = 0.15   PB = 0.1, SB = 0.25   PB = 0.02, SB = 0.75

Figure A2.4 Top: filter profiles and point spread functions for the Butterworth filter, with three different filter settings, where PB denotes "passband" and SB "stopband." (a) PB = 0.15, SB = 0.15, (b) PB = 0.1, SB = 0.25; and (c) PB = 0.02, SB = 0.25. Bottom: low-pass filtered version of test image.

The first filter (passband = stopband = 0.15) is nearly a sharp cutoff filter (top hat), causing "ringing" artifacts in the filtered image. This "ringing" is also seen as undulations in the outer lobes of the PSF.

The third filter (passband = 0.02, stopband = 0.75) passes more high frequencies, so that details are sharper. However, attenuating the low frequencies results in loss of contrast in some areas (e.g., the face). Note that, although the stopband can be specified beyond 0.5 and still determines the filter shape, the filter is cut off at the 0.5 (Nyquist) frequency (figure A2.4).

Appendix 3

Bibliography of Methods

Method	Reference
alignment	
2D	Radermacher, 1994; Lanzavecchia et al., 1996; Sigworth, 1998; Jensen, 2001; Joyeux and Penczek, 2002
2D, scaling	Aldroubi et al., 1992; Dryden et al., 1993
2D, statistical limitations	Saxton and Frank, 1977; Henderson, 1995
2D vs. 3D	*See under* orientation determination
3D	Penczek et al., 1992; Oefverstedt et al., 1997; Walz et al., 1997; Kozin and Svergun, 2000
3D, via canonical orientation	Lanzavecchia, et al., 1999b, 2001
3D, via spherical harmonics	Crowther, 1972; Kovacs and Wriggers, 2002
amplitude decay and correction	Conway and Steven, 1999; Schmid et al., 1999, 2004; Thurman-Commike et al., 1999; Gabashvili et al., 2000; Saad et al., 2001; Zhou and Chiu, 2003; Ludtke et al., 2004
antibody labeling	Boisset et al., 1992b, 1993, 1995; Koeck et al., 1996
charging effect	Henderson, 1992; Brink et al., 1998; Miyazawa et al., 1999; Gao et al., 2002
classification	van Heel et al., 1982b; Bretaudiere and Frank, 1986; Frank et al., 1988b, 1990
Bayesian	Samsó et al., 2002
bispectral projections	Marabini and Carazo, 1994b, 1996
case study	Falke et al., 2001a

(Continued)

Method	Reference
conformational heterogeneity	Saibil, 2000; Mellwig and Böttcher, 2001; Ranson et al., 2001; Valle et al., 2002; Heymann et al., 2003, 2004; Gao et al., 2004; Spahn et al., 2004a; White et al., 2004
hierarchical ascendant	Anderberg, 1973; Mezzich and Solomon, 1980; de Haas et al., 1999; Davis et al., 2002
self-organizing maps	Kohonen, 1990; Marabini and Carazo, 1994a; Fernández and Carazo, 1996; Kohonen, 1997, Wriggers et al., 1998: Pascual-Mantano et al., 2000; Ogura and Sato, 2001; Pascual-Montano et al., 2001; Radermacher et al., 2001; Gomez-Lorenzo et al., 2003; Ogura et al., 2003
SOMs, higher topology	Martinez and Schulten, 1994; Zuzan et al., 1997, 1998
3D	Penczek et al., 1996; Walz et al., 1997b; Winkler and Taylor, 1999; Pascual-Montano et al., 2001; Chen et al., 2002b; Liu et al., 2004
correlation functions	Frank, 1972b, 1973c, 1980; van Heel et al., 1992a
cross-validation	Brünger, 1992, 1993, 1997, Shaikh et al., 2003
cryo-EM	Steven, 1997; Conway and Steven, 1999; Frank et al., 1999b; Glaeser, 1999; Kühlbrandt and Williams, 1999; Baumeister and Steven, 2000; Mancini and Fuller, 2000; Ruprecht and Nield, 2001; Saibil, 2000b; Tao and Zhang, 2000; van Heel et al., 2000; Baker and Henderson, 2001; Ueno and Sato, 2001; Unger, 2001; Frank, 2002
comparison with X-ray	Penczek et al., 1999; Kalko et al., 2000; Harris et al., 2001; Roseman et al., 2001; Ludtke et al., 2004; Rossmann et al., 2005; Spahn et al., 2005
helium	Fujiyoshi, 1998; Grassucci et al., 2002
preparation	Taylor and Glaeser, 1976; Dubochet et al., 1982; Lepault et al., 1983; McDowall et al., 1983; Adrian et al., 1984; Wagenknecht et al., 1988b; Frederick et al., 2000; Braet et al., 2003; Al-Amoudi et al., 2004
cryo-negative staining	Adrian et al., 1998; de Carlo et al., 2002; Golas et al., 2003; Boehringer et al., 2004; Jurica et al., 2004
cryo-protection	Chiu et al., 1986
CTF	
correction	Frank and Penczek, 1995; Penczek et al., 1997; Grigorieff, 1998; Ludtke et al., 2001; Mouche et al., 2001; Roseman et al., 2001; Ludtke and Chiu, 2002; Sander et al., 2003; Sorzano et al., 2004
display	Jiang and Chiu, 2001
estimation	Zhou et al., 1996; Fernández et al., 1997; Mouche et al., 2001; Huang et al., 2003; Sander et al., 2003
data acquisition	Kisseberth et al., 1997; Downing and Hendrickson, 1999; Potter et al., 1999; Carragher et al., 2000; Faruqi and Subramaniam, 2000; Stewart et al., 2000; Rouiller et al., 2001; Potter et al., 2000; Zhang et al., 2000, 2003; Lei and Frank, 2005
CCD camera	Krivanek and Mooney, 1993; de Ruijter, 1995
database	
3D	Carazo et al., 1994; Marabini et al., 1996a; Carazo and Stelzer, 1999; Kalko et al., 2000

(*Continued*)

Method	Reference
denoising	
multiscale transformations	Stoschek and Hegerl, 1997b
nonlinear anisotropic diffusion	Frangakis and Hegerl, 1999, 2001; Frangakis et al., 2001
detection of macromolecules	Martín-Benito et al., 1997; Stoschek and Hegerl, 1997a; Böhm et al., 2000; Frangakis et al., 2002; Wagenknecht et al., 2002; Rath et al., 2003
difference maps/significance	Wagenknecht et al., 1994; Roseman et al., 1996
electron scattering (elastic and inelastic)	Angert et al., 1996, 2000
electron spectroscopic 3D imaging	Harauz and Ottensmeyer, 1984a,b; Andrews et al., 1985, 1987; Zabal et al., 1993; Bazett-Jones et al., 1996; Czarnota et al., 1996, 1997; Beniac et al., 1997
EM control	Kisseberth et al., 1997
energy filtering	Schröder, 1992; Gubbens and Krivanek, 1993; Koeck et al., 1996; Koster et al., 1997; Grimm et al., 1998; Angert et al., 2000
Ewald sphere, curvature	Frank, 1973b; DeRosier, 2000
expectation-maximization	Yin et al., 2002
field emission gun, coherence	Hewat and Neumann, 2002
fitting and docking	Rayment et al., 1993; Stewart et al., 1993; Agrawal et al., 1998; Wriggers et al., 1998, 1999; Volkmann and Hanein, 1999; Ishikawa et al., 2000; Kozin and Svergun, 2000; Roseman, 2000; Rossmann, 2000; Thouvenin and Hewat, 2000; Volkmann et al., 2000; Wriggers et al., 2000; Wriggers and Birmanns, 2001; Wriggers and Chacón, 2001; Chacón and Wriggers, 2002; Darst et al., 2002; Rath et al., 2003; Roseman, 2003; Wu et al., 2003
helix and fold hunter	Jiang et al., 2001; Chiu et al., 2002; Kong and Ma, 2003; Kong et al., 2004
low-resolution X-ray map	Murakami et al., 2002
real-space refinement	Chapman, 1995; Gao et al., 2003; Fabiola and Chapman, 2005
gold labeling	
nanogold, undecagold, colloidal gold	Hainfeld, 1987; Milligan et al., 1990; Wagenknecht et al., 1994; Hölzl et al., 2000; Büchel et al., 2001; Montesano-Roditis et al., 2001
handedness	Belnap et al., 1997; Rosenthal et al., 2003
helical reconstruction via single particles	Egelman, 2000; Schröder et al., 2002
helium temperature effects	Fujiyoshi, 1998; Grassucci et al., 2002
ice, scattering properties	Heide and Zeitler, 1985; Jenniskens and Blake, 1994
image-processing software	Crowther et al., 1996; Frank et al., 1996; Marabini et al., 1996b; Trus et al., 1996; Van Heel et al., 1996; Ludtke et al., 1999; Baxter et al., 2002; Liang et al., 2002
MSA/correspondence analysis	
molecules	van Heel and Frank, 1981; Frank et al., 1982; Frank and van Heel, 1982a,b; Frank, 1984a; Borland and van Heel, 1990

(*Continued*)

Method	Reference
single unit cells	Sherman et al., 1998
negative staining	Brenner and Horne, 1959; Harris and Horne, 1991; Hönger and Aebi, 1996; Massover and Marsh, 1997, 2000; Massover et al., 2001; Harris and Scheffler, 2002
cryo-negative staining	de Carlo et al., 2002; Golas et al., 2003; Boehringer et al., 2004; Jurica et al., 2004
double layer	Valentine et al., 1968; Tischendorf et al., 1974; Stöffler and Stoffler-Meilicke, 1983; Stöffler-Meilicke and Stöffler, 1988
methylamine tungstate	Stoops et al., 1991b; Kolodziej et al., 1997; Qazi et al., 1998, 1999; Kolodziej et al., 2000; Radermacher et al., 2001; Zhou et al., 2001b,c
normal-mode analysis (NMA)	Chacón et al., 2002; Tama and Brooks, 2002; Tama et al., 2002; 2003, 2004a,b; Brink et al., 2004; Wang et al., 2004; Ma, 2005
orientation determination for projections	
angular reconstitution (common lines)	Goncharov et al., 1987; van Heel, 1987a; Orlova and van Heel, 1994; Serysheva et al., 1995; Penczek et al., 1996; Bellon et al., 1998; Lindahl, 2001
common lines, viruses	Crowther et al., 1970, 1971; Baker and Cheng, 1996; Fuller et al., 1996
general moments	Salzman, 1990
heavy atom clusters	Jensen and Kornberg, 1998
MSA of sinogram lines	Bellon, 2001; Bellon et al., 2001, 2002
Radon transform	Lanzavecchia, 1999a; Lanzavecchia, 2002; Radermacher et al., 2001
reference-based	Cheng et al., 1994; Penczek et al., 1994; Baker and Cheng, 1996; Davis et al., 2002
particle selection	Stoschek and Hegerl, 1996; Shah and Stewart, 1998; Nicholson and Glaeser, 2001; Ogura and Sato, 2001, 2004; Zhu et al., 2002; Rath et al., 2003; Roseman, 2003; Ogura and Sato, 2004; Potter et al., 2004; Rath and Frank, 2004; Zhu et al., 2004
phase plates	Danev and Nagayama, 2001; Majorovits and Schröder, 2002a,b; Huang and Penczek, 2004;
projection alignment/ refinement	Harauz and Ottensmeyer, 1984a,b; Harauz and van Heel, 1986b; Cheng et al., 1994; Penczek et al., 1994; Yang et al., 2005
maximum-likelihood	Sigworth, 1998
radiation damage	Chiu et al., 1986; Zeitler, 1990; Henderson, 1996
Radon transform	
alignment	Radermacher, 1994; Radermacher et al., 2001
fast computation	Lanzavecchia and Bellon, 1998
reconstruction based on	Radermacher, 1994, 1997; Lanzavecchia et al., 1999
reconstruction	
ART with blobs	Marabini et al., 1998, 2002
artifacts from overabundance of projections	Boisset et al., 1998; Sorzano et al., 2001

(*Continued*)

Method	Reference
consistency measures	Sorzano et al., 2001
gridding	Penczek et al., 2004
random conical tilt	Radermacher et al., 1986a; Radermacher et al., 1987a,b; Radermacher, 1988; Lanzavecchia and Bellon, 1996
spherical reconstruction	Jiang et al., 2001
two-exposure	Lanzavecchia et al., 1999
weighted back-projection	Radermacher, 1988, 1992; Clark, 1992
resolution definition and estimation	Di Francia, 1955; Frank et al., 1970, 1972a, 1981a; van Heel and Hollenberg, 1980; Saxton and Baumeister, 1982; Kessel et al., 1985; van Heel and Harauz, 1986; Unser et al., 1987; van Heel, 1987a; Radermacher, 1988; Unser et al., 1989; de la Fraga et al., 1995; Böttcher et al., 1997; Orlova et al., 1997; Grigorieff, 1998, 2000; Malhotra et al., 1998; Stewart et al., 1999, 2000; Boisset and Mouche, 2000; Roseman et al., 2001; Penczek, 2002a; Rosenthal et al., 2003; Shaikh et al., 2003; Yang et al., 2003
rotation	Tosoni et al., 1996
rotational symmetry	Kocsis et al., 1995
segmentation	
eigenvector approach	Frangakis and Hegerl, 2002a,b
experimental (labeling)	Wagenknecht et al., 1994; Che et al., 1998
interactive	Li et al., 1997
nucleation and growth	Zheng and Kolb, 1994
RNA vs. protein	Spahn et al., 2000
watershed transform	Volkmann, 2002
sinograms	
alignment	Lanzavecchia et al., 1996
specimen grid preparation	Baumeister and Seredynski, 1976; Baumeister and Hahn, 1978; Toyoshima, 1989
specimen stage, He	Fujiyoshi et al., 1991
spot scanning	Downing, 1991; Downing and Glaeser, 1986; Soejima et al., 1993; Zhou et al., 1994
statistical tests	Trachtenberg and DeRosier, 1987; Roseman et al., 1996; Sharma et al., 1998, 2000; Samsó et al., 1999; Thomas et al., 2001
support films	Ermantraut et al., 1998
symmetry	Lanzavecchia et al., 2002
template matching	
2D	Stoschek and Hegerl, 1997a; Roseman, 2003; Huang and Penczek, 2004
3D	Böhm et al., 2000; Rath et al., 2003
time-resolved imaging	Frank, 2001b
caged compounds	Ménétret et al., 1991; Rapp, 1998
by classification	Heymann et al., 2003, 2004; Gao et al., 2004
snapshots, induced	Lata et al., 2000

(Continued)

Method	Reference
spray-freezing	White et al., 1998
tomography	
of molecules and cells	Fung et al., 1996; Koster et al., 1997; Walz et al., 1997b; Baumeister et al., 1999; McEwen and Marko, 1999, 2001; Böhm et al., 2000; Potter et al., 2002
spectroscopic	Midgley and Weyland, 2003; Möbus et al., 2003
without tilt	Metoz et al., 1997
topology-representing network	Wriggers et al., 1998
variance estimation/maps	
2D	Carrascosa and Steven, 1979; Frank et al., 1981a; Wagenknecht et al., 1988a; Boisset et al., 1990a; Wagenknecht et al., 1992, 1994; Liu et al., 2002
3D	Frank et al., 1992; Boisset et al., 1993b; Liu and Frank, 1995; Liu et al., 1995; Penczek, 2002b; Cantele et al., 2003
wavelet transform filtering	Frangakis et al., 2001

Appendix 4

Bibliography of Structures

Structure	Reference
α-latrotoxin, black widow spider	Orlova et al., 2000
α₂-macroglobulin, human	
chymotrypsin complex	Schroeter et al., 1992; Boissset et al., 1993a
native	Lambert et al., 1994b; Larquet et al., 1994a,b; Kolodziej et al., 1998
native + transformed	Schroeter et al., 1991; Boisset et al., 1992a, 1994a, 1996; Qazi et al., 1998, 1999
alcohol oxidase	Vonck and van Bruggen, 1990
Aquaporin-0	Zampighi et al., 2003
Arp 2/3 complex	
Acanthamoeba castellanii, *Saccaromyces cerevisiae*	Volkmann et al., 2001
ATPase, AAA p97	
bovine liver	Rouiller et al., 2000, 2001
rat liver	Zhang et al., 2000
ATPase, H^+ ($CF_0 F_1$)	
E. coli	Böttcher et al., 1998, 2000; Mellwig and Böttcher, 2001
Kalanchoë daigremontaniana	Domgali et al., 2002
Manduca sexta	Radermacher et al., 2001
ATP synthase	
bovine heart	Rubinstein et al., 2003
Ca^{2+} channel, voltage-gated L-type	Murata et al., 2001

(Continued)

337

Structure	Reference
calcium release channel/ ryanodine receptor	
review	Orlova et al., 1998; Samsó and Wagenknecht, 1998; Serysheva and Hamilton, 1998
cardiac	Sharma et al., 1998
skeletal	Wagenknecht et al., 1989a, 1997; Radermacher et al., 1994a,b; Serysheva et al., 1995; Orlova et al., 1996; Samsó et al., 1999; Serysheva et al., 1999, 2002; Pate et al., 2000; Baker et al., 2002; Samsó and Wagenknecht, 2002
type-3	Sharma et al., 2000; Liu et al., 2001
cGMP phosphodiesterase 6	Kajimura et al., 2002
chemotaxis receptor-signaling complex	Francis et al., 2004
clathrin	Vigers et al., 1986a,b
modeling	Musacchio et al., 1999
CRSP coactivator	Taatjes et al., 2002
CRSP–RNA polymerase II CTD	Näär et al., 2002
cyanide-degrading nitrilase	
Pseudomonas stutzeri	Sewell et al., 2003
cyanide dihydratase	
Bacillus pumilus	Jandhyala et al., 2003
cytoxin, 58 kDa cell-binding subunit	
Heliobacter pylori	Reyrat et al., 1999
DNA	
binding proteins, archaeal, RadA	Yang et al., 2001
polymerase, Klenow fragment	Ottensmeyer and Farrow, 1992
dynein, cytoplasmic	Samsó et al., 1998
fatty acid synthetase	Hoppe et al., 1974; Stoops et al., 1992a; Kolodziej et al., 1997; Brink et al., 2002; Brink et al., 2004
flagella	
cap, HAP2 decamer	
Salmonella enterica	Yonekura et al., 2000
motor	
Caulobacter crescentus	Stallmeyer et al., 1989a
S. enterica	Thomas et al., 1999, 2001
S. typhimurium	Stallmeyer et al., 1989b; Sosinski et al., 1992; Francis et al., 1994
head–tail connector/portal protein	
bacteriophage f 29	Tsuprun et al., 1994
bacteriophage SPP1	Dube et al., 1994
heat shock proteins/chaperonins	
aB-crystallin, human	Haley et al., 1998
archaeal	Phipps et al., 1993
CCT, cytosolic chaperonin	Llorca et al., 2000, 2001; McCormack et al., 2001
GroEL	Ludtke et al., 2001
GroEL–ATP	Langer et al., 1992; Saibil et al., 1993; White et al., 1997; Ranson et al., 2001; Roseman et al., 2001

(Continued)

Structure	Reference
GroEL–glutamine synthetase	Falke et al., 2001a,b
GroEL–substrate binding	Farr et al., 2000
GroEL/GroES	Chen et al., 1994; Llorca et al., 1994; Brinker et al., 2001
GroEL/GroES ATPase cycle	Roseman et al., 1996; Rye et al., 1999
Hsp26	Haslbeck et al., 1999
review	Saibil and Wood, 1993; Saibil, 2000
sHsps (overview)	Haley et al., 2000
TF55, archaeal, *Sulfolobus shibatae*	Schohn et al., 2000a
thermosome and TRiC (overview)	Gutsche et al., 1999
thermosome	
E. coli	Nitsch et al., 1998
Sulfolobus shibatae	Schohn et al., 2000a,b
Thermoplasma acidophilum	Nitsch et al., 1997
helicases	
bacteriophage T7	Egelman et al., 1995
DnaB	San Martin et al., 1995
RuvB branch migration protein, *E. coli*	Stasiak et al., 1994
hemaglutinin, influenza	Böttcher et al., 1999
hemocyanin	
chiton (*Lepidochiton sp.*)	Lambert et al., 1994a
cuttlefish (*Sepia officinalis*)	Lambert et al., 1995c
horseshoe crab (*Limulus polyphemus*)	van Heel et al., 1994
HtH1, *Haliotis tuberculata*	Meissner et al., 2000
KLH1, keyhole limpet, *M. crenulata*	Orlova et al., 1997; Mouche et al., 2004
octopus (*Octopus vulgaris*)	Lambert et al., 1994b
protobranch bivalve mollusc (*Nucula hanleyi*)	Lambert et al., 1995b
Roman snail (*Helix pomatia*)	Lambert et al., 1995a
scorpion (*Androctonus australis*)	Boisset et al., 1990b, 1992b, 1993b, 1994b, 1995
hemoglobin	
Eudistylia vancouverii	de Haas et al., 1996b
Lumbricus terrestris	Schatz et al., 1992, 1994, 1995; de Haas et al., 1997; Taveau et al., 1997, 1999; Mouche et al., 2001
M. decora (leech)	de Haas et al., 1996
Ophelia bicornis	Cejka et al., 1992
R. pachyptila (vent tube worm)	de Haas et al., 1996
Sabella spallanzanii	Lanzavecchia et al., 1999b
Sepia officinalis, Bentoctopus, Vampyroteuthis infernalis	Mouche et al., 1999; Boisset and Mouche, 2000
HIB pili	
Haemophilus influenza	Mu et al., 2002
inositol 1,4,5-triphosphate (IP3) receptor	Jiang et al., 2002; da Fonseca et al., 2003; Serysheva et al., 2003; Sato et al., 2004; Taylor et al., 2004
insulin receptor, insulin complex, human	Luo et al., 1999; Ottensmeyer et al., 2000

(*Continued*)

Structure	Reference
junctional pore complex	
P. uncinatum	Hoiczyk and Baumeister, 1998
low-density lipoproteins (LPL)	
human	Orlova et al., 1999
multimeric archaeal	
PEP-synthase (MAP)	
Staphylothermus marinus	Cicicopol et al., 1999
multisynthetase complex	Norcum and Boisset, 2002
NADH: ubiquinone	
oxidoreductase (Complex I)	
bovine	Grigorieff, 1998
E. coli	Böttcher et al., 2002
nuclear pore complex	
review	Pante and Aebi, 1994
Xenopus	Hinshaw et al., 1992; Akey and Radermacher, 1993
nucleosome	Harauz and Ottensmeyer, 1984a,b
peroxiredoxin-II	Harris et al., 2001
phosphoenolpyruvate synthase	
(PEP)	
Staphylothermus marinus	Harauz et al., 1996
phosphoribulokinase	
Chlamydomonas reinhardtii	Mouche et al., 2002
phosphorylase kinase, rabbit	Nadeau et al., 2002; Vénien-Bryan et al., 2002
photosystemIIsupercomplex,	Nield et al., 2000
spinach	
PsbH subunit	Büchel et al., 2001
pneumolysin	
Streptococcus pneumoniae	Gilbert et al., 1999
portal protein	
bacteriophage SPP1	Dube et al., 1993
protease	
ClpAP, *E. coli*	Kessel et al., 1995; Beuren et al., 1998; Ortega et al., 2000;
	Ishikawa et al., 2001
hClpXP, human	Kang et al., 2002
HsIVU, *E. coli*	Rohrwild et al., 1997; Ishikawa et al., 2000
tricorn, *T. acidophilum*	Walz et al., 1997
proteasome	
review	Zwickel et al., 2000
activation by 11S REG	Li et al., 2001
(PA28) homologs	
ClpQ, *E. coli* (20S homolog)	Kessel et al., 1996
20S, *Thermoplasma*	Hegerl et al., 1991; Wenzel and Baumeister, 1995;
acidophilum	Baumeister et al., 1998
26S, bovine–PA700 and	Adams et al., 1997
modulator protein	
26S, *Drosophila melanogaster*	Walz et al., 1998; Hölzl et al., 2000
26S, *Xenopus*	Walz et al., 1998
protein kinase	
CaM kinase Iia	Kolodziej et al., 2000
DNA-dependent	Chiu et al., 1998
pyruvate dehydrogenase	Stoops et al., 1992b, 1997; Zhou et al., 2001a,c;
complex (PDC)	Gu et al., 2003

(Continued)

Structure	Reference
RAD52 protein	Stasiak et al., 2000
RecA	Yu and Egelman, 1997
retrotransposon, Ty	
yeast	Al-Khayat et al., 1999
ribosome	
eukaryotic:	
40S	Verchoor et al., 1989; Srivastava et al., 1995
40S +eIF3	Srivastava et al., 1992a
80S, human, +CrPV IRES	Spahn et al., 2004c
80S, rabbit	Morgan et al., 2000
80S, rabbit, +HCV IRES	Spahn et al., 2001c
80S, rat liver	Dube et al., 1998
80S, wheat germ	Verschoor et al., 1996
80S, wheat germ, +SRP	Halic et al., 2004
80S, yeast, translating,	Beckmann et al., 1997, 2001; Ménétret et al., 2000;
+Sec61	Morgan et al., 2000; Spahn et al., 2001a
80S, yeast, +EF2	Gomez-Lorenzo et al., 2000; Spahn et al., 2004b
80S, yeast, +RACK1	Sengupta et al., 2004
prokaryotic:	
reviews	Verschoor et al., 1993; Frank, 1997, 1998, 2000, 2001a,b, 2003;
	Orlova, 2000; Stark, 2002
30S, *E. coli*	Knauer et al., 1983; Verschoor et al., 1984; Beniac and Harauz,
	1995; Lata et al., 1995
50S, *E. coli*	Öttl et al., 1983; Vogel and Provencher, 1988; Radermacher
	et al.,1987a,b, 1988, 1992b; Matadeen et al., 1999
50S −L7/L12	Carazo et al., 1988
50S +Fab	Srivastava et al., 1992b, 1995
50S, *E. coli*, rRNA modeling	Müller et al., 2000
50S, *E. coli* −5S rRNA	Radermacher et al., 1990
70S, *E. coli*	Carazo et al., 1989; Wagenknecht et al., 1989b; Frank et al., 1991;
	Penczek et al., 1992, 1994; 1995a,b; Öfverstedt et al., 1994;
	Stark et al., 1995
70S, *E. coli*, +tRNA	Agrawal et al., 1996, 2000; Stark et al., 1997
70S, *E. coli*, −L7/L12	Montesano-Roditis et al., 2001
70S, *E. coli*, −L11	Agrawal et al., 2001
70S, *E. coli*, −S1	Sengupta et al., 2001
70S, *E. coli*, +EF-G	Agrawal et al., 1998, 1999a, Frank and Agrawal, 2000;
	Stark et al., 2000; Wriggers et al., 2000; Valle et al., 2003a;
	Gao et al., 2004
70S, *E. coli*, +Tet(O)	Spahn et al., 2001b
70S, *E. coli*, +ternary	Stark et al., 1997, 2003; Agrawal et al., 2000a; Valle et al., 2002,
complex	2003c
70S, *E. coli*, +RF2	Klaholz et al., 2003; Rawat et al., 2003
70S, *E. coli*, translating,	Gilbert et al., 2004
+polypeptide	
70S, *E. coli*, +tmRNA	Valle et al., 2003b
RNA polymerase	
bacteriophage F6	de Haas et al., 1999
RNA polymerase II	
E. coli	Darst et al., 1991; Polyakov et al., 1998; Darst et al., 2002
holoenzyme	Asturias et al., 1999; Davis et al., 2002

(*Continued*)

Structure	Reference
yeast	Craighead et al., 2002
mediator	Asturias et al., 1999; Davis et al., 2002
RNA polymerase II	Murakami et al., 2002
holoenzyme DNA	
RNP particle	
recombinant influenza virus	Martín-Benito et al., 2001
RuvB branch migration protein	Chen et al., 2002a
signal recognition particle	Czarnota et al., 1994; Ottensmeyer et al., 1994
SRP54	
spliceosome	
B complex, human	Boehringer et al., 2004
C complex	Jurica et al., 2004
splicing factor SF3b	Golas et al., 2003
U1 snRNP	Stark et al., 2001
SV40 large T antigen hexamer	VanLoock et al., 2002
T antigen (TAG)	Gomez-Lorenzo et al., 2003
T7 bacteriophage gp4	VanLoock et al., 2001b
TEP1, telomerase/	Kickhoefer et al., 2001
vault-associated protein,	
mouse liver	
Tom40	Ahting et al., 2001
transcription factor	
preinitiation (review)	Nogales, 2000
CRSP coactivator, human	Taatjes et al., 2002
TFIID–IIA–IIB, human	Andel et al., 1999
TFIIH, human	Schultz et al., 2000
termination factor rho,	Yu et al., 2000
E. coli	
translin	VanLoock et al., 2001b
tripeptidyl peptidase II (TPPII)	Rochel et al., 2002b
tubulin rings, GDP-	Nicholson et al., 1999
VAT (valesine containing protein-like	
ATPase)	
Thermoplasma acidophilum	Coles et al., 1999; Rockel et al., 2002a
Vault	
rat liver/lung	Kong et al., 1999, 2000; Kickhoefer et al., 2001; Stewart et al., 2002

Appendix 5

Special Journal Issues on Image Processing Techniques

Topic	Journal and Issue	Editors
Advances in image processing for EM	*J. Elect. Microsc. Tech.* 9 (4) 1988	J. Frank and K. Kanaya
Three-dimensional image processing in microscopy	*Eur. J. Cell Biol.* 48 (Suppl. 25) 1989	H.R. Duncker and A. Kriete
Quantitative EM	*Ultramicroscopy* 46 (1–4) 1992	W. Baumeister and E. Zeitler
Image analysis software	*J. Struct. Biol.* 116 (1) 1996	U. Aebi, B. Carragher, and P.R. Smith
Electron tomography	*J. Struct. Biol.* 120 (3) 1997	A.J. Koster and D.A. Agard
Electron crystallography of biological macromolecules	*J. Struct. Biol.* 128 (1) 1999	E. Nogales and W. Chiu
Low-resolution phasing	*Acta Cryst.* D56 (10) 2000	J. Wilson, H. Saibil, and J. Grimes
Single-particle analysis	*J. Struct. Biol.* 133 (2&3) 2001	W. Chiu and S. Ludtke
Electron tomography	*J. Struct. Biol.* 138 (1&2) 2002	B.F. McEwen and A.J. Koster
Analytical methods and software tools for macromolecular microscopy	*J. Struct. Biol.* 144 (1&2) 2003	B. Carragher and P.A. Penczek
Automated particle selection for cryo-EM	*J. Struct. Biol.* 145 (1&2) 2004	C.S. Potter, Y. Zhu, and B. Carragher
Time-resolved imaging of macromolecular process and interactions	*J. Struct. Biol.* 147 (3) 2004	J. Frank and I. Schlichting
Macromolecular assemblies	*Structure* 13 (3) 2005	A. Sali and W. Chin

Note: The special issues are listed in chronological order, such that tabulation follows the steps in the development of methodology.

References

Abbott, A. (2002). The society of proteins. *Nature* **417**, 894–896.

Adams, G.M., Falke, S., Goldberg, A.L., Slaughter, C.A., DeMartino, G.N., and Gogol, E.P. (1997). Structural and functional effects of PA700 and modulator protein on proteasomes. *J. Mol. Biol.* **273**, 646–657.

Adiga, U., Maladi, R., Baxter, B., and Glaeser, R.M. (2004). A binary segmentation approach for boxing ribosome particles in cryo-EM micrographs. *J. Struct. Biol.* **145**, 142–151.

Adrian, M., Dubochet, J., Lepault, J., and McDowall, A.W. (1984). Cryoelectron microscopy of viruses. *Nature* **308**, 32–36.

Aebi, U., Smith, P.R., Dubochet, J., Henry, C., and Kellenberger, E. (1973). A study of the structure of the T-layer of *Bacillus brevis*. *J. Supramol. Struct.* **1**, 498–515.

Agrawal, R.K., and Frank, J. (1999). Structural studies of the translational apparatus. *Curr. Opin. Struct. Biol.* **9**, 215–221.

Agrawal, R.K., Heagle, A.B., and Frank, J. (2000a). Studies of elongation factor G-dependent tRNA translocation by three-dimensional cryo-electron microscopy. In: *The Ribosome: Structure, Function, Antibiotics and Cellular Interactions* (R.A. Garrett, S.R. Douthwate, A. Liljas, A.T. Matheson, P.B. Moore, and H.F. Noller, eds.). ASM Press, Washington, DC, pp. 53–62.

Agrawal, R.K., Heagle, A.B., Penczek, P., Grassucci, R.A., and Frank, J. (1999a). EF-G-dependent GTP hydrolysis induces translocation accompanied by large conformational changes in the 70S ribosome. *Nat. Struct. Biol.* **6**, 643–647.

Agrawal, R.K., Lata, K.R., and Frank, J. (1999b). Conformational variability in *E. coli* 70S ribosome as revealed by 3D cryo-electron microscopy. *Int. J. Biochem. Cell Biol.* **31**, 243–254.

Agrawal, R.K., Linde, J., Sengupta, J., Nierhaus, K.H., and Frank, J. (2001). Localization of L11 protein on the ribosome and elucidation of its involvement in EF-G-dependent translocation. *J. Mol. Biol.* **311**, 777–787.

345

Agrawal, R.K., Penczek, P., Grassucci, R.A., Burkhardt, N., Nierhaus, K.H., and Frank, J. (1999c). Effect of the buffer conditions on the position of the tRNAs on the 70S ribosome as visualized by cryo-electron microscopy. *J. Biol. Chem.* **13**, 8723–8729.

Agrawal, R.K., Penczek, P., Grassucci, R.A., Li, Y., Leith, A., Nierhaus, K.H., and Frank, J. (1996). Direct visualization of A-, P-, and E-site transfer RNA in the *Escherichia coli* ribosome. *Science* **271**, 1000–1002.

Agrawal, R.K., Penczek, P., Grassucci, R.A., and Frank, J. (1998). Visualization of elongation factor G on the *Escherichia coli* 70S ribosome: the mechanism of translocation. *Proc. Natl. Acad. Sci. USA* **95**, 6134–6138.

Agrawal, R.K., Spahn, C.M.T., Penczek, P., Grassucci, R.A., Nierhaus, K.H., and Frank, J. (2000b). Visualization of tRNA movements on the *Escherichia coli* 70S ribosome during the elongation cycle. *J. Cell Biol.* **150**, 447–459.

Ahting, U., Thieffry, M., Engelhardt, H., Hegerl, R., Neupert, W., and Nussberger, S. (2001). Tom40, the pore-forming component of the protein-conducting TOM channel in the outer membrane of mitochondria. *J. Cell Biol.* **153**, 1151–1160. [MSA of projection images.]

Akey, C.W., and Edelstein, S.J. (1983). Equivalence of the projected structure of thin catalase crystals preserved for electron microscopy by negative stain, glucose or embedding in the presence of tannic acid. *J. Mol. Biol.* **163**, 575–612.

Akey, C.W., and Radermacher, M. (1993). Architecture of the *Xenopus* nuclear pore complex revealed by three-dimensional cyro-electron microscopy. *J. Cell Biol.* **122**, 1–19.

Al-Ali, L. (1976). Translational alignment of differently defocused micro-graphs using cross-correlation. In: *Developments in Electron Microscopy & Analysis* (J.A. Venables, ed.). Academic Press, London.

Al-Ali, L., and Frank. J. (1980). Resolution estimation in electron microscopy. *Optik* **56**, 31–40.

Al-Amoudi, A., Chang, J.J., Leforestier, A., McDowall, A., Salamin, L.-A., Norlen, L.P.O., Richter, K., Sartori Blanc, N., Studer, D., and Dubochet, J. (2004). Cryo-electron microscopy of vitreous sections. *EMBO J.* **23**, 3583–3588.

Alberts, B. (1998). The cell as a collection of protein machines: preparing the next generation of molecular biologists. *Cell* **92**, 291–294.

Alberts, B., Johnson, A., Lewis, J., Raff, M., Roberts, K., and Walter, P. (2002). *Molecular Biology of the Cell*. Garland Publishing, New York.

Aldroubi, A., Trus, B.L., Unser, M., Booy, F.P., and Steven, A.C. (1992). Magnification mismatches between micrographs: corrective procedures and implications for structural analysis. *Ultramicroscopy* **46**, 175–188.

Al-Khayat, H.A., Bhella, D., Kenney, J.M., Roth, J.-F., Kingsman, A.J., Martin-Rendon, E., and Saibil, H.R. (1999). Yeast Ty retrotransposons assemble into virus-like particles whose T-numbers depend on the C-terminal length of the capsid protein. *J. Mol. Biol.* **292**, 65–73.

Allen, M.P., and Tildesley, D.J. (1987). *Computer Simulation of Liquids*. Clarendon, Oxford.

Amos, L.A., Henderson, R., and Unwin, P.N.T. (1982). Three-dimensional structure determination by electron microscopy of 2-dimensional crystals. *Prog. Biophys. Mol. Biol.* **39**, 183–231.

Andel, F., Ladurner, A.G., Inoyuye, C., Tjian, R., and Nogales, E. (1999). Three-dimensional structure of the human TFIID–IIA–IIB complex. *Science* **286**, 2153–2156.

Anderberg, M.R. (1973). *Cluster Analysis for Applications*. Academic Press, New York.

Andreasen, N.C., Arndt, S., Swayze, V., Cizadio, T., Flaum, M., O'Leary, D., Ehrhardt, J.C., and Yuh, W.T.C. (1994). Thalamic abnormalities in schizophrenia visualized through magnetic resonance image averaging. *Science* **266**, 294–298.

Andrews, D.W., Walter, P., and Ottensmeyer, F.P. (1985). Structure of the signal recognition particle by electron microscopy. *Proc. Natl. Acad. Sci. USA* **82**, 785–789.

Andrews, D.W., Yu, A.H.C., and Ottensmeyer, F.P. (1986). Automatic selection of molecular images from dark field electron micrographs. *Ultramicroscopy* **19**, 1–14.

Andrews, D.W., Walter, P., and Ottensmeyer, F.P. (1987). Evidence for an extended 7SL RNA structure in the signal recognition particle. *EMBO J.* **6**, 3471–3477.

Angert, I., Burmester, C., Dinges, C., Rose, H., and Schröder, R.R. (1996). Elastic and inelastic scattering cross-sections of amorphous layers of carbon and vitrified ice. *Ultramicroscopy* **63**, 181–192.

Angert, I., Majorovits, E., and Schröder, R.R. (2000). Zero-loss image formation and modified contrast transfer function theory in EFTEM. *Ultramicroscopy* **81**, 203–222.

Arad, T., Piefke, J., Weinstein, S., Gewitz, H.S., Yonath, A., and Wittmann, H.G. (1987). Three-dimensional image reconstruction from ordered arrays of 70S ribosomes. *Biochimie* **69**, 1001–1006.

Asturias, F.J., Jiang, Y.W., Myers, L.C., Gustafsson, C.M., and Kornberg, R.D. (1999). Conserved structures of mediator and RNA polymerase II holoenzyme. *Science* **283**, 985–987.

Asturias, F.J., Chung, W.-H., Kornberg, R.D., and Lorch, Y. (2002). Structural analysis of the RSC chromatin-remodeling complex. *Proc. Natl. Acad. Sci. USA* **99**, 13477–13480.

Auer, M. (2000). Three-dimensional electron cryo-microscopy as a powerful structural tool in molecular medicine. *J. Mol. Med.* **78**, 191–202.

Baker, M.L., Serysheva, I.I., Sencer, S., Wu, Y., Ludtke, S.J., Jiang, W., Hamilton, S.L., and Chiu, W. (2002). The skeletal muscle Ca^{2+} release channel has an oxidoreductase-like domain. *Proc. Natl. Acad. Sci. USA* **99**, 12155–12160.

Baker, T.S., and Cheng, R.H. (1996). A model-based approach for determining orientations of biological macromolecules imaged by cryoelectron microscopy. *J. Struct. Biol.* **116**, 120–130.

Baker, T.S., and Henderson, R. (2001). Electron cryomicroscopy. In: *International Tables for Crystallography. Vol. F: Crystallography of Biological Macromolecules* (M.G. Rossmann and E. Arnold, eds.), pp. 451–479. Kluwer Academic, Dordrecht.

Baker, T.S., and Johnson, J.E. (1996). Low resolution meets high: toward a continuum from cells to atoms. *Curr. Opin. Struct. Biol.* **6**, 585–594.

Baker, T.S., Olson, N.H., and Fuller, S.D. (1999). Adding the third dimension to virus life cycles: three-dimensional reconstruction of icosahedral viruses from cryo-electron micrographs. *Microbiol. Mol. Biol. Rev.* **63**, 862–922.

Ban, N., Freeborn, B., Nissen, P., Penczek, P., Grassucci, R.A., Sweet, R., Frank, J., Moore, P.B., and Steitz, T.A. (1998). A 9 Å resolution X-ray crystallographic map of the large ribosomal subunit. *Cell* **93**, 1105–1115.

Barth, M., Bryan, R.K., and Hegerl, R. (1989). Approximation of missing-cone data in 3D electron microscopy. *Ultramicroscopy* **31**, 365–378.

Baumeister, W., and Hahn, M. (1975). Relevance of three-dimensional reconstructions of stain distributions for structural analysis of biomolecules. *Hoppe-Seyler's Z. Physiol. Chem.* **356**, 1313–1316.

Baumeister, W., and Hahn, M. (1978). Specimen supports. In: *Principles and Techniques of Electron Microscopy: Biological Applications* (M. Hayat, ed.)., Vol. 8. Van Nostrand-Reinhold, New York, pp. 1–112.

Baumeister, W., and Herrmann, K.-H. (1990). High-resolution electron microscpopy in biology: sample preparation, image recording and processing. *Biophys. Elect. Microsc.* **5**, 109–131.

Baumeister, W., and Seredynski, J. (1976). Preparation of perforated films with predeterminable hole size distributions. *Micron* **7**, 49–54.

Baumeister, W., and Steven, A.C. (2000). Macromolecular electron microscopy in the era of structural genomics. *TIBS* **25**, 624–631.

Baumeister, W., Walz, J., Zühl, F., and Seemüller, E. (1998). The proteasome: paradigm of a self-compartmentalizing protease. *Cell* **92**, 367–380.

Baumeister, W., Grimm, R., and Walz, J. (1999). Electron tomography of molecules and cells. *Trends Cell Biol.* **9**, 81–85.

Baxter, W.T., Leith, A., and Frank, J. (2002). A scripting language approach to control software for cryo-electron microscopy. *Proc. IEEE Intl. Symp. Biomed. Imag.*, pp. 301–304.

Bazett-Jones, D.P., Mendez, E., Czarnota, G.J., Ottensmeyer, F.P., and Allfrey, V.G. (1996). Visualization and analysis of unfolded nucleosomes associated with transcribing chromatin. *Nucleic Acids Res.* **24**, 321–329.

Beckmann, R., Bubeck, D., Grassucci, R., Penczek, P., Verschoor, A., Blobel, G., and Frank, J. (1997). Alignment of conduits for the nascent polypeptide chain in the ribosome–Sec61 complex. *Science* **278**, 2123–2126.

Beckmann, R., Spahn, C.M.T., Eswar, N., Helmers, J., Penczek, P.A., Sali, A., Frank, J., and Blobel, G. (2001a). Architecture of the protein-conducting channel associated with the translating 80S ribosome. *Cell* **107**, 361–372.

Beckmann, R., Spahn, C.M.T., Frank, J., and Blobel, G. (2001b). The active 80S ribosome–Sec61 complex. *Cold Spring Harbor Symp. Quant. Biol.* **LXVI**, 543–554.

Bednar, J., Horowitz, R.A., Dubochet, J., and Woodcock, C.L. (1995). Chromatin conformation and self-induced compaction: three-dimensional structural information from cryoelectron microscopy. *J. Cell Biol.* **131**, 1365–1376.

Beer, M., and Moudrianakis, E.N. (1962). Determination of base sequence in nucleic acids with the electron microscope. III. Visibility of a marker. Chemistry and microscopy of guanine-labeled DNA. *Proc. Natl. Acad. Sci. USA* **48**, 409–416.

Beer, M., Frank, J., Hanszen, K.-J., Kellenberger, E., and Williams, R.C. (1975). The possibilities and prospects of obtaining high-resolution information (below 30 Å). on biological material using the electron microscope. *Quart. Rev. Biophys.* **7**, 211–238.

Bellon, P.L., Lanzavecchia, S., and Scatturin, V. (1998). A two exposures technique of electron tomography from projections with random orientation and a quasi-Boolean angular reconstitution. *Ultramicroscopy* **72**, 177–186.

Bellon, P.L., Cantele, F., and Lanzavecchia, S. (2001). Correspondence analysis of sinogram lines. Sinogram trajectories in factor space replace raw images in the orientation of projections of macromolecular assemblies. *Ultramicroscopy* **87**, 187–197.

Bellon, P.L., Cantele, F., de Carlo, S., and Lanzavecchia, S. (2002). A trajectory-based algorithm to determine and refine Euler angles of projections in three-dimensional electron microscopy, improvements and tests. *Ultramicroscopy* **93**, 111–121.

Belnap, D.M., Olson, N., and Baker, T.S. (1997). A method for establishing the handedness of biological macromolecules. *J. Struct. Biol.* **120**, 44–51.

Belnap, D.M., Kumar, A., Folk, J.T., Smith, T.J., and Baker, T.S. (1999). Low-resolution density maps from atomic models: how stepping "back" can be a step "forward." *J. Struct. Biol.* **125**, 166–175.

Beniac, D.R., and Harauz, G. (1995). Structures of small subunit ribosomal RNAs *in situ* from *Escherichia coli* and *Thertnomyces lanuginosus*. *Mol. Cell. Biochem.* **148**, 165–181.

Beniac, D.R., Czarnota, G.J., Rutherford, B.L., Ottensmeyer, F.P., and Harauz G. (1997a). The *in situ* architecture of *Escherichia coli* ribosomal RNA derived by electron spectroscopic imaging and three-dimensional reconstruction. *J. Micros.* **188**, 24–35.

Beniac, D.R., Czarnota, G.J., Rutherford, B.L., Ottensmeyer, F.P., and Harauz, G. (1997b). Three-dimensional architecture of *Thermomyces lanuginosus* small subunit ribosomal RNA. *Micron* **28**, 13–20.

Benzecri, J.P. (1969a). Statistical analysis as a tool to make pattens emerge from data. In: *Methodologies of Pattern Recognition* (S. Watanabe, ed.). Academic Press, New York.

Benzecri, J.P. (1969b). *L'Analyses des Donnees*, Vol. 1. La Taxinomie, Dunod, Paris.

Beroukhim, R., and Unwin, P.N.T. (1997). Distortion correction of tubular crystals: improvements in the acetylcholine receptor structure. *Ultramicroscopy* **70**, 57–81.

Berriman, J.A., and Unwin, P.N.T. (1994). Analysis of transient structures by cryo-electron microscopy combined with rapid mixing of spray droplets. *Ultramicroscopy* **56**, 241–252.

Beuren, F., Maurizi, M.R., Belnap, D.M., Kocsis, E., Booy, F.P., Kessel, M., and Steven, A.C. (1998). At sixes and sevens: characterization of the symmetry mismatch of the ClpAP chaperone-assisted protease. *J. Struct. Biol.* **123**, 248–259.

Bijlholt, M.M.C., van Heel, M.G., and Van Bruggen, E.F.J. (1982). Comparisons of 4 × 6-meric hemocyanins from three different arthropods using computer alignment and correspondence analysis. *J. Mol. Biol.* **161**, 139–153.

Billiald, P., Lamy, J., Taveau, J.C., Motta, G., and Lamy, J. (1988). Mapping of six epitopes in haemocyanin subunit Aa6 by immunoelectron microscopy. *Eur. J. Biochem.* **175**, 423–431.

Blechschmidt, B., Jahn, W., Hainfeld, J.F., Sprinzl, M., and Boublik, M. (1993). Visualization of a ternary complex of the *Escherichia coli* Phe-tRNA^{Phe} and Tu·GTP GTP from *Thermus thermophilus* by scanning electron microscopy. *J. Struct. Biol.* **110**, 84–89.

Boehringer, D., Makarov,. E.M., Sander, B., Makarova, O.V., Kastner, B., Lührmann, R., and Stark, H. (2004). Three-dimensional structure of a pre-catalytic human spliceosomal complex B. *Nat. Struct. Mol. Biol.* **11**, 463–468.

Boekema, E.J. (1991). Negative staining of integral membrane proteins. *Micron Microsc. Acta* **22**, 361–369.

Boekema, E.J., and Böttcher, B. (1992). The structure of ATP synthetase from chloroplasts. Conformational changes of CF, studied by electron microscopy. *Biochem. Biophys. Acta* **1098**, 131–143.

Boekema, E.J., and van Heel, M. (1989). Molecular shape of *Lumbricus terrestis* erythrocruorin studied by electron microscopy and image analysis. *Biochim. Biophys. Acta* **957**, 370–379.

Boekema, E.J., Berden, J.A., and van Heel, M.G. (1986). Structure of mitochondrial F1-ATPase studied by electron microscopy and image processing. *Biochim. Biophys. Acta* **851**, 353–360.

Böhm, J., Frangakis, A.S., Hegerl, R., Nickell, S., Typke, D., and Baumeister, W. (2000). Toward detecting and identifying macromolecules in a cellular context: template matching applied to electron tomograms. *Proc. Natl. Acad. Sci. USA* **97**, 14245–14250.

Böhm, J., Lambert, O., Frangakis, A.S., Letellier, L., Baumeister, W., and Rigaud, J.L. (2001). FhuA-mediated phage genome transfer into liposomes: a cryo-electron tomography study. *Curr. Biol.* **11**, 1168–1175.

Boier Martin, I.M., Marinescu, D.C., Lynch, R.E., and Baker, T.S. (1997). Identification of spherical virus particles in digitized images of entire electron micrographs. *J. Struct. Biol.* **120**, 146–157.

Boisset, N., and Mouche, F. (2000). *Sepia officinalis* hemocyanin: a refined 3D structure from field emission gun cryoelectron microscopy. *J. Mol. Biol.* **296**, 459–472.

Boisset, N., Frank, J., Taveau, J.C., Billiald, P., Motta, G., Lamy, J., Sizaret, P.Y., and Lamy, N. (1988). Intramolecular localization of epitopes within oligomeric protein by immunoelectron microscopy and image processing. *Proteins* **3**, 161–183.

Boisset, N., Taveau, J.C., Pochon, F., Tardieu, A., Barray, M., Lamy, J.N., and Delain, E. (1989). Image processing of proteinase- and methylamine-transformed human α$_2$-macroglobulin. *J. Biol. Chem.* **264**, 12046–12052.

Boisset, N., Taveau, J.-C., and Lamy, J.N. (1990a). An approach to the architecture of *Scutigera coleoptrata* haemocyanin by electron microscopy and image processing. *Biol. Cell* **68**, 73–84.

Boisset, N., Taveau, J.-C., Lamy, N., Wagenknecht, T., Radermacher, M., and Frank, J. (1990b). Three-dimensional reconstruction of native *Androctonus australis* hemocyanin. *J. Mol. Biol.* **216**, 743–760.

Boisset, N., Grassucei, R., Penczek, P., Delain, E., Pochon, F., Frank, J., and Lamy, J.N. (1992a). Three-dimensional reconstruction of a complex of human α$_2$-macroglobulin with monomaleimido nanogold (Au 1.4 nm) embedded in ice. *J. Struct. Biol.* **109**, 39–45.

Boisset, N., Grassucci, R., Motta, G., Lamy, J., Radermacher, M., Taveau, J.C., Liu, W., Frank, J., and Lamy, J.N. (1992b). Three-dimensional immunoelectron microscopy of *Androctonus australis* hemocyanin: the location of monoclonal Fab fragments specific for subunit Aa6. In: *Proceedings of the 10th European Congress on Electron Microscopy* (A. Rios, J.M. Arias, L. Megias-Megias, and A. Lopez-Galindo, eds.),

Vol. 1, pp. 407–408. Secratariado de Publicaciones de la Universidad Granada, Granada, Spain.

Boisset, N., Penczek, P., Pochon, F., Frank, J., and Lamy, J. (1993a). Three-dimensional architecture of human α_2-macroglobulin transformed with methylamine. *J. Mol. Biol.* **232**, 522–529.

Boisset, N., Radermacher, M., Grassucci, R., Taveau, J.-C., Liu, W., Lamy, J., Frank, J., and Lamy, J.N. (1993b). Three-dimensional immunoelectron microscopy of scorpion hemocyanin labeled with monoclonal Fab fragment. *J. Struct. Biol.* **111**, 234–244.

Boisset, N., Penczek, P., Pochon, F., Frank, J., and Lamy, J. (1994a). Three-dimensional reconstruction of human α_2-macroglobulin and refinement of the localization of thiol ester bonds with monomaleimido Nanogold. *Ann. NY Acad. Sci.* **737**, 229–244.

Boisset, N., Taveau, J.C., Penczek, P., and Frank, J. (1994b). Three-dimensional reconstruction in vitreous ice of *Androctonus australis* hemocyanin labelled with a monoclonal Fab fragment. In: *Proceedings of the 13th International Congress on Electron Microscopy (Paris)*, Vol. 1, pp. 527–528. Les Editions de Physiques, Les Ulis, France.

Boisset, N., Penczek, P., Taveau, J.-C., Lamy, J., Frank, J., and Lamy, J.N. (1995a). Three-dimensional reconstruction of *Androctonus australis* hemocyanin labeled with monoclonal Fab fragment. *J. Struct. Biol.* **115**, 16–29.

Boisset, N., Taveau, J.C., Pochon, F., and Lamy, J. (1996). Similar architectures of native and transformed human α_2-macroglobulin suggest the transformation mechanism. *J. Biol. Chem.* **271**, 25762–25769.

Boisset, N., Penczek, P.A., Taveau, J.-C., You, V., de Haas, F., and Lamy, J. (1998). Overabundant single-particle electron microscope views induce a three-dimensional reconstruction artifact. *Ultramicroscopy* **74**, 201–207.

Boisvert, D.C., Wang, J., Otwinowski, Z., Horwich A.L., and Sigler, P.B. (1996). The 2.4 Å crystal structure of the bacterial chaperonin GroEL complexed with ATP gamma S. *Nat. Struct. Biol.* **3**, 116–121.

Booy, F.P., and Pawley, J.B. (1992). Cryo-crinkling: what happens to carbon films on copper grids at low temperature. *Ultramicroscopy* **48**, 273–280.

Borland, L., and van Heel, M. (1990). Classification of image data in conjugate representation spaces. *J. Opt. Soc. Am.* **A7**, 601–610.

Born, M., and Wolf, E. (1975). *Principles of Optics*, 5th ed. Pergamon, Oxford.

Boroukhim, R., and Unwin, P.N.T. (1997). Distortion correction of tubular crystals: improvements in the acetylcholine receptor structure. *Ultramicroscopy* **70**, 57–81.

Böttcher, B., and Crowther, R.A. (1996). Difference imaging reveals ordered regions of RNA in turnip yellow mosaic virus. *Structure* **4**, 387–394.

Böttcher, B., Wynne, S.A., and Crowther, R.A. (1997). Determination of the fold of the core protein of hepatitis B virus by electron cryomicroscopy. *Nature* **36**, 88–91.

Böttcher, B., Schwarz, L., and Gräber, P. (1998). Direct indication for the existence of a double stalk in CF_0F_1. *J. Mol. Biol.* **281**, 757–762.

Böttcher, C., Ludwig, K., Herrmann, A., van Heel, M., and Stark, H. (1999). Structure of influenza haemagglutinin at neutral and at fusogenic pH by electron cryo-electron cryo-microscopy. *FEBS Lett.* **463**, 255–259.

Böttcher, B., Bertsche, I., Reuter, R., and Gräber, P. (2000). Direct visualisation of conformational changes in EF_0F_1 by electron microscopy. *J. Mol. Biol.* **296**, 449–457.

Böttcher, B., Scheide, D., Hesterberg, M., Nagel-Steger, L., and Friedrich, T. (2002). A novel, enzymatically active conformation of the *Escherichia coli* NADH:ubiquinone oxidoreductase (complex I). *J. Biol. Chem.* **277**, 17970–17977.

Boublik, M., Hellmann, W., and Kleinschmidt, A.K. (1977). Size and structure of *E. coli* ribosomes by electron microscopy. *Cytobiologie* **14**, 293–300.

Bracewell, R.N., and Riddle, A.C. (1967). Inversion of fan-beam scans in radio astronomy. *Astrophys. Soc.* **150**, 427–434.

Braet, F., Bomans, P.H.H., Wisse, E., and Frederick, P.M. (2003). The observation of intact hepatic endothelial cells by cryo-electron microscopy. *J. Microsc.* **212**, 175–185.

Braig, K., Simon, M., Furuya, F., Hainfeld, J. F., and Horwich, A.L. (1993). A polypeptide bound by chaperonin groEL is located within a central cavity. *Proc. Natl. Acad. Sci. USA* **90**, 3978–3982.

Bremer, A., Henn, C., Engel, A., Baumeister, W., and Aebi, U. (1992). Has negative staining still a place in biomacromolecular electron microscopy? *Ultramicroscopy* **46**, 85–111.

Brenner, S., and Horne, R.W. (1959). A negative staining method for high resolution electron microscopy of viruses. *Biochim. Biophys. Acta* **34**, 103–110.

Bretaudiere, J.P., and Frank, J. (1986). Reconstitution of molecule images analysed by correspondence analysis: A tool for structural interpretation. *J. Microsc.* **144**, 1–14.

Bricogne, G. (1976). Methods and programs for direct-space exploitation of geometric redundancies. *Acta Cryst.* **A32**, 832–847.

Brimacombe, R. (1995). Ribosomal RNA; a three-dimensional jigsaw puzzle. *Eur. J. Biochem.* **230**, 365–385.

Brink, J., and Chiu, W. (1994). Applications of a slow-scan CCD camera in protein electron crystallography. *J. Struct. Biol.* **113**, 23–34.

Brink, J., Chiu, W., and Dougherty, M. (1992). Computer-controlled spot-scan imaging of crotoxin complex crystals with 400 keV electrons at near-atomic resolution. *Ultramicroscopy* **46**, 229–240.

Brink, J., Sherman, M.B., Berriman, J., and Chiu, W. (1998). Evaluation of charging on macromolecules in electron cryomicroscopy. *Ultramicroscopy* **72**, 41–52.

Brink, J., Ludtke, S.J., Yang, C.-Y., Gu, Z.-W., Wakil, S.J., and Chiu, W. (2002). Quaternary structure of human fatty acid synthetase by electron cryomicroscopy. *Proc. Natl. Acad. Sci. USA* **99**, 138–143.

Brink, J., Ludtke, S.J., Kong, Y., Wakil, S.J., Ma, J., and Chiu, W. (2004). Experimental verification of conformational variations of human fatty acid synthetase as predicted by normal mode analysis. *Structure* **12**, 185–191.

Brinker, A., Pfeifer, G., Kerner, M.J., Naylor, D.J., Hartl, F.U., and Hayer-Hartl, M. (2001). Dual function of protein confinement in chaperonin-assisted protein folding. *Cell* **107**, 223–233.

Brooks, B.R., and Karplus, M. (1983). Harmonic dynamic of proteins: normal mode and fluctuations in bovine pancreatic trypsin inhibitor. *Proc. Natl. Acad. Sci. USA* **80**, 6571–6575.

Brünger, A. (1992). Free R value: a novel statistical quantity for assessing the accuracy of crystal structures. *Nature* **355**, 472–475.

Brünger, A. (1993). Assessment of phase accuracy by cross validation: the Free R value. Methods and applications. *Acta Cryst.* **D49**, 24–36.

Brünger, A.T. (1997). Free R value: cross-validation in crystallography. *Methods Enzymol.* **222**, 366–396.

Büchel, C., Morris, E., Orlova, E., and Barber, J. (2001). Localization of the PsbH subunit in Photosystem II: a new approach using labeling of His-tags with a Ni^{2+}–NTA gold cluster and single particle analysis. *J. Mol. Biol.* **312**, 371–379.

Bullough, P., and Henderson, R. (1987). Use of spotscan procedure for recording low-dose micrographs of beam-sensitive specimens. *Ultramicroscopy* **21**, 223–230.

Burge, R.E., and Scott, R.F. (1975). Electron microscope calibration by astigmatic images. *Optik* **43**, 503–507.

Burgess, S., Walker, M.L., Sakakibara, H., Knight, P.J., and Oiwa, K. (2003). Dynein structure and power stroke. *Nature* **421**, 715–718.

Burley, S.K. (2000). An overview over structural genomics. *Nat. Struct. Biol.* **7**, 932–934.

Butt, H.-J., Wang, D.N., Hansma, P.K., and Kühlbrandt, W. (1991). Effect of surface roughness of carbon films on high-resolution electron diffraction of protein crystals. *Ultramicroscopy* **36**, 307–318.

Campbell, I.D. (2002). The march of structural biology. *Nat. Rev.: Mol. Cell Biol.* **3**, 377–381.

Cantele, F., Lanzavecchia, S., and Bellon, P.L. (2003). The variance of icosahedral virus models is a key indicator in the structure determination: a model-free reconstruction of viruses, suitable for refractory particles. *J. Struct. Biol.* **141**, 84–92.

Carazo, J.M. (1992). The fidelity of 3D reconstruction from incomplete data and the use of restoration methods. In: *Electron Tomography* (J. Frank, ed.), pp. 117–166. Plenum Press, New York.

Carazo, J.M., and Carrascosa, J.L. (1987a). Restoration of direct Fourier three-dimensional reconstructions of crystalline specimens by the method of convex projections. *J. Microsc.* **145**, 159–177.

Carazo, J.M., and Carrascosa, J.L. (1987b). Information recovery in missing angular data cases: An approach by the convex projections method in three dimensions. *J. Microsc.* **145**, 23–43.

Carazo, J.M., and Frank. J. (1988). Three-dimensional matching of macromolecular structures obtained from electron microscopy: An application to the 70S and 50S *E. coli* ribosomal particles. *Ultramicroscopy* **25**, 13–22.

Carazo, J.M., and Stelzer, E.H.K. (1999). The BioImage Database Project: organizing multidimensional biological images in an object-relational database. *J. Struct. Biol.* **125**, 97–102.

Carazo, J.M., Wagenknecht, T., Radermacher, M., Mandiyan, V., Boublik, M., and Frank, J. (1988). Three-dimensional structure of 50S *E. coli* ribosomal subunits depleted of proteins L7/LI2. *J. Mol. Biol.* **201**, 393–404.

Carazo, J.M., Wagenknecht, T., and Frank, J. (1989). Variations of the three-dimensional structure of *Escherichia coli* ribosome in the range of overlap views. *Biophys. J.* **55**, 465–477.

Carazo, J.M., Rivera, F.F., Zapata, E.L., Radermacher, M., and Frank, K. (1990). Fuzzy sets-based classification of electron microscopy images of biological macromolecules with an application to ribosomal particles. *J. Microsc.* **157**, 187–203.

Carazo, J.M., Benavides, I., Rivera, F.F., and Zapata, E. (1992). Identification, classification, and 3D reconstruction on hypercube computers. *Ultramicroscopy* **40**, 13–32.

Carazo, J.M., Marabini, R., Vaquerizo, C., and Frank, J. (1994). Towards a data-bank of three-dimensional macromolecular structures: A WWW-based prototype. In: *Proceedings of the 13th International Congress on Electron Microscopy (Paris)*, Vol. 1, pp. 519–520. Les Editions de Physiques, Les Ulis, France.

Carragher, B., and Smith, P.R. (1996). Advances in computational image processing for microscopy. *J. Struct. Biol.* **116**, 2–8.

Carragher, B., Kisseberth, N., Kriegman, D., Milligan, R.A., Potter C.S., Pulokas, J., and Reilein, A. (2000). Leginon: an automated system for acquisition of images from vitreous ice specimens. *J. Struct. Biol.* **132**, 33–45.

Carrascosa, J.L., and Steven, A.C. (1979). A procedure for evaluation of significant structural differences between related arrays of protein molecules. *Micron* **9**, 199–206.

Carson, M. (1997). Ribbons. *Methods Enzymol.* **277**, 493–505.

Cate, J.H., Yusupov, M.M., Yusupova, G.Z., Earnest, T.N., and Noller, H.N. (1999). X-ray structures of 70S ribosome functional complexes. *Science* **285**, 2095–2104.

Cejka, Z. Kleinz, J., Santini, C., Hegerl, R., and Ghiretti Magaldi, A. (1992). The molecular architecture of the extracellular hemoglobin of *Ophelia bicornis:* analysis of individual molecules. *J. Struct. Biol.* **109**, 52–60.

Chacón, P., and Wriggers, W. (2002). Multi-resolution contour-based fitting of macromolecular structures. *J. Mol. Biol.* **317**, 375–384.

Chacón, P., Tama, F., and Wriggers, W. (2002). Mega-dalton biomolecular motion captured from electron microscopy reconstructions. *J. Mol. Biol.* **317**, 485–492.

Chapman, M. (1995). Restrained real-space macromolecular refinement using a new resolution-dependent electron-density function. *Acta Cryst.* **A51**, 69–80.

Che, Z., Olson, N.H., Leippe, D., Lee, W.-M., Mosser, A.G., Rueckert, R.R., Baker, T.S., and Smith, T.J. (1998). Antibody-mediated neutralization of human rhinovirus 14 explored by means of cryoelectron microscopy and X-ray crystallography of virus–Fab complexes. *J. Virol.* **72**, 4610–4622.

Chen, L.F., Winkler, H., Reedy, M.C., and Taylor, K.A. (2002b). Molecular modeling of averaged rigor cross-bridges from tomograms of insect flight muscle. *J. Struct. Biol.* **138**, 92–104.

Chen, S., Roseman, A.M., Hunter, A., Wood, S.P., Burston, S.G., Ranson N., Clarke, A.Z.R., and Saibil, H.R. (1994). Location of a folding protein and shape changes in GroEl–GroES complexes imaged by cryo-electron microscopy. *Nature* **371**, 261–264.

Chen, S., Roseman, A.M., and Saibil, H.R. (1998). Electron microscopy of chaperonins. *Meth. Enzymol.* **290**, 242–253.

Chen, Y.-J., Yu, X., and Egelman, E.H. (2002a). The hexameric ring structure of the *Escherichia coli* RuvB branch migration protein. *J. Mol. Biol.* **319**, 587–591.

Cheng, R.H., Reddy, V.S., Olson, N.H., Fisher, A.J., Baker, T.S., and Johnson, J.E. (1994). Functional implications of quasi-equivalence in a $T = 3$ icosohedral animal virus established by cryo-electron microscopy and X-ray crystallography. *Structure* **2**, 271–282.

Chiu, C.Y., Cary, R.B., Chen, D.J., Peterson, S.R., and Stewart, P.L. (1998). Cryo-EM imaging of the catalytic subunit of the DNA-dependent protein kinase. *J. Mol. Biol.* **284**, 1075–1081.

Chiu, W. (1986). Electron microscopy of frozen, hydrated biological specimens. *Annu. Rev. Biophys. Biophys. Chem.* **15**, 237–257.

Chiu, W. (1993). What does electron cryomicroscopy provide that X-ray crystallography and NMR spectroscopy cannot? *Annu. Rev. Biophys. Biomol. Struct.* **22**, 233–255.

Chiu, W., Downing, K.H., Dubochet, J., Glaeser, R.M., Heide, H.G., Knapek, E., Kopf, D.A., Lamvik, M.K., Lepault, J., Robertson, J.D., Zeitler, E., and Zemlin, F. (1986). Cryoprotection in electron microscopy. *J. Microsc.* **141**, 385–391.

Chiu, W., Burnett, R., and Garcea, R. (1997). *Structural Biology of Viruses.* Oxford University Press, New York.

Chiu, W., McGough, A., Sherman, M.B., and Schmid, M.F. (1999). High-resolution electron cryomicroscopy of macromolecular assemblies. *Trends Cell Biol.* **9**, 154–159.

Chiu, W., Baker, M.L., Jiang, W., and Zhou, Z.H. (2002). Deriving folds of macro-molecular complexes through electron microscopy and bioinformatics approaches. *Curr. Opin. Struct. Biol.* **12**, 263–269.

Cicicopol, C., Peters, J., Lupas, A., Cejka, Z., Müller, S.A., Golbik, R., Pfeifer, G., Lilie, H., Engel, A., and Baumeister, W. (1999). Novel molecular architecture of the multimeric archaeal PEP–synthetase homologue (MAPS) from *Staphylothermus marinus*. *J. Mol. Biol.* **290**, 347–361.

Clark, R. (1992). Towards a complete description of the three-dimensional filtered backprojection. *Phys. Med. Biol.* **37**, 646–660.

Coles, M., Diercks, T., Lierman, J., Gröger, A., Rockel, B., Baumeister, W., Koretke, K.K., Lupas, A., Peters, J., and Kessler, H. (1999). The solution structure of VAT-N reveals a "missing link" in the evolution of complex enzymes from a simple $\beta\alpha\beta\beta$ element. *Curr. Biol.* **9**, 1158–1168.

Conway, J.F., and Steven, A.C. (1999). Methods for reconstructing density maps of "single" particles from cryoelectron micrographs to subnanometer resolution. *J. Struct. Biol.* **128**, 106–118.

Conway, J.F., Trus, B.L., Booy, F.P., Newcomb, W.W., Brown, J.C., and Steven, A.C. (1993). The effects of radiation damage on the structure of frozen-hydrated HSV-1 capsids. *J. Struct. Biol.* **111**, 222–233.

Conway, J.F., Cheng, N., Zlotnick, A., Wingfield, P.T., Stahl, S.J., and Steven, A.C. (1997). Visualization of a 4-helix bundle in the hepatitis B virus capsid by cryo-electron microscopy. *Nature* **386**, 91–94.

Cooley, J.W., and Tukey, J.W. (1965). An algorithm for the machine calculation of complex Fourier series. *Math. Comput.* **19**, 297–301.

Craighead, J.L., Chang, W., and Asturias, F.J. (2002). Structure of yeast RNA polymerase II in solution: implications for enzyme regulation and interaction with promoter DNA. *Structure* **18**, 1117–1125.

Crowther, R.A. (1971). Procedures for three-dimensional reconstruction of spherical viruses by Fourier synthesis from electron micrographs. *Phils. Trans. R. Soc. Lond. B* **261**, 221–230.

Crowther, R.A. (1972). The fast rotation function. In: *The Molecular Replacement Method* (M.G. Rossmann, ed.), pp. 173–178. Gordon & Breach, New York.

Crowther, R.A. (1976). The interpretation of images reconstructed from electron micrographs of biological particles. In: *Proceedings of the Third John Innes Symposium* (R. Markham and R.W. Horne, eds.), pp. 15–25. North-Holland, Amsterdam.

Crowther, R.A., and Amos, L.A. (1971). Three-dimensional image reconstructions of some small spherical viruses. *Cold Spring Harbor Symp. Quant. Biol.* **36**, 489–494.

Crowther, R.A., DeRosier, D.J., and Klug, A. (1970). The reconstruction of a three-dimensional structure from projections and its application to electron microscopy. *Proc. R. Soc. Lond.* **317**, 319–340.

Crowther, R.A., Henderson, R., and Smith, J.M (1996). MRC image processing programs. *J. Struct. Biol.* **116**, 9–16.

Cruickshank, D.W.J. (1959). *International Tables for X-Ray Crystallography*, Vol. 11, pp. 84–98. Kynoch Press, Birmingham, UK.

Crum, J., Gruys, K.J., and Frey, T.G. (1994). Electron microscopy of cytochrome *c* oxidase crystals: labeling of subunit III with a monomaleimide undecagold cluster compound. *Biochemistry* **33**, 13719–13726.

Cyrklaft, M., and Kühlbrandt, W. (1994). High-resolution electron microscopy of biological specimens in cubic ice. *Ultramicroscopy* **55**, 141–153.

Cyrklaff, M., Adrian, M., and Dubochet, J. (1990). Evaporation during preparation of unsupported thin vitrified aqueous layers for cryo-electron microscopy. *J. Electron Microsc.* **16**, 351–355.

Czarnota, G.J., and Ottensmeyer, F.P. (1996). Structural states of the nucleosome. *J. Biol. Chem.*, **271**, 3677–3683.

Czarnota, G.J., Andrews, D.W., Farrow, N.A., and Ottensmeyer, F.P. (1994). A structure for the signal sequence binding protein SRP54: 3D reconstruction from STEM images of single molecules. *J. Struct. Biol.* **113**, 35–46.

Czarnota, G.J., Bazett-Jones, D.P., Mendez, E., Allfrey, V.G., and Ottensmeyer, F.P. (1997). High resolution microanalysis and three-dimensional nucleosome structure associated with transcribing chromatin. *Micron* **28**, 419–431.

Danev, R., and Nagayama, K. (2001).Transmission electron microscopy with Zernicke plate. *Ultramicroscopy* **88**, 243–252.

da Fonseca, P.C., Morris, S.A., Nerou, E.P., Taylor, C.W., and Morris, E.P. (2003). Domain organization of the type 1 inositol 1,4,5-trisphosphate receptor as revealed by single-particle analysis. *Prac. Natl. Acad. Sci. USA* **100**, 3936–3941.

Darst, S.A., Edwards, A.M., Kubalek, E.W., and Kornberg, R.D. (1991). Three-dimensional structure of yeast RNA polymerase II at 16 Å resolution. *Cell* **66**, 121–128.

Darst, S.A., Opalka, N. Chacon, P., Polykov, A., Richter, C., Zhang, G., and Wriggers, W. (2002). Conformational flexibility of bacterial RNA polymerase. *Proc. Natl. Acad. Sci. USA* **99**, 429–431.

Davies, C., and White, S.W. (2000). Electrons and X-rays gang up on the ribosome. *Structure* **8**, 41–45.

Davis, J.A., Takagi, Y., Kornberg, R.D., and Asturias, F.J. (2002). Structure of the yeast RNA polymerase II holoenzyme: mediator conformations and polymerase interaction. *Mol. Cell* **10**, 409–415.

De-Alarcon, P.A., Pascual-Montano, A., Gupta, A., and Carazo, J.M. (2002). Modeling shape and topology of low-resolution density maps of biological macromolecules. *Biophys. J.* **83**, 619–632.

de Carlo, S., El-Bez, C., Alvarez-Rua, C., Borge, J., and Dubochet, J. (2002). Cryo-negative staining reduces electron-beam sensitivity of vitrified biological particles. *J. Struct. Biol.* **138**, 216–226.

de Haas, F., and van Bruggen, E.F.J. (1994). The interhexameric contacts in the four-hexameric hemocyanin from the tarantula *Eurypelma californicum*. *J. Mol. Biol.* **237**, 464–478.

de Haas, F., Bijlholt, M.M.C., and van Bruggen, E.F.J. (1991). An electron microscopic study of two-hexameric hemocyanins from the crab *Cancerpagurus* and the tarantula

Eurypelma califomicum: determination of their quaternary structure using image processing and simulation models based on X-ray diffraction data. *J. Struct. Biol.* **107**, 86–94.

de Haas, F., van Bremen, J.F.L., Boekema, E.J., and Keegstra, W. (1993). Comparative electron microscopy and image analysis of oxy- and deoxy-hemocyanin from the spiny lobster *Panulirus interruptus*. *Ultramicroscopy* **49**, 426–435.

de Haas, F., Boisset, N., Taveau, J.C., Lambert, O., Vinogradov, S.N., and Lamy, J.N. (1996a). Three-dimensional reconstruction of *Macrobdella decora* (leech) hemoglobin by cryoelectron microscopy. *Biophys. J.* **70**, 1973–1984.

de Haas, F., Taveau, J.-C., Boisset, N., Lambert, O., Vinogradov, S.N., and Lamy, J.N. (1996b). Three-dimensional reconstruction of the chlorocruorin of the polychaete annelid *Eudistylia vancouverii*. *J. Mol. Biol.* **255**, 140–153.

de Haas, F., Zal, F., Lallier, F.H., Toulmond, A., and Lamy, J. (1996c). Three-dimensional reconstruction of the hexagonal bilayer hemoglobin of the hydrothermal vent tube worm *Riftia pachyptila* by cryoelectron microscopy. *Proteins* **26**, 241–256.

de Haas, F., Kuchumov, A., Taveau, J.-C., Boisset, N., Vinogradov, S.N., and Lamy, J.N. (1997). Three-dimensional reconstruction of native and reassembled *Lumbricus terrestris* extracellular hemoglobin. Location of the monomeric globin chains. *Biochemistry* **36**, 7330–7338.

de Haas, F., Paatero, A.O., Mindich, L., Bamford, D.H., and Fuller, S.D. (1999). A symmetry mismatch at the site of RNA packaging in the polymerase complex of dsRNA bacteriophage F6. *J. Mol. Biol.* **294**, 357–372.

de Jong, A.F., and van Dyck, D. (1993). Ultimate resolution and information in electron microscopy. II. The information limit of transmission electron microscopes. *Ultramicroscopy* **49**, 66–80.

de la Fraga, L., Dopazo, J., and Carazo, J.M. (1995). Confidence limits for resolution estimation in image averaging by random subsampling. *Ultramicroscopy* **60**, 385–391.

Dengler, J. (1989). A multi-resolution approach to the 3D reconstruction from an electron microscope tilt series solving the alignment problem without gold particles. *Ultramicroscopy* **30**, 337–348.

DeRosier, D.J. (1997). Who needs crystals anyway? *Nature* **38**, 26–27.

DeRosier, D.J. (2000). Correction of high-resolution data for curvature of the Ewald sphere. *Ultramicroscopy* **81**, 83–98.

DeRosier, D.J., and Harrison, S.C. (1997). Macromolecular assemblages, sizing things up. *Curr. Opin. Struct. Biol.* **7**, 237–238.

DeRosier, D., and Klug, A. (1968). Reconstruction of 3-dimensional structures from electron micrographs. *Nature* **217**, 130–134.

DeRosier, D.J., and Moore, P.B. (1970). Reconstruction of three-dimensional images from electron micrographs of structures with helical symmetry. *J. Mol. Biol.* **52**, 355–369.

de Ruijter, W.J. (1995). Imaging properties and applications of slow-scan charge-coupled device cameras suitable for electron microscopy. *Micron* **26**, 247–275.

Diday, E. (1971). La methode des nuees dynamiques. *Rev. Stat. Appl.* **19**, 19–34.

Dierksen, K., Typke, D., Hegerl, R., Koster, A.J., and Baumeister, W. (1992). Towards automatic tomography. *Ultramicroscopy* **40**, 71–87.

Dierksen, K., Typke, D., Hegerl, R., and Baumeister, W. (1993). Towards automatic tomography. II. Implementation of autofocus and low-dose procedures. *Ultramicroscopy* **49**, 109–120.

Di Francia, T. (1955). Resolution power and information. *J. Opt. Soc. Am.* **45**, 497–501.

Domgall, I., Venzke, D., Lüttke, U., Ratajezak, R., and Böttcher, B. (2002). Three-dimensional map of a plant V-ATPase based on electron microscopy. *J. Biol. Chem.* **277**, 13115–13121.

Dover, S.D., Elliot, A., and Kernagham, A.K. (1980). Three-dimensional reconstruction from images of tilted specimens: the paramyosin filament. *J. Microsc.* **122**, 23–33.

Downing, K.H. (1991). Spot scan imaging in TEM. *Science* **251**, 53–59.

Downing, K.H. (1992). Automatic focus correction for spot-scan imaging of tilted specimens. *Ultramicroscopy* **46**, 199–206.

Downing, K.H., and Glaeser, R.M. (1986). Improvement in high resolution image quality of radiation-sensitive specimens achieved with reduced spot size of the electron beam. *Ultramicroscopy* **20**, 269–278.

Downing, K.H., and Grano, D.A. (1982). Analysis of photographic emulsions for electron microscopy of two-dimensional crystalline specimens. *Ultramicroscopy* **7**, 381–404.

Downing, K.H., and Hendrickson, F.M. (1999). Performance of a 2k CCD camera designed for electron crystallography at 400 kV. *Ultramicroscopy* **75**, 215–233.

Downing, K.H., Koster, A.J., and Typke, D. (1992). Overview of computer-aided electron microscopy. *Ultramicroscopy* **46**, 189–197.

Dryden, K.A., Wang, G., Yeager, M., Nibert, M.L., Coombs, K.M., Furlong, D.B., Fields, B.N., and Baker, T.S. (1993). Early steps in reovirus infection are associated with dramatic changes in supramolecular structure and protein conformation: analysis of virions and subviral particles by cryoelectron microscopy and image reconstruction. *J. Cell Biol.* **122**, 1023–1041.

Dube, P., Tavares, P., Lurz, R., and van Heel, M. (1993). The portal protein of bacteriophage SPP1: a DNA pump with 13-fold symmetry. *EMBO J.* **12**, 1303–1309.

Dube, P., Tavares, P., Orlova, E.V., Zemlin, F., and van Heel, M. (1994). Three-dimensional structure of portal protein from bacteriophage SPP1. In: *Proceedings of the International Congress on Electron Microscopy (Paris)*, Vol. 3, pp. 533–534. Les Editions de Physiques, Les Ulis, France.

Dube, P., Bacher, G., Stark, H., Müller, F., Zemlin, F., van Heel, M., and Brimacombe. R. (1998a). Correlation of the expansion segments in mammalian rRNA with the fine structure of the 80S ribosome: a cryoelectron microscope reconstruction of the rabbit reticulocyte ribosome at 21 Å resolution. *J. Mol. Biol.* **279**, 403–421.

Dube, P., Wieske, M., Stark, H., Schatz, M., Stahl, J., Zemlin, F., Lutsch, G., and van Heel, M. (1998b). The 80S rat liver ribosome at 25 Å resolution by electron cryomicroscopy and angular reconstitution. *Structure* **6**, 389–399.

Dubochet, J., Lepault, J., Freeman, R., Berriman, J.A., and Homo, J.-C. (1982). Electron microscopy of frozen water and aqueous solutions. *J. Microsc.* **128**, 219–237.

Dubochet, J., Adrian, M., Lepault, J., and McDowall, A.W. (1985). Cryo-electron microscopy of vitrified biological specimens. *Trends Biochem Sci.* **10**, 143–146.

Dubochet, J., Adrian, Chang, J.-J., Homo, J.-C., Lepault, J., McDowall, A.W., and Schultz, P. (1988). Cryo-electron microscopy of vitrified specimens. *Quart. Rev. Biophys.* **21**, 129–228.

Duda, R., and Hart, P. (1973). *Pattern Classification and Analysis*, pp. 10–43, 189–260. John Wiley, New York.

Editorial, unsigned (2000). RNA–protein machines rule. *Nature Struct. Biol.* **7**, 817.

Egelman, E. (1986). An algorithm for straightening images of curved filamentous structures. *Ultramicroscopy* **19**, 367–374.

Egelman, E.H. (2000). A robust algorithm for the reconstruction of helical filaments using single-particle methods. *Ultramicroscopy* **85**, 225–234.

Egelman, E.H., Yu, X., Wild, R., Hingorani, M.M., and Patel, S.S. (1995). T7 helicase/primase proteins form rings around single-stranded DNA that suggest a general structure for hexameric helicases. *Proc. Natl. Acad. Sci. USA* **92**, 3869–3873.

Egerton, R.F. (1975). Inelastic scattering of 80 keV electrons in amorphous ice. *Phil. Mag.* **31**, 199–215.

Erickson, H.P., and Klug, A. (1970). The Fourier transform of an electron micrograph: effects of defocussing and aberrations, and implications for the use of underfocus contrast enhancement. *Ber. Bunsenges. Phys. Chem.* **74**, 1129–1137.

Erickson, H.P., and Klug, A. (1971). Measurement and compensation of defocusing and aberrations by Fourier processing of electron micrographs. *Phil. Trans.* **B261**, 105–118.

Ermantraut, E., Wohlfart, K., and Tichelaar, W. (1998). Perforated support foils with pre-defined hole size, shape, and arrangement. *Ultramicroscopy* **74**, 75–81.

Fabiola, F., and Chapman, M.S. (2005). Fitting of high-resolution structures into electron microscopy reconstruction images. *Structure* **13**, 389–400.

Falke, S., Fisher, M.T., and Gogol, E.P. (2001a). Classification and reconstruction of a heterogeneous set of electron microscopic images: a case study of GroEL-substrate complexes. *J. Struct. Biol.* **133**, 203–213.

Falke, S., Fisher, M.T., and Gogol, E.P. (2001b). Structural changes in GroEL effected by binding a denatured protein substrate. *J. Mol. Biol.* **308**, 569–577.

Fan, G., Mercurio, P., Young, S., and Ellisman, M.H. (1993). Telemicroscopy. *Ultramicroscopy* **52**, 499–503.

Farr, G.W., Furtak, K., Rowland, M.B., Ranson, N.A., Saibil, H.R., Kirchhausen, T., and Horvich, A.L. (2000). Multivalent binding of nonnative substrate proteins by the chaperonin GroEl. *Cell* **100**, 561–573.

Farrow, N.A., and Ottensmeyer, F.P. (1989). Maximum entropy methods and dark field microscopy images. *Ultramicroscopy* **31**, 275–284.

Farrow, N.A., and Ottensmeyer, F.P. (1992). *A-posteriori* determination of relative projection directions of arbitrarily oriented macromolecules. *J. Opt. Soc. Am.* **A9**, 1749–1760.

Faruqi, A.R., and Subramaniam. (2000). CCD detectors in high-resolution biological electron microscopy. *Quart. Rev. Biol.* **33**, 1–27.

Fernández, J.J., and Carazo, J.M. (1996). Analysis of structural variability within two-dimensional biological crystals by a combination of patch averaging techniques and self-organizing maps. *Ultramicroscopy* **65**, 81–93.

Fernández, J.J., Sanjurjo, J.R., and Carazo, J.M. (1997). A spectral estimation approach to contrast transfer function detection in electron microscopy. *Ultramicroscopy* **68**, 267–295.

Francis, N.R., Sosinsky, G.E., Thomas, D., and DeRosier, D.J. (1994). Isolation, characterization and structure of bacterial flagellar motors containing the switch complex. *J. Mol. Biol.* **235**, 1261–1270.

Francis, N.R., Wolanin, P.M., Stock, J.B., DeRosier, D.J., and Tomas, D.R. (2004). Three-dimensional structure and organization of a receptor/signaling complex. *Proc. Natl. Acad. Sci. USA* **101**, 17480–17485.

Frangakis, A.S., and Hegerl, R. (1999). Nonlinear anisotropic diffusion in three-dimensional electron microscopy. In: *Lecture Notes in Computer Science*, Vol. 1682, Springer-Verlag, Berlin, pp. 386–397.

Frangakis, A.S., and Hegerl, R. (2001). Noise reduction in electron tomographic reconstructions using nonlinear anisotropic diffusion. *J. Struct. Biol.* **135**, 239–250.

Frangakis, A.S. and Hegerl, R. (2002a). Segmentation of biomedical images with eigenvectors. *Proc. IEEE Int. Symp. Biol. Imag.* pp. 90–93.

Frangakis, A.S., and Hegerl, R. (2002b). Segmentation of two- and three-dimensional data from electron microscopy using eigenvector analysis. *J. Struct. Biol.* **138**, 105–113.

Frangakis, A.S., Stoschek, A., and Hegerl, R. (2001). Wavelet transform filtering and nonlinear anisotropic diffusion assessed for signal reconstruction performance on multidimensional biomedical data. *IEEE Trans. Biomed. Eng.* **2**, 213–222.

Frangakis, A.S., Boehm, J., Forster, F., Nickell, S., Nicastro, D., Typke, D., Hegerl, R., and Baumeister, W. (2002). Identification of macromolecular complexes in cryoelectron tomograms of phantom cells. *Proc. Natl. Acad. Sci. USA* **99**, 14153–14158.

Frank, J. (1969). Nachweis von Objektbewegungen im licht-optischen Diffraktogramm von elektronenmikroskopischen Aufnahmen. *Optik* **30**, 171–180.

Frank, J. (1972a). Observation of the relative phases of electron microscopic phase contrast zones with the aid of the optical diffractometer. *Optik* **35**, 608–612.

Frank, J. (1972b). Two-dimensional correlation functions in electron microscope image analysis. In: *Proceedings of the Fifth European Congress on Electron Microscopy, Manchester*, pp. 622–623. The Institute of Physics, London.

Frank, J. (1972c). A study on heavy/light atom discrimination in bright field electron microscopy using the computer. *Biophys. J.* **12**, 484–511.

Frank, J. (1973a). The envelope of electron microscopic transfer functions for partially coherent illumination. *Optik* **38**, 519–536.

Frank, J. (1973b). Use of anomalous scattering for element discrimination. In: *Image Processing and Computer-Aided Design in Electron Optics* (P.W. Hawkes, ed.), pp. 196–211. Academic Press, San Diego.

Frank, J. (1973c). Computer processing of electron micrographs. In: *Advanced Techniques in Biological Electron Microscopy* (J.K. Koehler, ed.), pp. 215–274. Springer-Verlag, Berlin.

Frank, J. (1975a). Averaging of low exposure electron micrographs of nonperiodic objects. *Ultramicroscopy* **1**, 159–162.

Frank, J. (1975b). Partial coherence and efficient use of electrons in bright field electron microscopy. *Optik* **43**, 103–109.

Frank, J. (1976). Determination of source size and energy spread from electron micrographs using the method of Young's fringes. *Optik* **44**, 379–391.

Frank, J. (1979). Image analysis in electron microscopy. *J. Microsc.* **117**, 25–38.

Frank, J. (1980). The role of correlation techniques in computer image processing. In: *Computer Processing of Electron Microscope Images* (P.W. Hawkes, ed.). Springer-Verlag, Berlin, pp. 187–222.

Frank, J. (1982). New methods for averaging non-periodic objects and distorted crystals in biologic electron microscopy. *Optik* **63**, 67–89.

Frank, J. (1984a). The role of multivariate statistical analysis in solving the architecture of the *Limulus polyphemus* hemocyanin molecule. *Ultramicroscopy* **13**, 153–164.

Frank, J. (1984b). Recent advances of image processing in the structural analysis of biological macromolecules. In: *Proceedings of the 8th European Congress on Electron Microscopy*, Vol. 2, pp. 1307–1316. Electron Microscopy Foundation, Program Committee, Budapest.

Frank, J. (1985). Image analysis of single molecules. *Electron Microsc. Rev.* **7**, 53–74.

Frank, J. (1989b). Image analysis of single macromolecules. *Electron Microsc. Rev.* **2**, 53–74.

Frank, J. (1990). Classification of macromolecular assemblies studied as single particles. *Quart. Rev. Biophys.* **23**, 281–329.

Frank, J. (1992a). Three-dimensional reconstruction at the molecular level. *Microsc. Microanal. Microstruct.* **3**, 45–54.

Frank, J. (ed.). (1992b). *Electron Tomography*. Plenum, New York.

Frank, J. (1995). Approaches to large-scale structures. *Curr. Opin. Struct. Biol.* **5**, 194–201.

Frank, J. (1996). *Three-Dimensional Electron Microscopy of Macromolecular Assemblies*, 1st ed. Academic Press, San Diego.

Frank, J. (1997). The ribosome at higher resolution—the donut takes shape. *Curr. Opin. Struct. Biol.* **7**, 266–272.

Frank, J. (1998). The ribosome—structure and functional ligand-binding experiments using cryo-electron microscopy. *J. Struct. Biol.* **124**, 142–150.

Frank, J. (2000). The ribosome—a macromolecular machine *par excellence*. *Chem. Biol.* **7**, R133–R141.

Frank, J. (2001a). Cryo-electron microscopy as an investigative tool: the ribosome as an example. *BioEssays* **23**, 725–732.

Frank, J. (2001b). Ribosomal dynamics explored by cryo-electron microscopy. *Methods* **25**, 309–315.

Frank, J. (2002). Single-particle imaging of macromolecules by cryo-electron microscopy. *Ann. Rev. Biophys. Biomol. Struct.* **31**, 303–319.

Frank, J. (2003). Electron microscopy of functional ribosome complexes. *Biopolymers* **68**, 223–233.

Frank, J., and Agrawal, R.K. (2001). Ratchet-like movements between the two ribosomal subunits: their implications in elongation factor recognition and tRNA translocation. *Cold Spring Harbor Symp. Quant. Biol.* **LXVI**, 67–75.

Frank, J. and Agrawal, R.K. (1998). The movement of tRNA through the ribosome. *Biophys. J.* **74**, 589–594.

Frank. J. and Agrawal, R.K. (2000). A ratchet-like intersubunit reorganization of the ribosome during translocation. *Nature* **406**, 318–322.

Frank, J., and Al-Ali, L. (1975). Signal-to-noise ratio of electron micrographs obtained by cross-correlation. *Nature* **256**, 376–378.

Frank, J., and Goldfarb, W. (1980). Methods of averaging of single molecules and lattice fragments. In: *Electron Microscopy at Molecular Dimensions. State of the Art and Strategies for the Future* (W. Baumeister and W. Vogell, eds.), pp. 154–160. Springer-Verlag, Berlin.

Frank, J., and Penczek, P. (1995). On the correction of the contrast transfer function in biological electron microscopy. *Optik* **98**, 125–129.

Frank, J., and Radermacher, M. (1986). Three-dimensional reconstruction of non-periodic macromolecular assemblies from electron micrographs. In: *Advanced Techniques in Biological Electron Microscopy* (J.K. Koehler, ed.), Vol. 3, pp. 1–72. Springer-Verlag, Berlin.

Frank, J., and Radermacher, M. (1992). Three-dimensional reconstruction of single particles negatively stained or in vitreous ice. *Ultramicroscopy* **46**, 241–262.

Frank, J., and van Heel, M. (1982a). Correspondence analysis of aligned images of biological particles. *J. Mol. Biol.* **161**, 134–137.

Frank, J., and van Heel, M. (1982b). Averaging techniques and correspondence analysis. In: *Proceedings of the 10th International Congress on Electron Microscopy*, Vol. 1, pp. 107–114. Deutsche Gesellschaft für Elektronenmikroskopie, Frankfurt (Main).

Frank, J., and Verschoor, A. (1984). Masks for pre-screening of molecule projections. *J. Mol. Biol.* **178**, 696–698.

Frank, J., and Wagenknecht, T. (1984). Automatic selection of molecular images from electron micrographs. *Ultramicroscopy* **12**, 169–176.

Frank, J., Bussler, P., Langer, R., and Hoppe, W. (1970). Einige Erfahrungen mit der rechnerischen Analyse und Synthese von elektronenmikroskopischen Bildern hoher Auflösung. *Ber. Bunsenges. Phys. Chem.* **74**, 1105–1115.

Frank, J., Goldfarb, W., Eisenberg, D., and Baker, T.S. (1978a). Reconstruction of glutamine synthetase using computer averaging. *Ultramicroscopy* **3**, 283–290.

Frank, J., McFarlane, S.C., and Downing, K.H. (1978b). A note on the effect of illumination aperture and defocus spread in bright field electron microscopy. *Optik* **52**, 49–60.

Frank, J., Verschoor, A., and Boublik, M. (1981a). Computer averaging of electron micrographs of 40S ribosomal subunits. *Science* **214**, 1353–1355.

Frank, J., Shimkin, B., and Dowse, H. (1981b). SPIDER-A modular software system for electron image processing. *Ultramicroscopy* **6**, 343–358.

Frank, J., Verschoor, A., and Boublik, M. (1982). Multivariate statistical analysis of ribosome electron micrographs. *J. Mol. Biol.* **161**, 107–137.

Frank, J., Verschoor, A., and Wagenknecht, T. (1985). Computer processing of electron microscopic images of single macromolecules. In: *New Methodologies in Studies of Protein Configuration* (T.T.W., ed.), pp. 36–89. Van Nostrand-Reinhold, New York.

Frank, J., Radermacher, M., Wagenknecht, T., and Verschoor, A. (1986). A new method for three-dimensional reconstruction of single macromolecules using low dose electron micrographs. *Ann. NY Acad. Sci.* **483**, 77–87.

Frank, J., Radermacher, M., Wagenknecht, T., and Verschoor, A. (1988a). Studying ribosome structure by electron microscopy and computer-image processing. *Methods Enzymol.* **164**, 3–35.

Frank, J., Bretaudiere, J. P., Carazo, J.M., Verschoor, A., and Wagenknecht, T. (1988b). Classification of images of biomolecular assemblies: A study of ribosomes and ribosomal subunits of *Escherichia coli*. *J. Microsc.* **150**, 99–115.

Frank, J., Chiu, W., and Degn, L. (1988c). The characterization of structural variations within a crystal field. *Ultramicroscopy* **26**, 345–360.

Frank, J., Penczek, P., Grassucci, R., and Srivastava, S. (1991). Three-dimensional reconstruction of the 70S *E. coli* ribosome in ice: the distribution of ribosomal RNA. *J. Cell Biol.* **115**, 597–605.

Frank, J., Penczek, P., and Liu, W. (1992). Alignment, classification, and three-dimensional reconstruction of single particles embedded in ice. In: *Scanning Microscopy Supplement 6: Proceedings of the Tenth Pfefferkorn Conference, Cambridge University, England, September 1992* (P.W. Hawkes, ed.), pp. 11–22. Scanning Microscopy International, Chicago.

Frank, J., Chiu, W., and Henderson, R. (1993). Flopping polypeptide chains and Suleika's subtle imperfections: analysis of variations in the electron micrograph of a purple membrane crystal. *Ultramicroscopy* **49**, 387–396.

Frank, J., Zhu, J., Penczek, P., Li, Y., Srivastava, S., Verschoor, A., Radermacher, M., Grassucci, R., Lata, R.K., and Agrawal, R.K. (1995a). A model of protein synthesis based on cryo-electron microscopy of the *E. coli* ribosome. *Nature* **376**, 441–444.

Frank, J., Verschoor, A., Li, Y., Zhu, J., Lata, R.K., Radermacher, M., Penczek, P., Grassucci, R., Agrawal, R.K., and Srivastava, S. (1995b). A model of the translational apparatus based on three-dimensional reconstruction of the *E. coli* ribosome. *Biochem. and Cell Biol.* **73**, 757–765.

Frank, J., Radermacher, M., Penczek, P., Zhu, J., Li, Y., Ladjadj, M., and Leith, A. (1996). SPIDER and WEB: processing and visualization of images in 3D electron microscopy and related fields. *J. Struct. Biol.* **116**, 190–199.

Frank, J., Heagle, A.B. and Agrawal, R.K. (1999a). Animation of the dynamic events of the elongation cycle based on cryoelectron microscopy of functional complexes of the ribosome. *J. Struct. Biol.* **128**, 15–18.

Frank, J., Penczek, P., Agrawal, R.K., Grassucci, R.A., and Heagle, A.B. (2000a). Three-dimensional cryo-electron microscopy of ribosomes. *Methods Enzymol.* **317**, 276–291.

Frank, J., Penczek, P., Grassucci, R.A., Heagle, A., Spahn, C.M.T., and Agrawal, R.K. (2000b). Cryo-electron microscopy of the translational apparatus: experimental evidence for the paths of mRNA, tRNA, and the polypeptide chain. In: *The Ribosome: Structure, Function, Antibiotics and Cellular Interactions* (R.A Garrett, S.R. Douthwate, A. Liljas, A.T. Matheson, P.B. Moore, and H.F. Noller, eds.), pp. 45–51. ASM Press, Washington, DC.

Frank, J., Wagenknecht, T., McEwen, B.F., Marko, M., Hsieh, C.-E., and Mannella, C. (2002). Three-dimensional imaging of biological complexity. *J. Struct. Biol.* **138**, 85–91.

Franzini-Armstrong, C. (1994). Unraveling the ryanodine receptor. *Biophys. J.* **67**, 2135–2136.

Fraser, R.D.B., Furlong, D.B., Trus, B.L., Nibert, M.L., Fields, B.N., and Steven, A.C. (1990). Molecular structure of the cell-attachment protein of reovirus: correlation averaging of computer-processed electron micrographs with sequence-based predictions. *J. Virol.* **64**, 2990–3000.

Frederick, P.M., Bomans, P., Franssen, V., and Laeven, P. (2000). A vitrification robot for time resolved cryo-electron microscopy. *Proceedings of 12th European Congress on Electron Microscopy* (S. Cech and R. Janisch, eds.), Vol. I, pp. B383–B384. Reklamni Atelier Kupa Brno.

Fujiyoshi,Y. (1998). The structural study of membrane proteins by electron crystallography. *Adv. Biophys.* **35**, 25–80.

Fujiyoshi, Y. (1999). Molecular structure of proton pump revealed with electron crystallography. *FASEB J.* **13**, S191–S194.

Fujiyoshi, Y., Mizusaki, T., Morikawa, K., Yamegishi, H., Aoki, Y., Kihara, H., and Harada, Y. (1991). Development of a superfluid helium stage for high-resolution electron microscopy. *Ultramicroscopy* **38**, 241–251.

Fukunaka, K., and Koontz, W.L.G. (1970). Application of the Karhunen–Loève expansion to feature selection and ordering. *IEEE Trans. Comp.* **C-19**, 311–318.

Fuller, S.D., Butcher, S.J., Cheng, R.H., and Baker, T.S. (1996). Three-dimensional reconstruction of isocahedral particles—the uncommon line. *J. Struct. Biol.* **116**, 48–55.

Fung, J.C., Liu, W., de Ruijter, J., Chen, H., Abbey, C.K., Sedat, J.W., and Agard, D.A. (1996). Toward fully automated high-resolution electron tomography. *J. Struct. Biol.* **116**, 181–189.

Furcinitti, P.S., van Oostrum, J., and Burnett, R.M. (1989). Adenovirus polypeptide IX revealed as capsid cement by difference images from electron microscopy and crystallography. *EMBO J.* **8**, 3563–3570.

Gabashvili, I.S., Agrawal, R.K., Grassucci, R., Squires, C.L., Dahlberg, A.E., and Frank, J. (1999a). Major rearrangements in the 70S ribosomal 3D structure caused by a conformational switch in 16S ribosomal RNA. *EMBO J.* **18**, 6501–6507.

Gabashvili, I.S., Agrawal, R.K., Grassucci, R.A., and Frank, J. (1999b). Structure and structural variations of the *E. coli* 30S ribosomal subunit as revealed by three-dimensional cryo-electron microscopy. *J. Mol. Biol.* **5**, 1285–1291.

Gabashvili, I.S., Agrawal, R.K., Spahn, C.M.T., Grassucci R.A., Frank, J., and Penczek, P. (2000). Solution structure of the *E. coli* 70S ribosome at 11.5 Å resolution. *Cell* **100**, 537–549.

Gabashvili, I.S., Gregory, S.T., Valle, M., Grassucci, R., Worbs, M., Wahl, M.C., Dahlberg, A.E., and Frank, J. (2001). The polypeptide tunnel system in the ribosome and its gating in erythromycin resistance mutants of L4 and L22. *Mol. Cell* **8**, 181–188.

Galton, F.J. (1878). Composite portraits. *Nature* **18**, 97–100.

Gao, H., and Frank, J. (2005). Molding atomic structures into intermediate-resolution cryo-EM density maps of ribosomal complexes using real-space refinement. *Structure* **13**, 401–406.

Gao, H. Spahn, C.M.T., Grassucci, R.A., and Frank, J. (2002). An assay for local quality in cryo-electron micrographs of single particles. *Ultramicroscopy* **93**, 169–178.

Gao, H., Sengupta, J., Valle, M., Korostelev, A., Eswar, N., Stagg, S.M., Van Roey, P., Agrawal, R.K., Harvey, S.C., Sali, A., Chapman, M.S., and Frank, J. (2003). Study of structural dynamics of the *E. coli* 70S ribosome using real-space refinement. *Cell* **113**, 789–801.

Gao, H., Valle, M., Ehrenberg, M., and Frank, J. (2004). Dynamics of EF-G interaction with the ribosome explored by classification of a heterogeneous cryo-EM dataset. *J. Struct. Biol.* **147**, 283–290.

Gavin, A.C., Bösche, M., Krause, R., Grandi, P., Marzioch, M., Bauer, A., Schultz, J., Rick, J.M., Michon, A.-M., Cruciat, C.-M., Remor, M., Höfert, C., Scheider, M., Brajeovic, M., Ruffner, H., Merino, A., Klein, K., Hudak, M., Dickson, D., Rudi, T., Gnau, V., Bauch, A., Bastuck, S., Huhse, B., Leutwein, C., Heurtier, M.-A., Copley, R.R., Edelmann, A., Querfurth, E., Rybin, V., Drewes, G., Ralda, M., Bouwmeester, T., Bork, P., Seraphin, B., Kuster, B., Neubauer, G., and Superti-Furga, G. (2002). Functional organization of the yeast proteome by systematic analysis of protein complexes. *Nature* **415**, 141–147.

Gerchberg, R. W. (1974). Super resolution through energy reduction. *Opt. Acta* **21**, 709–720.

Gerchberg, R.W., and Saxton, W. O. (1971). A practical algorithm for the determination of phase from image and diffraction plane pictures. *Optik* **34**, 275–284.

Gilbert, P.F.C. (1972). The reconstruction of a three-dimensional structure from projections and its application to electron microscopy. II. Direct methods. *Proc. R. Soc. London B* **182**, 89–117.

Gilbert, R.J.C., Jimenez, J.L., Chen, S., Tickle, I.J., Rossjohn, J., Parker, M., Andrew, P.W., and Saibil, H.R. (1999). Two structural transitions in membrane pore formation by pneumolysin, the pore-forming toxin of *Streptococcus pneumoniae*. *Cell* **87**, 647–655.

Gilbert, R.J.C., Fucini, P., Connell, S., Fuller, S.D., Nierhaus, K.H., Robinson, C.V., Dobson, C.M., and Stuart, D.I. (2004). Three-dimensional structures of translating ribosomes by cryo-EM. *Mol. Cell* **14**, 57–66.

Glaeser, R.M. (1971). Limitations to significant information in biological electron microscopy as a result of radiation damage. *J. Ultrastruct. Res.* **36**, 466–482.

Glaeser, R.M. (1985). Electron crystallography of biological macromolecules. *Am. Rev. Phys. Chem.* **36**, 243–275.

Glaeser, R.M. (1992a). Specimen flatness of thin crystalline arrays—influence of the substrate. *Ultramicroscopy* **46**, 33–43.

Glaeser, R.M. (1992b). Cooling-induced wrinkling of thin crystals of biological macromolecules can be prevented by using molybdenum grids. In: *Proceedings of the 50th Annual Meeting, EMSA*, pp. 520–521. San Francisco, Inc., San Francisco.

Glaeser. R.M. (1999). Electron crystallography: present excitement, a nod to the past, anticipating the future. *J. Struct. Biol.* **128**, 3–14.

Glaeser, R.M., and Downing, K.H. (1992). Assessment of resolution in biological electron microscopy. *Utramicroscopy* **47**, 256–265.

Glaeser, R.M., and Taylor, K.A. (1978). Radiation damage relative to transmission electron microscopy of biological specimens at low temperature: a review. *J. Microsc.* **112**, 127–138.

Glaeser, R.M., Kuo, I., and Budinger, T.F. (1971). Method for processing of periodic images at reduced levels of electron radiation. In: *Proceedings of the 29th Annual Meeting, EMSA*, pp. 466–467. Boston.

Gogol, E.P., Johnston, E., Aggeler, R., and Capaldi, R.A. (1990). Ligand-dependent structural variations in *Escherichia coli* F1–ATPase revealed by cryoelectron microscopy. *Proc. Natl. Acad. Sci. USA* **87**, 9585–9589.

Golas, M.M., Sander, B., Will, C.L., Lührmann, R., and Stark, H. (2003). Molecular architecture of the multiprotein splicing factor SF3b. *Science* **300**, 980–984.

Gomez-Lorenzo, M.G., Spahn, C.M.T., Agrawal, R.K., Grassucci, R.A., Penczek, P., Chakrabartty, K., Ballesta, J.P.G., Lavandera, J.L., Garcia-Bustos, J.F., and Frank, J. (2000). Three-dimensional cryo-electron microscopy localization of EF2 in the *Saccharomyces cerevisiae* 80S ribosome at 17.5 Å resolution. *EMBO J.* **19**, 2710–2718.

Gomez-Lorenzo, M.G., Valle, M., Frank, J., Gruss, C., Sorzano, C.O.S., Chen, X.S., Donate, L.E., and Carazo, J.M. (2003). Large T antigen on the simian virus 40 origin of replication: a 3D snapshot prior to DNA repliucation. *EMBO J.* **22**, 6205–6213.

Goncharov, A.B., Vainshtein, B.K., Ryskin, A.I., and Vagin, A.A. (1987). Three-dimensional reconstruction of arbitrarily oriented particles from their electron photomicrographs. *Sov. Phys. Crystallogr.* **32**, 504–509.

Gonzales, R.C. and Woods, R.E. (1993). *Digital Image Processing*. Addison-Wesley, Reading, Mass.

Goodman, J.W. (1968). *Introduction to Fourier Optics*. McGraw-Hill, New York.

Goodsell, D.S., and Olson, A.J. (1993). Soluble proteins: size, shape and function. *TIBS* **18**, 65–68.

Gordon, R., Bender, R., and Herman, G.T. (1970). Algebraic reconstruction techniques. *J. Theor. Biol.* **29**, 471–482.

Grassucci, R.A., Liu, Z., Wagenknecht, T., and Frank, J. (2002). HHMI Tecnai F30 helium microscope: initial results and observations. *Microsc. Microanal.* **8** (Suppl. 2), 854CD-855CD.

Greenacre, M.J. (1984). *Theory and Applications of Correspondence Analysis*. Academic Press, London/Orlando.

Grigorieff, N. (1998). Three-dimensional structure of bovine NADH: ubiquinone oxidoreductase (complex I) at 22 Å in ice. *J. Mol. Biol.* **277**, 1033–1046.

Grigorieff, N. (2000). Resolution measurement in structures derived from single particles. *Acta Cryst.* **D56**, 1270–1277.

Grimm, R., Typke, D., and Baumeister, W. (1998). Improving image quality by zero-loss energy filtering: quantitative assessment by means of image cross-correlation. *J. Microsc.* **190**, 339–349.

Grüber, G., Radermacher, M., Ruiz, T., Godovac-Zimmermann, J., Canas, B., Kleine-Kohlbrecher, D., Huss, M., Harvey, W.R., and Wieczorek, H. (2000). Three-dimensional structure and subunit topology of the V_1 ATPase from *Manduca sexta* midgut. *Biochemistry* **39**, 8609–8616.

Grünewald, K., Medalia, O., Gross, A., Steven, A.C., and Baumeister, W. (2003). Prospects of electron cryotomography to visualize macromolecular complexes inside cellular compartments: implications of crowding. *Biophys. Chem.* **100**, 577–591.

Gu, Yingqi, Zhou, Z.H., McCarthy, D.B., Reed, L.J., and Stoops, J.K. (2003). 3D electron microscopy reveals the variable deposition and protein dynamics of the peripheral pyruvate dehydrogenase component about the core. *PNAS* **100**, 7015–7020.

Gubbens, A.J., and Krivanek, O.J. (1993). Applications of a post-column filter in biology and materials science. *Ultramicroscopy* **51**, 146–159.

Guckenberger, R. (1982). Determination of a common origin in the micrographs of tilt series in three-dimensional electron microscopy. *Ultramicroscopy* **9**, 167–174.

Gutsche, I., Essen, L.-O., and Baumeister, W. (1999). Group II chaperonins: new TRIC(k)s and turns of a protein folding machine. *J. Mol. Biol.* **293**, 295–312.

Haase, J. (1970a). Zusammenstellung der Koeffizienten für die Anpassung komplexer atomarer Streufaktoren für schnelle Elektronen durch Polynome. 1. Hartee–Fock– Fall *Z. Naturforsch.* **A25**, 936–945.

Haase, J. (1970b). Zusammenstellung der Koeffizienten für die Anpassung komplexer atomarer Streufaktoren für schnelle Elektronen durch Polynome. 2. Thomas–Fermi–Dirac–Fall *Z. Naturforsch.* **A25**, 1219–1235.

Hainfeld, J.F. (1987). A small gold-conjugated antibody label: improved resolution for electron microscopy. *Science* **236**, 450–453.

Hainfeld, J.F. (1992). Site-specific cluster labels. *Ultramicroscopy* **46**, 135–144.

Hainfeld, J.F., and Furuya, F.R. (1992). A 1.4 nm gold cluster covalently attached to antibodies improves immunolabelling. *J. Histochem. Cytochem.* **40**, 177–184.

Haley, D.A., Horwitz, J., and Stewart, P.L. (1998). The small heat-shock protein, αB-crystallin, has a variable quaternary structure. *J. Mol. Biol.* **277**, 27–35.

Haley, D.A., Bova, M.P., Huang, Q.-L., Mchaourab, H.S., and Stewart, P.L. (2000). Small heat-shock protein structures reveal a continuum from symmetry to variable assemblies. *J. Mol. Biol.* **298**, 261–272.

Halic, M., Becker, T., Pool, M.R., Spahn, C.M.T., Grassucci, R.A., Frank, J., and Beckmann, R. (2004). Structure of the signal recognition particle interacting with the elongation arrested ribosome. *Nature* **427**, 808–814.

Halic, M., Becker, T., Frank, D., Spahn, C.M.T., and Beckmann, R. (2005). Localization and dymanic behavior of ribosomal protein L30e. *Nat. Strut. Mol. Biol.* **12**, 467–468.

Hanein, D., Volkmann, N., Goldsmith, S., Michon, A.-M., Lehman, W., Craig, R., DeRosier, D., Almo, S., and Matsudaira, P. (1998). An atomic model of fimbrin binding to F-actin and its implications for filament crosslinking and regulation. *Nat. Struct. Biol.* **5**, 787–792.

Hänicke, W. (1981). *Mathematische Methoden zur Aufbereitung elektronenmikroskopischer Bilder*. Thesis, Universität Göttingen.

Hänicke, W., Frank, J., and Zingsheim, H.P. (1984). Statistical significance of molecule projections by single particle averaging. *J. Microsc.* **133**, 223–238.

Hanszen, K.-J. (1971). The optical transfer theory of the electron microscope: fundamental principles and applications. *Adv. Opt. Microsc.* **4**, 1–84.

Hanszen, K.-J., and Trepte, L. (1971a). The contrast transfer of the electron microscope with partial coherent illumination. A. The ring condenser. *Optik* **33**, 166–181.

Hanszen, K.-J., and Trepte, L. (1971b). B. Disc-shaped source. *Optik* **33**, 182–198.

Harauz, G. (1990). Representation of rotations by unit quaternions. *Ultramicroscopy* **33**, 209–213.

Harauz, G., and Chiu, D.K.Y. (1991). Covering events in eigenimages of biomolecules. *Ultramicroscopy* **38**, 307–317.

Harauz, G., and Chiu, D.K.Y. (1993). Complementary applications of correspondence analysis and event covering of noisy image sequences. *Optik* **95**, 1–8.

Harauz, G., and Fong-Lochovsky, A. (1989). Automatic selection of macromolecules from electron micrographs by component labelling and symbolic processing. *Ultramicroscopy* **31**, 333–344.

Harauz, G., and Ottensmeyer, F.P. (1984a). Direct three-dimensional reconstruction for macromolecular complexes from electron micrographs. *Ultramicroscopy* **12**, 309–320.

Harauz, G., and Ottensmeyer, F.P. (1984b). Nucleosome reconstruction via phosphorus mapping. *Science* **226**, 936–940.

Harauz, G., and van Heel, M. (1986a). Exact filters for general geometry three-dimensional reconstruction. *Optik* **73**, 146–156.

Harauz, G., and van Heel, M. (1986b). Direct 3D reconstruction from projections with initially unknown angles. In: *Pattern Recognition in Practice II* (E.S. Gelsema and L. N. Kanal, eds.), pp. 279–288. North-Holland, Elsevier, Amsterdam.

Harauz, G., Stöffler-Meilicke, M., and van Heel, M. (1987). Characteristic views of prokaryotic 50S ribosomal subunits. *J. Mol. Evol.* **26**, 347–357.

Harauz, G., Boekema, E., and van Heel, M. (1988). Statistical image analysis of electron micrographs of ribosomal subunits. *Methods Enzymol.* **164**, 35–49.

Harauz, G., Chiu, D.K.Y., MacAulay, C., and Palcic, B. (1994). Probabilistic inference in computer-aided screening for cervical cancer: An event covering approach to information extraction and decision rule formulation. *Anal. Cell. Pathol.* **6**, 37–50.

Harauz, G., Cicicopoli, C., Hegerl, R., Cejka, Z., Goldie, K., Santarius, U., Engel, A., and Baumeister, W. (1996). Structural studies on the 2.25-MDa homomultimeric phospoenolpyruvate synthase from *Staphylothermus marinus*. *J. Struct. Biol.* **116**, 290–301.

Harris, J.R., and Scheffler, D. (2002). Routine preparation of air-dried negatively stained and unstained specimens on holey carbon support films: a review of applications. *Micron* **33**, 461–480.

Harris, J.R., Schröder, E., Isupov, M.N., Scheffler, D., Kristensen, P., Littlechild, J.A., Vagin, A.A., and Meissner, U. (2001). Comparison of the decameric structure of peroxiredoxin-II by transmission electron microscopy and X-ray crystallography. *Biochem. Biophys. Acta* **1547**, 221–234.

Harris, R., and Horne, R. (1991). Negative staining. In: *Electron Microscopy in Biology* (J. R. Harris, ed.), pp. 203–228. IRL Press at Oxford University Press, Oxford.

Haslbeck, M., Walke, S., Stromer, T., Ehrnsperger, M., White, H.E., Chen, S., Saibil, H.R., and Buchner, J. (1999). Hsp26: a temperature-regulated chaperone. *EMBO J.* **18**, 6744–6751.

Hawkes, P.W. (1980). Image processing based on the linear transfer theory of image formation. In: *Computer Processing of Electron Microscope Images*, pp. 1–33. Springer-Verlag, Berlin.

Hawkes, P.W. (1992). The electron microscope as a structure projector. In: *Electron Tomography* (J. Frank, ed.), pp. 17–38. Plenum Press, New York.

Hawkes, P.W. (1993). Reflections on the algebraic manipulation of sets of electron images or spectra. *Optik* **93**, 149–154.

Hawkes, P.W., and Kasper, E. (1994). *Principles of Electron Optics, Vol. 3: Wave Optics.* Academic Press, London.

Hayward, S.B., and Glaeser, R. (1979). Radiation damage of purple membrane at low temperature. *Ultramicroscopy* **4**, 201–210.

Hegerl, R. (1992). A brief survey of software packages for image processing in biological electron microscopy. *Ultramicroscopy* **47**, 417–423.

Hegerl, R. (1996). The EM package: a platform for image processing in biological electron microscopy. *J. Struct. Biol.* **116**, 30–34.

Hegerl, R., and Altbauer, A. (1982). The "EM" program system. *Ultramicroscopy* **9**, 109–116.

Hegerl, R., and Hoppe, W. (1976). Influence of electron noise on three-dimensional image reconstruction. *Z. Naturforsch.* **31a**, 1717–1721.

Hegerl, R., Pfeifer, G., Pilhier, G., Dahlmann, B., and Baumeister, W. (1991). The three-dimensional structure of proteosomes from *Thermoplasma acidophilum* as determined by electron microscopy using random conical tilting. *FEBS Lett.* **283**, 117–121.

Heide, H.-G., and Zeitler, E. (1985). The physical behavior of solid water at low temperature and the embedding of electron microscopical specimens. *Ultramicroscoscopy* **16**, 151–160.

Henderson, R. (1992). Image contrast in high-resolution electron microscopy of biological specimens: TMV in ice. *Ultramicroscopy* **46**, 1–18.

Henderson, R. (1995). The potential and limitation of neutrons, electrons, and X-rays for atomic resolution microscopy of unstained biological molecules. *Quart. Rev. Biophys.* **28**, 171–193.

Henderson, R., and Glaeser, R.M. (1985). Quantitative analysis of image contrast in electron micrographs of beam-sensitive crystals. *Ultramicroscopy* **16**, 139–150.

Henderson, R., and Unwin, P.N.T. (1975). Three-dimensional model of purple membrane obtained by electron microscopy. *Nature* **257**, 28–32.

Henderson, R., Baldwin, J.M., Downing, K.H., and Zemlin, F. (1986). Structure of purple membrane from *Halobacterium holobium*: Recording, measurement and evaluation of electron micrographs at 3.5 Å resolution. *Ultramicroscopy* **19**, 147–178.

Henderson, R., Baldwin, J.M., Ceska, T.A., Zemlin, F., Beckmann, E., and Downing, K. H. (1990). Model of the structure of bacteriorhodopsin based on high-resolution electron cryomicroscopy. *J. Mol. Biol.* **213**, 899–929.

Herman, G.T. (1980). *Image Reconstruction from Projections: The Fundamentals of Computerized Tomography.* Academic Press, New York.

Hewat, E.A. and Neumann, E. (2002). Characterization of the performance of a 200-kV field emission gun for cryo-electron microscopy of biological molecules. *J. Struct. Biol.* **139**, 60–64.

Heymann, J.B., Cheng, N., Newcomb, W.W., Trus, B.L., Brown, J.C., and Steven, A.C. (2003). Dynamics of herpes simplex virus capsid maturation visualized by time-lapse cryo-electron microscopy. *Nature Struct. Biol.* **10**, 334–341.

Heymann, J.B., Conway, J.F., and Steven, A.C. (2004). Molecular dynamics of protein complexes from four-dimensional cryo-electron microscopy. *J. Struct. Biol.* **147**, 291–301.

Hinshaw, J.E., Carragher, B.O., and Milligan, R.A. (1992). Architecture and design of the nuclear pore complex. *Cell* **69**, 1133–1141.

Ho, Y., et al. (45 coauthors) (2002). Systematic identification of protein complexes in *Saccaromyces cerevisiae* by mass spectroscopy. *Nature* **415**, 141–183.

Hoiczyk, E., and Baumeister, W. (1998). The junctional pore complex, a prokaryotic secretion organelle, is the molecular motor underlying gliding motility in cyanobacteria. *Curr. Biol.* **8**, 1161–1168.

Hollander, M., and Wolfe, D.A. (1973). *Nonparametric Statistical Methods.* John Wiley, New York.

Hölzl, H., Kapelari, B., Kellermann, J., Seemüller, E., Sümegi, M., Udvardy, A., Medalia, O., Sperling, J., Müller, S.A., Engel, A., and Baumeister, W. (2000). The regulatory complex of *Drosophila melanogaster* 26S proteasomes: subunit composition and localization of a deubiquitylating enzyme. *J. Cell Biol.* **150**, 119–129.

Hönger, A., and Aebi, U. (1996). 3-D reconstructions from ice-embedded and negatively stained biomacromolecular assemblies: a critical comparison. *J. Struct. Biol.* **117**, 99–116.

Hoppe, W. (1961). Ein neuer Weg zur Erhöhung des Auflösungsvermögens des Elektronenmikroskops. *Naturwissenschaften* **48**, 736–737.

Hoppe, W. (1969). Das Endlichkeitspostulat und das Interpolationstheorem der dreidimensionalen elektronenmikroskopischen Analyse aperiodischer Strukturen. *Optik* **29**, 617–621.

Hoppe, W. (1972). Drei-dimensional abbildende Elektronenmikroskope. *Z. Naturforsch.* **27a**, 919–929.

Hoppe, W. (1974). Towards three-dimensional "electron microscopy" at atomic resolution. *Naturwissenschaften* **61**, 239–249.

Hoppe, W. (1981). Three-dimensional electron microscopy. *Annu. Rev. Biophys. Bioeng.* **10**, 563–592.

Hoppe, W. (1983). Electron diffraction with the transmission microscope as a phase-determining diffractometer—from spatial frequency filtering to the three dimensional Structure analysis of ribosomes. *Angew. Chem. Int., Ed. Engl.* **22**, 456–485.

Hoppe, W., and Hegerl, R. (1980). Three-dimensional structure determination by electron microscopy (nonperiodic specimens). In: *Computer Processing of Electron Microscope Images* (P.W. Hawkes, ed.), pp. 127–185. Springer-Verlag, Berlin.

Hoppe, W., and Hegerl, R. (1981). Some remarks concerning the influence of electron noise on 3D reconstruction. *Ultramicroscopy* **6**, 205–206.

Hoppe, W., Langer, R., Knesch, G., and Poppe, C. (1968). Protein-Kristallstrukturanalyse mit Elektronenstrahlen. *Naturwissenschaften* **55**, 333–336.

Hoppe, W., Langer, R., Frank, J., and Feltynowski, A. (1969). Bilddifferenzverfahren in der Elektronenmikroskopie. *Naturwissenschaften* **56**, 267–272.

Hoppe, W., Gassmann, J., Hunsmann, N., Schramm, H.J., and Sturm, M. (1974). Three-dimensional reconstruction of individual negatively stained fatty-acid synthetase molecules from tilt series in the electron microscope. *Hoppe-Seyler's Z. Physiol. Chem.* **355**, 1483–1487.

Hoppe, W., Schramm, H.J., Sturm, M., Hunsmann, N., and Gassmann, J. (1986). Three-dimensional electron microscopy of individual biological objects. 1. Methods. *Z. Naturforsch.* **A31**, 645–655.

Hu, J.J., and Li, F.H. (1991). Maximum entropy image deconvolution in high resolution electron microscopy. *Ultramicroscopy* **35**, 339–350.

Huang, Z., Baldwin, P.R., Mullapudi, S., and Penczek, P.A. (2003). Automated determination of parameters describing power spectra of micrograph images in electron microscopy. *J. Struct. Biol.* **144**, 79–94.

Hunt, B.R. (1973). The application of constraint least squares estimation to image restoration by digital computer. *IEEE Trans. Comput.* **22**, 805–812.

Hutchinson, G., Tichelaar, W., Weiss, H., and Leonard, K. (1990). Electron microscopic characterization of helical filaments formed by subunits I and II (core proteins) of ubiquinol: Cytochrome *c* reductase from *Neurospora mitochondria*. *J. Struct. Biol.* **103**, 75–88.

Ishikawa, T., Maurizi, M.R., Belnap, D.M., and Steven, A.C. (2000). Docking of components in ATP-dependent HslVU protease. *Nature* **408**, 667–668.

Ishikawa, T., Beuron, F., Kessel, M., Wickner, S., Maurizi, M.R., and Steven, A.C. (2001). Translocation pathway of protein substrates in ClpAP protease. *Proc. Natl. Acad. Sci. USA* **98**, 4328–4333.

Jandhyala, D., Berman, M.N., Meyers, P.R., Sewell, B.T., Willson, R.C., and Benedik, M.J. (2003). CynD, the cyanide dihydratase from *Bacillus pumilus*: gene cloning and structural studies. *Appl. Environ. Microbiol.* **69**, 4794–4805.

Jap, B. (1989). Molecular design of PhoE porin and its functional consequences. *J. Mol. Biol.* **205**, 407–419.

Jap, B. (1991). Structural architecture of an outer membrane channel as determined by electron crystallography. *Nature* **350**, 167–170.

Jap, B.K., Zulauf, M., Scheybani, T., Hefti, A., Baumeister, W., Aebi, U., and Engel, A. (1992). 2D crystallization: From art to science. *Ultramicroscopy* **46**, 45–84.

Jeng, T.W., Crowther, R.A., Stubbs, G., and Chiu, W. (1989). Visualization of alpha-helices in TMV by cryo-electron microscopy. *J. Mol. Biol.* **205**, 251–257.

Jenkins, G.M., and Watts, D.G. (1968). *Spectral Analysis and Its Applications*. Holden-Day, Oakland, CA.

Jenniskens, P., and Blake, D.F. (1994). Structural transitions in amorphous ice and astrophysical implications. *Science* **265**, 753–756.

Jensen, G.J. (2001). Alignment error envelopes for single particle analysis. *J. Struct. Biol.* **133**, 143–155.

Jensen, G.J., and Kornberg, R.D. (1998). Single-particle selection and alignment with heavy atom cluster–antibody conjugates. *Proc. Natl. Acad. Sci. USA* **95**, 9262–9267.

Jiang,W., and Chiu, W. (2001). Web-based simulation for contrast transfer function and envelope functions. *Microsc. Microanal.* **7**, 329–334.

Jiang, Q.-X., Chester, D.W., and Sigworth, F.J. (2001a). Spherical reconstruction: a method for structure determination of membrane proteins from cryo-EM images. *J. Struct. Biol.* **133**, 119–131.

Jiang, W., Baker, M.I., Ludtke, S.I., and Chiu, W. (2001b). Bridging the information gap: computational tools for intermediate resolution structure interpretation. *J. Mol. Biol.* **308**, 1033–1044.

Jiang, Q.-X., Thrower, E.C., Chester, D.W., Ehrlich, B.E., and Sigworth, F.J. (2002). Three-dimensional structure of the type 1 inositol 1,4,5-triphosphate receptor at 24 Å resolution. *EMBO J.* **21**, 3575–3581.

Jiménez, L.J., Guijarro, J.I., Orlova, E., Zurdo, J., Dodson, C.M., Sunde, M., and Saibil, H.R. (1999). Cryo-electron microscopy structure of an SH3 amyloid fibril and model of the molecular packing. *EMBO J.* **18**, 815–821.

Jiménez, L.J., Tennent,G., Pepys, M., and Saibil, H. (2001). Structural diversity of *ex vivo* amyloid fibrils studied by cryo-electron microscopy. *J. Mol. Biol.* **311**, 241–247.

Jiménez, L.J., Nettleton, E.J., Bouchard, M., Robinson, C.V., and Dodson, C.M. (2002). The protofilament structure of insulin amyloid fibrils. *Proc. Natl. Acad. Sci. USA* **99**, 9196–9201.

Johansen, B.V. (1975). Optical diffractometry. In: *Principles and Techniques of Electron Microscopy: Biological Applications* (M.A. Hayat, ed.), Vol. 5, pp. 114–173. Van Nostrand-Reinhold, New York.

Jones, T.A. (1978). A graphics model building and refinement system for macromolecules. *J. Appl. Crystallogr.* **11**, 268–272.

Jones, T.A., Zou, J.-Y., Cowan, S.W., and Kjeldgaard, M. (1991). Improved methods for building protein models in electron density maps and the location of errors in these models. *Acta Crystallogr.* **A47**, 110–119.

Joyeux, L., and Penczek, P. (2002). Efficiency of 2D alignment methods. *Ultramicroscopy* **92**, 33–46.

Jurica, M.S., Sousa, D., Moore, M.J., and Grigorieff, N. (2004). Three-dimensional structure of C complex spliceosomes by electron microscopy. *Nat. Struct. Mol. Biol.* **11**, 265–269.

Kajimura, N., Yamazaki, M., Morikawa, K., Yamazaki, A., and Mayanagi, K. (2002). Three-dimensional structure of non-activated cGMP phosphodiesterase 6 and comparison of its image with those of activated forms. *J. Struct. Biol.* **139**, 27–38.

Kalko, S.G., Chagoyen, M., Jiménez-Lozano, N., Verdaguer, N., Fita, N., and Carazo, J.M. (2000). The need for a shared database infrastructure: combining X-ray crystallography and electron microscopy. *Eur. Biophys. J.* **29**, 457–462.

Kam, Z. (1980). The reconstruction of structure from electron micrographs of randomly oriented particles. *J. Theor. Biol.* **82**, 15–39.

Kang, S.G., Ortega, J., Singh, S.K., Wang, N., Huang, N., Steven, A.C., and Maurizi, M.R. (2002). Functional proteolytic complexes of the human mitochondrial ATP-dependent protease, hClpXP. *J. Biol. Chem.* **277**, 21095–21102.

Karczmarz, S. (1937). Angenährte Auflösung von Systemen linearer Gleichungen. *Bull. Acad. Pol. Sci. Leu.* **A35**, 355–357.

Kellenberger, E., and Kistler, J. (1979). The physics of specimen preparation. In: *Advances in Structure Research by Diffraction Methods* (W. Hoppe and R. Mason, eds.), Vol. 3, pp. 49–79. Vieweg, Wiesbaden.

Kellenberger, E., Häner, M., and Wurtz, M. (1982). The wrapping phenomenon in air-dried and negatively stained preparations. *Ultramicroscopy* **9**, 139–150.

Kessel, M., Radermacher, M., and Frank, J. (1985). The structure of the stalk surface layer of a brine pond microorganism: correlation averaging applied to a double layered lattice structure. *J. Microsc.* **139**, 63–74.

Kessel, M., Maurizi, M.R., Kim, B., Kocsis, E., Trus, B.L., Singh, S.K., and Steven, A.C. (1995). Homology in structural organization between *E. coli* ClpAP protease and the eukaryotic 26S proteasome. *J. Mol. Biol.* **250**, 587–594.

Kessel, M., Wu, W.-F., Gottesman, S., Kocsis, E., Steven, A.C., and Maurizi, M. (1996). Six-fold rotational symmetry of ClpQ, the *E. coli* homolog of the 20S proteasome, and its ATP-dependent activator, ClpY. *FEBS Lett.* **398**, 274–278.

Kickhoefer, V.A., Liu, Y., Kong, L.B., Snow, B.E., Stewart, P.L., Harrington, L., and Rome, L.H. (2001). The telomerase/vault-associated protein TEP1 is required for vault RNA stability and its association with the vault particle. *J. Cell Biol.* **152**, 157–164.

Kinder, E., and Súffert, F. (1943). Über den Feinban schillernder Schmetterlingsschuppen vom Morpho-Typ. *Biol Zentr.* **63**, 268–288.

Kirkland, E.J. (1984). Improved high resolution image processing of bright field electron micrographs. *Ultramicroscopy* **15**, 151–172.

Kirkland, E.J., Siegel, B.M., Uyeda, N., and Fujiyoshi, Y. (1980). Digital reconstruction of bright field phase contrast images from high resolution electron micrographs. *Ultramicroscopy* **5**, 479–503.

Kisseberth, N., Whittaker, M., Weber, D., Potter, C.S., and Carragher, B. (1997). emScope: a tool kit for control and automation of a remote electron microscope. *J. Struct. Biol.* **120**, 309–319.

Klaholz, B.P., Pape, T., Zavialov, A.V., Myasnikov, A.G., Orlova, E.V., Vestergaard, B., Ehrenberg, M., and van Heel, M. (2003). Structure of the *Escherichia coli* ribosomal termination complex with release factor 2. *Nature* **421**, 90–93.

Klug, A. (1983). From macromolecules to biological assemblies (Nobel lecture). *Angew. Chem.* **22**, 565–636.

Klug, A., and Berger, J.E. (1964). An optical method for the analysis of periodicities in electron micrographs, and some observations on the mechanism of negative staining. *J. Mol. Biol.* **10**, 565–569.

Klug, A., and Crowther, R.A. (1972). Three-dimensional image reconstruction from the viewpoint of information theory. *Nature* **228**, 435–440.

Klug, A., and DeRosier, D.J. (1966). Optical filtering of electron micrographs: Reconstruction of one-sided images. *Nature* **212**, 29–32.

Knauer, V., Hegerl, R., and Hoppe, W. (1983). Three-dimensional reconstruction and averaging of 30S ribosomal subunits of *Escherichia coli* from electron micrographs. *J. Mol. Biol.* **163**, 409–430.

Kocsis, E., Cerritelli, M.E., Trus, B.L., Cheng, N., and Steven, A.C. (1995). Improved methods for determination of rotational symmetries in macromolecules. *Ultramicroscopy* **60**, 219–228.

Koeck, P.J.B., Schröder, R.R., Haider, M., and Leonard, K.R. (1996). Unconventional immuno double labeling by energy filtered transmission electron microscopy. *Ultramicroscopy* **62**, 65–78.

Kohonen, T. (1990). The self-organizing map. *Proc. IEEE* **78**, 1464–1480.

Kohonen, T. (1997). *Self-Organizing Maps*, 2nd ed. Springer-Verlag, Berlin.

Koller, T., Beer, M., Müller, M., and Múhlethaler, M. (1971). Electron microscopy of selectively stained molecules. *Cytobiologie* **4**, 369–408.

Kolodziej, S.J., Penczek, P., and Stoops, J.K. (1997). Utility of Butvar support film and methylamine tungstate stain in three-dimensional electron microscopy: agreement between stain and frozen-hydrated reconstructions. *J. Struct. Biol.* **120**, 158–167.

Kolodziej, S.J., Kluppenberg, H.U., Nolasco, N., Ehses, W., Strickland, D.K., and Stoops, J.K. (1998). Three-dimensional structure of the human plasmin α_2-macroglobulin complex. *J. Struct. Biol.* **123**, 124–133.

Kolodziej, S.J., Hudman, A., Waxham, M.N., and Stoops, J.K. (2000). Three-dimensional reconstructions of CaM kinase IIα and truncated CaM kinase IIα reveal an unique organization for its structural core and functional domains. *J. Biol. Chem.* **275**, 14354–14359.

Kong, L.B., Siva, A.C., Rome, L.H., and Stewart, P.L. (1999). Structure of the vault, a ubiquitous cellular component. *Structure* **7**, 371–379.

Kong, L.B., Siva, A.C., Kickhoefer, V.A., Rome, L.H., and Stewart, P.L. (2000). RNA location and modeling of a WD40 repeat domain within the vault. *RNA* **6**, 890–900.

Kong, Y., and Ma, J. (2003). A structural-informatics approach for mining β-sheets: locating sheets in intermediate-resolution density maps. *J. Mol. Biol.* **332**, 399–413.

Kong, Y., Zhang, X., Baker, T.S., and Ma, J. (2004). A structural-informatics approach for tracing β-sheets: building pseudo-C traces for β-strands in intermediate-resolution density maps. *J. Mol. Biol.* **339**, 117–130.

Kornberg, R., and Darst, S.A. (1991). Two-dimensional crystals of proteins on liquid layers. *Curr. Opin. Struct. Biol.* **1**, 642–646.

Koster, A.J., de Ruijter, W.J., van Den Bos, A., and van Der Mast, K.D. (1989). Autotuning of a TEM using minimum electron dose. *Ultramicroscopy* **27**, 251–272.

Koster, A.J., Typke, D., and de Jong, M.J.C. (1990). Fast and accurate autotuning of a TEM for high resolution and low dose electron microscopy. In: *Proceedings of the XII International Congress for Electron Microscopy*, pp. 114–115. San Francisco Press, San Francisco.

Koster, A.J., Chen, J.W., Sedat, J.W., and Agard, D.A. (1992). Automated microscopy for electron tomography. *Ultramicroscopy* **46**, 207–227.

Koster, A.J., Grimm, R., Typke, D., Hegerl, R., Stoschek, A., Walz, J., and Baumeister, W. (1997). Perspectives of molecular and cellular electron tomography. *J. Struct. Biol.* **120**, 276–308.

Kostyuchenko, V.A., Leiman, P.G., Chipman, P.R., Kanamaru, S., van Raaij, M.J., Arisaka, F., Mesyanzhinov, V.V., and Rossmann, M.G. (2003). Three-dimensional structure of bacteriophage T4 baseplate. *Nat. Struct. Biol.* **10**, 688–693.

Kovacs, J.A., and Wriggers, W. (2002). Fast rotational matching. *Acta Cryst.* **D58**, 1371–1376.

Kozin, M.B., and Svergun, D.I. (2000). Automated matching of high- and low-resolution structural models. *J. Appl. Cryst.* **34**, 33–41.

Krakow, W., Downing, K.H., and Siegel, B.M. (1974). The use of tilted specimens to obtain the contrast transfer characteristics of an electron microscopy imaging system. *Optik* **40**, 1–13.

Krivanek, O.L., and Ahn, C. (1986). Energy-filtered imaging with quadrupole lenses. In: *Electron Microscopy 1986, Proceedings of the XI International Congress on Electron Microscopy* (T. Imura, S. Maruse, and T. Suzuki, eds.), Vol. 1, pp. 519–520. The Japanese Society of Electron Microscopy, Tokyo.

Krivanek, O.L., and Mooney, P.E. (1993). Applications of slow-scan CCD cameras in transmission electron microscopy. *Ultramicroscopy* **49**, 95–108.

Krivanek, O.L., Gubbens, A.J., and Dellby, N. (1991). Developments in EELS instrumentation for spectroscopy and imaging. *Microsc. Microanal. Microstruct.* **2**, 315–332.

Kübler, O., Hahn, M., and Serendynski, J. (1978). Optical and digital spatial frequency filtering of electron micrographs. I. Theoretical considerations. *Optik* **51**, 171–188. II. Experimental results. *Optik* **51**, 235–256.

Kühlbrandt, W. (1982). Discrimination of protein and nucleic acids by electron microscopy using contrast variation. *Ultramicroscopy* **7**, 221–232.

Kühlbrandt, W., and Downing, K.H. (1989). Two-dimensional structure of plant light-harvesting complex at 3.7 Å resolution by electron crystallography. *J. Mol. Biol.* **207**, 823–828. [Note that the actual title states "37 Å" due to a misprint.]

Kühlbrandt, W. and Unwin, P.N.T. (1982). Distribution of RNA and portein in crystalline eukaryotic ribosomes. *J. Mol. Biol.* **156**, 431–448.

Kühlbrandt, W., and Wang, D.N. (1991). 3-dimensional structure of plant light harvesting complex determined by electron crystallography. *Nature* **350**, 130–134.

Kühlbrandt, W., and Williams, K.A. (1999). Analysis of macromolecular structure and dynamics by electron cryo-microscopy. *Curr. Opin. Chem. Biol.* **3**, 537–543.

Kühlbrandt, W., Wang, D.N., and Fujiyoshi, Y. (1994). Atomic model of plant light-harvesting complex by electron crystallography. *Nature* **367**, 614–621.

Kunath, W., Weiss, K., Sack-Kongehl, H., Kessel, M., and Zeitler, E. (1984). Time-resolved low-dose microscopy of glutamine synthetase molecules. *Ultramicroscopy* **13**, 241–252.

Kuo, I., and Glaeser, R.M. (1975). Development of methodology for low exposure, high resolution electron microscopy of biological specimens. *Ultramicroscopy* **1**, 53–66.

Lake, J. (1971). Biological structures. In: *Optical Transforms* (H. Lipson, ed.), p. 174. Academic Press, London.

Lake, J.A. (1976). Ribosome structure determined by electron microscopy of *Escherichia coli* small subunits, large subunits, and monomeric ribosomes. *J. Mol. Biol.* **105**, 131–139.

Lambert, O., Boisset, N., Taveau, J.-C., and Lamy, J.N. (1994a). Three-dimensional reconstruction from frozen-hydrated specimen of the chiton *Lepidochiton sp.* hemocyanin. *J. Mol. Biol.* **244**, 640–647.

Lambert, O., Boisset, N., Penczek, P., Lamy, J., Taveau, J.C., Frank, J., and Lamy, J.N. (1994b). Quaternary structure of *Octopus vulgaris* hemocyanin: three-dimensional reconstruction from frozen-hydrated specimens and intramolecular location of functional units Ove and Ovb. *J. Mol. Biol.* **238**, 75–87.

Lambert, O., Boisset, N., Pochon, F., Delain, E., and Lamy, J.N. (1994c). An approach to the intramolecular localization of the thiol ester bonds in the internal cavity of human α_2-macroglobulin based on correspondence analysis. *J. Struct. Biol.* **112**, 148–159.

Lambert, O., Boisset, N., Taveau, J.-C., Preaux, G., and Lamy, J.N. (1995a). Three-dimensional reconstruction of the aD- and Bc-hemocyanins of *Helix pomatia* from frozen-hydrated specimens. *J. Mol. Biol.* **248**, 431–448.

Lambert, O., Taveau, J.-C., Boisset, N., and Lamy, J.N. (1995b). Three-dimensional reconstruction of the hemocyanin of the protobranch bivalve mollusc *Nucula hanleyi* from frozen-hydrated specimens. *Arch. Biochem. Biophys.* **319**, 231–243.

Lambert, O., Boisset, N., Taveau, J.-C., and Lamy, J.N. (1995c). Three-dimensional reconstruction of *Sepia officinalis* hemocyanin from frozen-hydrated specimens. *Arch. Biochem. Biophys.* **316**, 950–959.

Lamy, J. (1987). Intramolecular localization of antigenic determinants by molecular immunoelectron microscopy. In: *Biological Organization: Macromolecular Interactions at High Resolution* (R.M. Burnett and H.J. Vogel, eds.), pp. 153–191. Academic Press, San Diego.

Lamy, J., Sizaret, P.-Y., Frank, J., Verschoor, A., Feldmann, R., and Bonaventura, J. (1982). Architecture of *Limulus polyphemus* hemocyanins. *Biochemistry* **21**, 6825–6833.

Lamy, J., Lamy, J.N., Billiald, P., Sizaret, P.-Y., Cave, G., Frank, J., and Motta, G. (1985). Approach to the direct intramolecular localization of antigenic determinants in *Androctonus australis* hemocyanin with monoclonal antibodies by molecular immunoelectron microscopy. *Biochemistry* **24**, 5532–5542.

Lamy, J., Billiald, P., Taveau, J.-C., Boisset, N., Motta, G., and Lamy, J.N. (1990). Topological mapping of 13 epitopes on a subunit of *Androctonus australis* hemocyanin. *J. Struct. Biol.* **103**, 64–74.

Lamy, J., Gielens, C., Lambert, O., Taveau, J.C., Motta, G., Loncke, P., De Geest, N., Preaux, G., and Lamy, J.N. (1993a). Further approaches to the quaternary structure of *Octopus* hemocyanin: a model based on immunoelectron microscopy and image processing. *Arch. Biochem. Biophys.* **305**, 17–29.

Lamy, J., Frank, J., and Lamy, J.N. (1993b). Three-dimensional immunoelectron microscopy of scorpion hemocyanin labeled with monoclonal Fab fragment. *J. Struct. Biol.* **111**, 234–244.

Langer, R., and Hoppe, W. (1966). Die Erhöhung der Auflösung und Kontrast im Elektronenmikroskop mit Zonenkorrekturplatten. *Optik* **24**, 470–489.

Langer, R., Frank, J., Feltynowski, A., and Hoppe, W. (1970). Anwendung des Bilddifferenzverfahrens auf die Untersuchung von Strukturänderungen dünner Kohlefolien bei Elektronenbestrahlung. *Ber. Bunsenges. Phys. Chem.* **74**, 1120–1126.

Langer, T., Pfeifer, G., Martin, J., Baumeister, W., and Hartl, F.-U. (1992). Chaperonin-mediated protein folding: GroES binds to one end of the GroL cylinder, which accommodates the protein substrate within its central cavity. *EMBO J.* **11**, 4757–4765.

Langmore, J., and Smith, M. (1992). Quantitative energy-filtered microscopy of biological molecules in ice. *Ultramicroscopy* **46**, 349–373.

Lanio, S. (1986). High-resolution imaging magnetic filter with simple structure. *Optik* **73**, 99–107.

Lanzavecchia, S., and Bellon, P.L. (1994). A moving window Shannon reconstruction algorithm for image interpolation. *J. Vis. Commun. Image Repres.* **5,** 255–264.

Lanzavecchia, S., and Bellon, P.L. (1996). Electron tomography in conical tilt geometry. The accuracy of a direct Fourier method (DFM) and the suppression of non-tomographic noise. *Ultramicroscopy* **63,** 247–261.

Lanzavecchia, S., and Bellon, P.L. (1998). Fast computation of 3D Radon transform via a direct Fourier method. *Bioinformatics* **14,** 212–216.

Lanzavecchia, S., Bellon, P.L., and Scatturin, V. (1993). SPARK, a kernel of software programs for spatial reconstruction in electron microscopy. *J. Microsc.* **171,** 255–266.

Lanzavecchia, S., Tosoni, L., and Bellon, P.L. (1996). Fast sinogram computation and the sinogram-based alignment of images. *Cabios* **12,** 531–537.

Lanzavecchia, S., Bellon, P.L., Lupetti, P., Dallai, R., Rappuoli, R., and Telford, J.L. (1998). Three-dimensional reconstruction of metal replicas of the *Heliobacter pylori* vacuolating cytoxin. *J. Struct. Biol.* **121,** 9–18.

Lanzavecchia, S., Bellon, P.L., and Radermacher, M. (1999a). Fast and accurate three-dimensional reconstruction from projections with random orientations via Radon transforms. *J. Struct. Biol.* **128,** 152–164.

Lanzavecchia, S., Wade, R.H., Ghiretti Magaldi, A., Tognon, G., and Bellon, P.L. (1999b). A two exposure technique for ice embedded samples successfully reconstructs the chlorocruorin pigment of *Sabella spallanzanii* at 2.2 nm resolution. *J. Struct. Biol.* **127,** 53–63.

Lanzavecchia, S., Cantele, F., and Luigi, P. (2001). Alignment of 3D structures of macromolecular assemblies. *Bioinformatics* **17,** 58–62.

Lanzavecchia, S., Cantele, F., Radermacher, M., and Bellon, P.L. (2002). Symmetry embedding in the reconstruction of macromolecular assemblies via discrete Radon transform. *J. Struct. Biol.* **17,** 259–272.

Larquet, E., Boisset, N., Pochon, F., and Lamy, J. (1994a). Three-dimensional cryoelectron microscopy of native human α_2-macroglobulin. In: *Proceedings of the 13th International Congress on Electron Microscopy (Paris)*, Vol. 3A, pp. 529–530. Les Editions de Physiques, Les Ulis, France.

Larquet, E., Boisset, N., Pochon, F., and Lamy, J. (1994b). Architecture of native human α_2-macroglobulin studies by cryoelectron microscopy and three-dimensional reconstruction. *J. Struct. Biol.* **113,** 87–98.

Lata, K.R., Penczek, P., and Frank, J. (1994). Automatic particle picking from electron micrographs. In: *Proceedings of the 52nd Annual Meeting MSA (New Orleans)* (G. W. Bailey and A.J. Garratt-Reed, eds.), pp. 122–123. San Francisco Press, San Francisco.

Lata, K.R., Penczek, P., and Frank, J. (1995). Automated particle picking from electron micrographs. *Ultramicroscopy* **58,** 381–391.

Lata, K.R., Agrawal, R.K., Penczek, P., Grassucci, R., Zhu, J., and Frank, J. (1996). Three-dimensional reconstruction of the *Escherichia coli* 30S ribosomal subunit in ice. *J. Mol. Biol.* **262,** 43–52.

Lata, K.R., Conway, J.F., Cheng, N., Duda, R.L., Hendrix, R.W., Wikoff, W.R., Johnson, J.E., Tsurata, H., and Steven, A.C. (2000). Maturation dynamics of a viral capsid: visualization of transitional intermediate states. *Cell* **100,** 253–263.

Lawrence, M.C., Jaffer, M.A., and Sewell, B.T. (1989). The application of the maximum entropy method to electron microscope tomography. *Ultramicroscopy* **31,** 285–301.

Lebart, L., Morineau, A., and Tabard, N. (1977). *Techniques de la Description Statistique.* Dunod, Paris.

Lebart, L., Maurineau, A., and Warwick, K.M. (1984). *Multivariate Descriptive Statistical Analysis.* John Wiley, New York.

Lei, J., and Frank, J. (2005). Automated acquisition of cryo-electron micrographs for single-particle reconstruction on an FEI Tecnai electron microscope. *J. Struct. Biol.* **150,** 69–80.

Leith, A. (1992). Computer visualization of volume data in electron tomography. In: *Electron Tomography* (J. Frank, ed.), pp. 215–236. Plenum Press, New York.

Lenz, F. (1971). *Electron Microscopy in Material Science* (U. Valdré, ed.). Academic Press, New York.

Leonard, K.R., and Lake, J.A. (1979). Ribosome structure: hand determination by electron microscopy of 30S subunits. *J. Mol. Biol.* **129**, 155–163.

Lepault, J., and Pitt, T. (1984). Projected structure of unstained, frozen-hydrated T-layer of *Bacillus brevis*. *EMBO. J.* **3**, 101–105.

Lepault, J., Booy, F.P., and Dubochet, J. (1983). Electron microscopy of frozen biological suspensions. *J. Microsc.* **129**, 89–102.

Li, J., Gao, X, Nazif, T., Joss, L., Bogyo, M., Steven, A.C., and Rechsteiner, M. (2001). Proteasome activation by 11S REG (PA28) homologs: lysine 188 substitutions convert the pattern of proteasome activation by REGγ to that of REGs α and β. *EMBO J.* **20**, 3359–3369.

Li, Y., Leith, A., and Frank, J. (1997). Tinkerbell—a tool for interactive segmentation of 3D data. *J. Struct. Biol.* **120**, 266–275.

Liang, Y., Ke, E.Y., and Zhou, Z.H. (2002). IMIRS: a high-resolution 3D reconstruction package integrated with a relational image database. *J. Struct. Biol.* **137**, 292–304.

Lim, V., Venclovas, C., Spirin, A., Brimacombe, R., Mitchell, P., and Müller, F. (1992). How are tRNAs and MRNA arranged in the ribosome? An attempt to correlate the stereochemistry of the tRNA—mRNA interaction with constraints imposed by the ribosomal topography. *Nucleic Acids Res.* **20**, 2627–2637.

Lindahl, M. (2001). Strul—a method for 3D alignment of single-particle projections based on common line correlation in Fourier space. *Ultramicroscopy* **87**, 165–175.

Linfoot, E.H. (1964). *Fourier Methods in Optical Image Evaluation*. The Focal Press, London.

Liu, W. (1991). 3-D variance of weighted back-projection reconstruction and its application to the detection of 3-D particle conformational changes. In: *Proceedings of the 49th Annual Meeting, EMSA* (G.W. Bailey, ed.), pp. 542–543. San Francisco Press, San Francisco.

Liu, W. (1993). *Three-Dimensional Variance of Weighted Back Projection*. Ph.D. thesis, State University of New York at Albany.

Liu, W., and Frank, J. (1995). Estimation of variance distribution in three-dimensional reconstruction. I. Theory. *J. Opt. Soc. Am.* **12**, 2615–2627.

Liu, W., Boisset, N., and Frank, J. (1995). Estimation of variance distribution in three-dimensional reconstruction. II. Applications. *J. Opt. Soc. Am.* **12**, 2628–2635.

Liu, Z., Zhang, J., Sharma, M.R., Li, P., Chen, S.R.W., and Wagenknecht, T. (2001). Three-dimensional reconstruction of the recombinant type 3 ryanodine receptor and localization of its amino terminus. *Proc. Natl. Acad. Sci. USA* **98**, 6104–6109.

Liu, Z., Zhang, J., Li, P., Chen, S.R., and Wagenknecht, T. (2002). Three-dimensional reconstruction of the recombinant type 2 ryanodine receptor and localization of its divergent region 1. *J. Biol. Chem.* **277**, 46712–46719.

Liu, J., Reedy, M.C., Goldman, Y.E., Franzini-Armstrong, C., Sasaki, H., Tregear, R.T., Lucaveche, C., Winkler, H. Baumann, B.A.J., Squire, J.M., Irving, T.C., Reedy, M.K., and Taylor, K.A. (2004). Electron tomography of fast frozen, stretched rigor fibers reveals elastic distortions in the myosin crossbridges. *J. Struct. Biol.* **147**, 268–282.

Llorca, O., Marco, S., Carrascosa, L., and Valpuesta, J.M. (1994). The formation of symmetrical GroEL–GroES complexes in the presence of ATP. *FEBS Lett.* **345**, 181–186.

Llorca, O., Martin-Benito, J., Ritco-Vonsovici, M., Grantham, J., Hynes, G.M., Willison, K.R., Carrascosa, J.L., and Valpuesta, J.M. (2000). Eukaryotic chaperonin CCT stabilizes actin and tubulin folding intermediates in open quasi-native onformations. *EMBO J.* **19**, 5971–5979.

Llorca, O., Martin-Benito, J., Grantham, J., Ritco-Vonsovici, M., Willison, K.R., Carrascosa, J.L., and Valpuesta, J.M. (2001). The "sequential allosteric ring" mechanism in the eukaryotic chaperonin-assisted folding of actin and tubulin. *EMBO J.* **20**, 4065–4075.

Ludtke, Z.J., Baldwin, P.R., and Chiu, W. (1999). EMAN: semiautomated software for high-resolution single-particle reconstructions. *J. Struct. Biol.* **128**, 82–97.

Ludtke, S.L., and Chiu, W. (2002). Image restoration in sets of noisy electron micrographs. *Proc. IEEE Intl. Symp. Biomed. Imag.* pp. 745–748.

Ludtke, S.L., Jakana, J., Song, J.-L., Chuang, D.T., and Chiu, W. (2001). A 11.5 Å single particle reconstruction of GroEl using EMAN. *J. Mol. Biol.* **314**, 253–262.

Ludtke, S.L., Chen, D.H., Song, J.-L., Chuang, D.T., and Chiu, W. (2004). Seeing GroEL at 6 Å resolution by single particle electron cryomicroscopy. *Structure* **12**, 1129–1136.

Luo, R.Z.-T., Beniac, D.R., Fernandez, A., Yip, C.C., and Ottensmeyer, F.P. (1999). Quaternary structure of the insulin–insulin receptor complex. *Science* **285**, 1077–1080.

Lutsch, G., Pleissner, K.-P., Wangermann, G., and Noll, F. (1977). Studies on the structure of animal ribosomes. VIII. Application of a digital image processing method to the enhancement of electron micrographs of small ribosomal subunits. *Acta Biol. Med. Germ.* **36**, K-59.

Ma, J. (2005). Usefulness and limitations of normal mode analysis in modeling dynamics of biomolecular complexes. *Structure* **13**, 373–380.

Majorovits, E., and Schröder, R.R. (2002a). Enhancing contrast of weak phase objects using a Zernike-type phase plate in phase contrast TEM. *Microsc. Microanal.* **8** (Suppl. 2), 864CD–865CD.

Majorovits, E., and Schröder, R.R. (2002b). Improved information recovery in phase contrast EM for non-two-fold symmetric Boersch phase plate geometry. *Microsc. Microanal.* **8** (Suppl. 2), 540CD–541CD.

Malhotra, A., and Harvey, S.C. (1994). A quantitative model of the *Escherichia coli* 16S RNA in the 30S ribosomal subunit. *J. Mol. Biol.* **240**, 308–340.

Malhotra, A., Penczek, P., Agrawal, R.K., Gabashvili, I.S., Grassucci, R.A., Jünemann, R., Burkhardt, N., Nierhaus, K.H., and Frank, J. (1998). *Escherichia coli* 70S ribosome at 15 Å resolution by cryo-electron microscopy: localization of fMet–tRNA$_f$^{Met} and fitting of L1 protein. *J. Mol. Biol.* **280**, 103–116.

Mancini, E.J., and Fuller, S.D. (2000). Supplanting crystallography or supplementing microscopy? A combined approach to the study of an enveloped virus. *Acta Cryst.* **D56**, 1278–1287.

Mannella, C., Marko, M., Penczek, P., Barnard, D., and Frank, J. (1994). The internal compartmentation of rat-liver mitochondria: tomographic study using the high-voltage electron microscope. *Microsc. Res. Tech.* **27**, 278–283.

Marabini, R., and Carazo, J.M. (1994a). Pattern recognition and classification of images of biological macromolecules using artificial neural networks. *Biophys. J.* **66**, 1804–1814.

Marabini, R., and Carazo, J.M. (1994b). Practical issues on invariant image averaging using the bispectrum. *Signal Process.* **40**, 119–128.

Marabini, R., and Carazo, J.M. (1996). On a new computationally fast image invariant based on bispectral projections. *Patt. Recog. Lett.* **17**, 959–967.

Marabini, R. Vaquerizo, C., Fernandez, J.J., Carazo, J.M., Ladjadj, M., Odesanya, O., and Frank, J. (1994). On a prototype for a new distributed data base of volume data obtained by 3D imaging. In: *Visualization in Biomedical Computing* (R.A. Robb, ed.), pp. 466–472. Proc. SPIE 2359.

Marabini, R., Vaquerizo, C., Fernández, J.J., Carazo, J.M., Engel, A., and Frank, J. (1996a). Proposal for a new distributed database of macromolecular and subcellular structures from different areas of microscopy. *J. Struct. Biol.* **116**, 161–166.

Marabini, R., Masegosa, M., San Martin, M.C., Marco, S., Fernández, J.J., de la Fraga, L.G., Vaquerizo, C., and Carazo, J.M. (1996b). Xmipp: an image processing package for electron microscopy. *J. Struct. Biol.* **116**, 237–240.

Marabini, R., Vaquerizo, C., Fernandez, J.J., Carazo, J.M., Engel, A., and Frank, J. (1996c). Proposal for a new distributed database of macromolecular and subcellular structures from different areas of microscopy. *J. Struct. Biol.* **116**, 161–165.

Marabini, R., Herman, G.T., and Carazo, J.M. (1998). 3D reconstruction in electron microscopy using ART with smooth spherically symmetric volume elements (blobs). *Ultramicroscopy* **72**, 53–65.

Marabini, R., Herman, G.T., Sorzano, C.O.S., and Carazo, J.M. (2002). Volume constraints in 3D tomography applied to electron microscopy. *Proc. IEEE Intl. Symp. Biomed. Imag.*, pp. 641–644.

Markham, R., Frey, S., and Hills, G.J. (1963). Methods for the enhancement of image detail and accentuation of structure in electron microscopy. *Virology* **22**, 88–102.

Markham, R., Hitchborn, J.H., Hills, G., and Frey, S. (1964). The anatomy of the tobacco mosaic virus. *Virology* **22**, 342–359.

Martín-Benito, J., Area, E., Ortega, J., Llorca, O., Valpuesta, J.M., Carrascosa, J.L., and Ortín, J. (2001). Three-dimensional reconstruction of a recombinant influenza virus ribonucleoprotein particle. *EMBO Rep.* **2**, 313–317.

Martinez, T., and Schulten, K. (1994). Topology representing networks. *Neural Networks* **7**, 507–522.

Massover, W.M., and Marsh, P. (1997). Unconventional negative stains: heavy metals are not required for negative staining. *Ultramicroscopy* **69**, 139–150.

Massover, W.M., and Marsh, P. (2000). Light atom derivatives of structure-preserving sugars are unconventional negative stains. *Ultramicroscopy* **85**, 107–121.

Massover, W.M., Lai, P.F., and Marsh, P. (2001). Negative staining permits 4.0 Å resolution with low-dose electron diffraction of catalase crystals. *Ultramicroscopy* **90**, 7–12.

Matadeen, R., Patwardhan, A., Gowen, B., Orlova, E.V., Müller, F., Brimacombe, R., and van Heel, M. (1999). The *Escherichia coli* large ribosomal subunit at 7.5 Å resolution. *Structure* **7**, 1575–1583.

McCormack, E.A., Llorca, O., Carrascosa, J.L., Valpuesta, J.-M., and Willison, K.R. (2001). Point mutations in a hinge linking the small and large domains of β-actin result in trapped folding intermediates bound to cytosolic chaperonin CCT. *J. Struct. Biol.* **135**, 198–204.

McCutcheon, J.P., Agrawal, R.K., Philips, S.M., Grassucci, R.A., Gerchman, S.E., Clemons, W.M., Ramakrishnan, V., and Frank, J. (1999). Location of translational initiation factor IF3 on the small ribosomal subunit. *Proc. Natl. Acad. Sci. USA* **96**, 4301–4306.

McDowall, A.W., Chang, J.J., Freeman, R., Lepault, J., Walter, C.A., and Dubochet, J. (1983). Electron microscopy of frozen-hydrated sections of vitreous ice and vitrified biological samples. *J. Microsc.* **131**, 1–9.

McEwen, B., and Marko, M. (1999). Three-dimensional transmission electron microscopy and its application to mitosis research. In: *Mitosis and Meiosis.* (C. Rieder, ed.), *Methods in Cell Biology*, Vol. 61, pp. 81–111. Academic Press, San Francisco.

McEwen, B.F. and Frank, J. (2001). Electron tomographic (and other approaches) for imaging molecular machines. *Curr. Opin. Neurobiol.* **11**, 594–600.

McEwen, B.F., and Marko, M. (2001). The emergence of tomography as an important tool for investigating cellular ultrastructure. *J. Histochem. Cytochem.* **49**, 553–563.

McIntosh, J.R. (2001). Electron microscopy of cells: a beginning for a new century. *J. Cell Biol.* **153**, F25–F32.

Meissner, U., Dube, P., Harris, J.R., Stark, H., and Markl, J. (2000). Structure of a molluscan hemocyanin dodecamer (HtH1 from *Haliotis tuberculata*) at 12 Å resolution by cryoelectron microscopy. *J. Mol. Biol.* **298**, 21–34.

Mellwig, C., and Böttcher, B. (2001). Dealing with particles in different conformational states by electron microscopy and image processing. *J. Struct. Biol.* **133**, 214–220.

Ménétret, J.-F., Hofmann, W., Schröder, R.R., Rapp, G., and Goody, R.S. (1991). Time-resolved cryo-electron microscopic study of the dissociation of actomyosin induced by photolysis of photolabile nucteotides. *J. Mol. Biol.* **219**, 139–144.

Ménétret, J.-F., Neuhof, A., Morgan, D.G., Plath, K., Radermacher, M., Rapoport, T.A., and Akey, C.W. (2000). The structure of ribosome–channel complexes engaged in protein translocation. *Mol. Cell* **6**, 1219–1232.

Metoz, F., Arnal, I., and Wade, R.H. (1997). Tomography without tilt: three-dimensional imaging of microtubule/motor complexes. *J. Struct. Biol.* **118**, 159–168.

Mezzich, J.E., and Solomon, H. (1980). *Taxonomy and Behavioral Science*. Academic Press, London.

Midgley, P.A., and Weyland, M. (2003). 3D electron microscopy in the physical sciences: the development of Z-contrast and EFTEM tomography. *Ultramicroscopy* **96**, 413–431.

Milligan, R.A., and Flicker, P.F. (1987). Structural relationships of actin, myosin, and tropomyosin revealed by cryo-electron microscopy. *J. Cell Biol.* **105**, 29–39.

Milligan, R.A., and Unwin, P.N.T. (1986). Location of exit channel for nascent protein in 80S ribosomes. *Nature* **319**, 693–695.

Milligan, R.A., Whittaker, M., and Safer, D. (1990). Molecular structure of F-actin and location of surface binding sites. *Nature* **348**, 217–221.

Mindell, J.A. and Grigorieff, N. (2003). Accurate determination of local defocus and specimen tilt in electron microscopy. *J. Struct. Biol.* **142**, 334–337.

Miyazawa, A., Fujioshi, Y., Stowell, M., and Unwin, N. (1999). Nicotinic acetylcholine receptor at 4.6 Å resolution: transverse tunnels in the channel well. *J. Mol. Biol.* **288**, 765–786.

Miyazawa, A., Fujioshi, Y., and Unwin, N. (2003). Structure and gating mechanism of the acetylcholine receptor pore. *Nature* **423**, 949–955.

Möbus, G., and Rühle, M. (1993). A new procedure for the determination of the chromatic contrast transfer envelope of electron microscopes. *Optik* **93**, 108–118.

Möbus, G., Doole, R.C., and Inkson, B.J. (2003). Spectroscopic electron tomography. *Ultramicroscopy* **96**, 433–451.

Montesano-Roditis, L., Glitz, D.G., Traut, R.B., and Stewart, P.L. (2001). Cryo-electron microscopic localization of protein L7/L12 within the *Escherichia coli* 70S ribosome by difference mapping and Nanogold labeling. *J. Biol. Chem.* **276**, 14117–14123.

Moody, M.F. (1967). Structure of the sheath of bacteriophage T4. 1. The structure of the contracted sheath and polysheath. *J. Mol. Biol.* **25**, 167–200.

Moody, M.F. (1990). Image analysis in electron microscopy. In: *Biophysical Electron Microscopy* (P.W. Hawkes and U. Valdrè, eds.), pp. 145–287. Academic Press, London.

Moore, P.B. (1995). Ribosomes seen through a glass less darkly. *Structure* **3**, 851–852.

Moore, P.B., and Steitz, T.A. (2003). The structural basis of large ribosomal subunit function. *Ann. Rev. Biochem.* **72**, 813–850.

Morgan, D.G., Owen, C., Melanson, L.A., and DeRosier, D.J. (1995). Structure of bacterial flagellar filaments at 11 Å resolution: packing of the alpha-helices. *J. Mol. Biol.* **249**, 88–110.

Morgan, D.G., Ménétret, J.F., Radermacher, M., Neuhof, A., Akey, I.V., Rapaport, T.A., and Akey, C.W. (2000). A comparison of the yeast and rabbit 80S ribosome reveals the topology of the nascent chain exit tunnel, intersubunit bridges, and mammalian rRNA expansion segments. *J. Mol. Biol.* **301**, 301–321.

Mouche, F., Boisset, N., Lamy, J., Zal, F., and Lamy, J.N. (1999). Structural comparison of cephalopodan hemocyanins: phylogenetic significance. *J. Struct. Biol.* **127**, 199–212.

Mouche, F., Boisset, N., and Penczek, P. (2001). *Lumbricus terrestris* hemoglobin—the architecture of linker chains and structural variation of the central toroid. *J. Struct. Biol.* **133**, 176–192.

Mouche, F., Gontero, B., Callebaut, I., Mornon, J.P., and Boisset, N. (2002). Striking conformational change suspected within phosphoribulokinase dimer induced by interaction with GAPdH. *J. Biol. Chem.* **277**, 6743–6749.

Mouche, F., Zhu, Y., Pulokas, J., Potter, C.S., and Carragher, B. (2004). Automated three-dimensional reconstruction of keyhole limpet hemocyanin type I. *J. Struct. Biol.* **144**, 301–312.

Mu, X.-Q., Egelman, E.H., and Bullitt, E. (2002). Structure and function of Hib pili from *Haemophilus influenzae* type b. *J. Bact.* **184**, 4868–4874.

Mullapudi, S., Pullan, L., Bishop, O.T., Khalil, H., Stoops, J.K., Beckmann, R., Kloetzel, P.M., Kruger, E., and Penczek, P.A. (2004). Rearrangement of the 16S precursor subunits is essential for the formation of the active 20S proteasome. *Biophys. J.* **87**, 4098–4105.

Müller, F., and Brimacombe, R. (1997). A new model for the three-dimensional folding of *Escherichia coli* 16S ribosomal RNA. I. Fitting the RNA to a 3D electron microscopic map at 20 Å. *J. Mol. Biol.* **271**, 534–544.

Müller, F., Sommer, I., Baranov, P., Matadeen, R., Stoldt, M., Wühnert, P., Görlach, M., van Heel, M., and Brimacombe, R. (2000). The 3D arrangement of the 23S and 5S RNA in the *E. coli* ribosomal subunit based on cryo-electron microscopic reconstruction at 7.5 Å resolution. *J. Mol. Biol.* **298**, 35–59.

Murakami, K.S., Masuda, S., Campbell, E.A., Muzzin, O., and Darst, S.A. (2002). Structural basis of transcription initiation: an RNA polymerase holoenzyme–DNA complex. *Science* **296**, 1285–1290.

Murata, K., Odahara, N., Kuniyasu, A., Sato, Y., Nakayama, H., and Nagayama, K. (2001). Asymmetric arrangement of auxiliary subunits of skeletal muscle voltage-gated L-type Ca^{2+} channel. *Biochem. Biophys. Res. Commun.* **282**, 284–291.

Musacchio, A., Smith, C.J., Roseman, A.M., Harrison, S.C., Kirchhausen, T., and Pearse, B.M.F. (1999). Functional organization of clathrin in coats: combining electron cryomicroscopy and X-ray crystallography. *Mol. Cell* **3**, 761–770.

Näär, A.M., Taatjes, D.J., Zhai, W., Nogales, E., and Tjian, R. (2002). Human CRSP interacts with RNA polymerase II CTD and adopts a specific CTD-bound conformation. *Genes Devel.* **16**, 1339–1344.

Nadeau, O.W., Carlson, G.M., and Gogol, E.P. (2002). A Ca^{2+}-dependent global conformational change in the 3D structure of phosphorylase kinase obtained from electron microscopy. *Structure* **10**, 23–32.

Nagano, K. and Hrel, M. (1987). Approaches to a three-dimensional model of *E. coli* ribosome. *Progr. Biophys. Mol. Biol.* **48**, 67–101.

Namba, K., Caspar, D.L.D., and Stubbs, G. (1988). Enhancement and simplification of macromolecular images. *Biophys. J.* **53**, 469–475.

Nathan, R. (1970). Computer enhancement of electron micrographs. In: *Proceedings of the 28th Annual Meeting, EMSA*, pp. 28–29. Houston, TX.

Natterer, F. (1986). *The Mathematics of Computerized Tomography*. John Wiley, Stuttgart.

Nicholls, A., Bharadwaj, R., and Honig, B. (1991). Protein folding and association: insights from the interfacial and thermodynamic properties of hydrocarbons. *Proteins* **11**, 281–296.

Nicholson, W.V., and Glaeser, R.M. (2001). Automatic particle detection in electron microscopy. *J. Struct. Biol.* **133**, 90–101.

Nicholson, W.N., Lee, M., Downing K.H., and Nogales, E. (1999). Cryo-electron microscopy of GDP–tubulin rings. *Cell Biochem. Biophys.* **31**, 175–183.

Nield, J., Orlova, E.V., Morris, E.P., Gowen, B., and van Heel, M., and Barber, J. (2000). 3D map of the plant photosystem II supercomplex obtained by cryoelectron microscopy and single particle analysis. *Nat. Struct. Biol.* **7**, 44–47.

Nitsch, M., Klumpp, M., Lupas, A., and Baumeister, W. (1997). The thermosome: alternating α- and β-subunits within the chaperonin of the archeon *Thermoplasma acidophilum*. *J. Mol. Biol.* **267**, 142–149.

Nitsch, M., Walz, J., Typke, D, Klumpp, M., Essen, L.-O., and Baumeister, W. (1998). Group II chaperonin in an open conformation examined by electron tomography. *Nat. Struct. Biol.* **5**, 855–857.

Nogales, E. (2000). Recent structural insights into transcriptional preinitiation complexes. *J. Cell Sci.* **113**, 4391–4397.

Nogales, E., and Grigorieff, N. (2001). Molecular machines: putting the pieces together. *J. Cell Biol.* **152**, F1–F10.

Nogales, E., Wolf, S.G., and Downing, K.H. (1998). Structure of the αβ tubulin dimer by electron crystallography. *Nature* **391**, 199–203.

Norcum, M., and Boisset, N. (2002). Three-dimensional architecture of the eukaryotic multisynthetase complex from negatively stained and cryoelectron micrographs. *FEBS Lett.* **512**, 298–302.

Oakes, M.I., Scheiman, A., Atha, T., Shankweiler, G., and Lake, J.A. (1990). Ribosome structure: three-dimensional locations of rRNA and proteins. In: *The Ribosome, Structure, Function, and Evolution* (W. Hill, A. Dahlberg, R.A. Garrett, P.B. Moore, D. Schlessinger, and J.R. Warner, eds.), pp. 180–193. American Society for Microbiology, Washington, D.C.

O'Brien, J. (1991). Cartoon. *The New Yorker*, Feb. 25, p. 37.

Öfverstedt, L.-G., Zhang, K., Tapio, S., Skoglund, U., and Isakson, L.A. (1994). Starvation *in vivo* for aminoacyl-tRNA increases the spatial separation between the two ribosomal subunits. *Cell* **79**, 629–638.

Öfverstedt, L.-G., Zhang, K., Bricogne, G., Isaksson, L.A., and Skoglund, U. (1997). Automated correlation and averaging of three-dimensional reconstructions obtained by electron tomography. *J. Struct. Biol.* **120**, 329–342.

Ogura, T., and Sato, C. (2001). An automatic particle pickup method using a neural network applicable to low-contrast electron micrographs. *J. Struct. Biol.* **136**, 227–238.

Ogura, T. and Sato, C. (2003). Automatic particle pickup method using a neural network has high accuracy by applying an initial weight derived from eigenimages: a new reference free method for single-particle analysis. *J. Struct. Biol.* **145**, 63–75.

Ogura, T., Iwasaki, K., and Sato, C. (2003). Topology representing network enables highly accurate classification of protein images taken by cryo electron-microscope mithout masking. *J. Struct. Biol.* **143**, 185–200.

Ohi, M., Li, Y., Cheng, Y., and Walz, T. (2004). Negative staining and image classification—powerful tools in modern elctron microscopy. *Biol. Proced. Online* **6**, 23–34.

O'Keefe, M.A. (1992). "Resolution" in high-resolution electron microscopy. *Ultramicroscopy* **47**, 282–297.

O'Neill, E.L. (1969). *Introduction to Statistical Optics*. Addison-Wesley, Reading, Mass.

Oostergetel, G.T., Keegstra, W., and Brisson, A. (1998). Automation of specimen selection and data acquisition for protein crystallography. *Ultramicroscopy* **40**, 47–49.

Orlova, E., and van Heel, M. (1994). Angular reconstitution of macromolecules with arbitrary point-group symmetry. In: *Proceedings of the 13th International Congress on Electron Microscopy (Paris)*, Vol. 1, pp. 507–508. Les Editions de Physiques, Les Ulis, France.

Orlova, E.V. (2000). Structural analysis of non-crystalline macromolecules: the ribosome. *Acta Cryst.* **D56**, 1253–1258.

Orlova, E.V., Serysheva, I.I., van Heel, M., Hamilton, S.L., and Chiu, W. (1996). Two structural configurations of the skeletal muscle calcium release channel. *Nat. Struct. Biol.* **3**, 547–552.

Orlova, E.V., Dube, P., Harris, J.R., Beckman, E., Zemlin, F., Markl, J., and van Heel, M. (1997). Structure of keyhole limpet hemocyanin type 1 (KLH1) at 15 Å resolution by electron cryomicroscopy and angular reconstitution. *J. Mol. Biol.* **271**, 417–437.

Orlova, E.V., Serysheva, I.I., Hamilton, S.L., Chiu, W., and van Heel, M. (1998). The skeletal muscle calcium-release channel visualized by electron cryomicroscopy and angular reconstitution. In: *Ryanodine Receptors* (R. Sitsapesan and A. Williams, eds.), pp. 23–46. J.B. Lippincott Co., Philadelphia, PA.

Orlova, E.V., Sherman, M.B., Chiu, W., Mowri, H., Smith, L.C., and Gotto, A.M. (1999). Three-dimensional structure of low density lipoproteins by electron cryomicroscopy. *Proc. Natl. Acad. Sci. USA* **96**, 8420–8425.

Orlova, E.V., Atiqur Rahman, M., Gowen, B., Volynski, K.E., Ashon, A.C., Manser, C., van Heel, M., and Ushkaryov, Y.A. (2000). Structure of α-latrotoxin oligomers reveals that divalent cation-dependent tetramers form membrane pores. *Nat. Struct. Biol.* **7**, 48–53.

Ortega, J., Singh, S.K., Ishikawa, T., Maurizi, M.R., and Steven, A.C. (2000). Visualization of substrate binding and translocation by the ATP-dependent protease, ClpXP. *Cell* **6**, 1509–1514.

Ottensmeyer, F.P., and Farrow, N.A. (1992). Three-dimensional reconstruction from dark-field electron micrographs of macromolecules at random unknown angles. In: *Proceedings of the 50th Annual Meeting, EMSA* (G.W. Bailey, J. Bentley, and J.A. Small, eds.), pp. 1058–1059. San Francisco Press, San Francisco.

Ottensmeyer, F.P., Schmidt, E.E., Jack, T., and Powell, J. (1972). Molecular architecture: The optical treatment of dark field electron micrographs of atoms. *J. Ultrastruct. Res.* **40**, 546–555.

Ottensmeyer, F.P., Andrew, J.W., Bazett-Jones, D.P., Chan. A.S., and Hewitt, J. (1977). Signal to noise enhancement in dark field electron micrographs of vasopressin: Filtering of arrays of images in reciprocal space. *J. Microsc.* **109**, 259–268.

Ottensmeyer, F.P., Czarnota, G.J., Andrews, D.W., and Farrow, N.A. (1994). Three-dimensional reconstruction of the 54 kDa signal recognition protein SRP54 from STEM darkfield images of the molecule at random orientations. In: *Proceedings of the 13th International Congress on Electron Microscopy (Paris)*, Vol. 1, pp. 509–510. Les Editions de Physiques, Les Ulis, France.

Ottensmeyer, F.P., Beniac, D.R., Luo, R.Z.-T., and Yip, C.C. (2000). Mechanism of transmembrane signaling: insulin binding and the insulin receptor. *Biochemistry* **39**, 12103–12112.

Öttl, H., Hegerl, R., and Hoppe, W. (1983). Three-dimensional reconstruction and averaging of 50S ribosomal subunits of *Escherichia coli* from electron micrographs. *J. Mol. Biol.* **163**, 431–450.

Pante, N., and Aebi, U. (1994). Towards understanding the three-dimensional structure of the nuclear pore complex at the molecular level. *Curr. Opin. Struct. Biol.* **4**, 187–196.

Pascual-Montano, A., Bárcena, M., Merelo, J.J., and Carazo, J.-M. (2000). Mapping and fuzzy classification of macromolecular images using self-organizing neural networks. *Ultramicroscopy* **84**, 85–99.

Pascual-Montano, A., Donate, L.E., Valle, M., Barcena, M., Pascual-Marqui, R.D., and Carazo, J.M. (2001). A novel neural network technique for analysis and classification of EM single-particle images. *J. Struct. Biol.* **133**, 233–245.

Pascual-Montano, A., Taylor, K.A., Winkler, H., Pascual-Marqui, R.D., and Carazo, J.-M. (2002). Quantitative self-organizing maps for clustering electron tomograms. *J. Struct. Biol.* **138**, 114–122.

Pate, P., Mochca-Morales, J., Wu, Y., Zhang, J.Z., Rodney, G.G., Serysheva, I.I., Williams, B.Y., Anderson, M.E., and Hamilton, S.L. (2000). Determinants for calmodulin binding on voltage-dependent Ca^{2+} channels. *J. Biol. Chem.* **275**, 39786–39792.

Pavelcik, F., Zelinka, J., and Otwinowski, Z. (2002). Methodology and applications of automatic electron-density map interpretation by six-dimensional rotational and translational search for molecular fragments. *Acta Cryst.* **D58**, 275–283.

Penczek, P., Radermacher, M., and Frank, J. (1992). Three-dimensional reconstruction of single particles embedded in ice. *Ultramicroscopy* **40**, 33–53.

Penczek, P., Grassucci, R.A., and Frank, J. (1994). The ribosome at improved resolution: New techniques for merging and orientation refinement in 3D cryoelectron microscopy of biological particles. *Ultramicroscopy* **53**, 251–270.

Penczek, P.A. (2002a). Three-dimensional spectral signal-to-noise ratio for a class of reconstruction algorithms. *J. Struct. Biol.* **138**, 34–46.

Penczek, P.A. (2002b). Variance in three-dimensional reconstructions from projections. *Proc. IEEE Intl. Symp. Biomed. Imag.* pp. 749–752.

Penczek, P., Ban, N., Grassucci, R.A., Agrawal, R.K., and Frank, J. (1999). *Haloarcula marismortui* 50S subunit—complementarity of electron microscopy and X-ray crystallographic information. *J. Struct. Biol.* **128**, 44–50.

Penczek, P., Renka, R., and Schomberg, H. (2004). Gridding-based direct Fourier inversion of the three-dimensional ray transform. *J. Opt. Soc. Am.* **21**, 499–509.

Penczek, P.A., Zhu, J., and Frank, J. (1996). A common-lines based method for determining orientations for $N > 3$ particle projections simultaneously. *Ultramicroscopy* **63**, 205–218.

Penczek, P.A., Zhu, J., Schröder, R., and Frank, J. (1997). Three-dimensional reconstruction with contrast transfer compensation from defocus series. *Scanning Microscopy*, **11**, 147–154.

Phipps, B.M., Typke, D., Hegerl, R., Volker, S., Hoffmann, A., Stetter, K.O., and Baumeister, W. (1993). Structure of a molecular chaperone from a thermophilic archaebacterium. *Nature* **361**, 475–477.

Plitzko, J.M., Frangakis, A.S., Nickell, S., Foerster, F., Gross, A., and Baumeister, W. (2002). *In vivo veritas*: electron cryotomography of cells. *Trends Biotechnol.* **20**, S40–S44.

Polyakov, A., Richter, C., Malhotra, A., Koulich, D., Borukhov, S., and Darst, S. (1998). Visualization of the binding site for the transcript cleavage factor GreB on *Escherichia coli* RNA polymerase. *J. Mol. Biol.* **281**, 465–473.

Potter, C.S., Chu, H., Frey, B., Green, C., Kisseberth, N., Madden, T.J., Miller, K.L., Nahrstedt, K., Pulokas, J., Reilein, A., Tcheng, D., Weber, D., and Carragher, B. (1999). Leginon: a system for fully automated acquisition of 1000 micrographs a day. *Ultramicroscopy* **77**, 153–161.

Potter, C.S., Fellmann, D., Milligan, R.A., Pulokas, J., Suloway, C., Zhu, Y., and Carragher, B. (eds.) (2002). Automated cryo-electron microscopy. *Proc. IEEE Intl. Symp. Biomed. Imag.*, pp. 741–744.

Provencher, S.W., and Vogel, R.H. (1983). Regularization techniques for inverse problems in molecular biology. In: *Progress in Scientific Computing* (S. Abarbanel, R. Glowinski, G. Golub, and H.-O. Kreiss, eds.). Birkhauser, Boston.

Provencher, S.W., and Vogel, R.W. (1988). Three-dimensional reconstruction from electron micrographs of disordered specimens, 1. Method. *Ultramicroscopy* **25**, 209–222.

Qazi, U., Gettins, P.G.W., and Stoops, J.K. (1998). On the structural changes of native human α_2-macroglobulin upon proteinase entrapment: three-dimensional structure of the half-transformed molecule. *J. Biol. Chem.* **273**, 8987–8993.

Qazi, U., Gettins, P.G.W., Strickland, D.K., and Stoops, J.K. (1999). Structural details of proteinase entrapment by human α_2-macroglobulin emerge from three-dimensional reconstructions of Fab-labeled native, half-transformed molecules. *J. Biol. Chem.* **274**, 8137–8142.

Radermacher, M. (1980). *Dreidimensionale Rekonstruktion bei kegelförmiger Kippung im Elektronenmikroskop.* Thesis, Technical University, Munich.

Radermacher, M. (1988). The three-dimensional reconstruction of single particles from random and non-random tilt series. *J. Electron Microsc. Tech.* **9**, 359–394.

Radermacher, M. (1991). Three-dimensional reconstruction of single particles in electron microscopy. In: *Image Analysis in Biology* (D.-P. Häder, ed.), pp. 219–249. CRC Press, Boca Raton, FL.

Radermacher, M. (1992). Weighted back-projection methods. In: *Electron Tomography* (J. Frank, ed.). Plenum Press, New York, pp. 91–115.

Radermacher, M. (1994). Three-dimensional reconstruction from random projections: Orientational alignment via Radon transforms. *Ultramicroscopy* **53**, 121–136.

Radermacher, M. (1997). *Scanning microscopy*, Scanning Microscopy International, AMF O'Hare (Chicago), IL, Vol. 11, 169–176.

Radermacher, M., and Frank, J. (1984). Representation of objects reconstructed in 3D by surfaces of equal density. *J. Microsc.* **136**, 77–85.

Radermacher, M., and Frank, J. (1985). Use of nonlinear mapping in multivariate statistical analysis of molecule projections. *Ultramicroscopy* **17**, 117–126.

Radermacher, M., Wagenknecht, T., Verschoor, A., and Frank, J. (1986a). A new 3-D reconstruction scheme applied to the 50S ribosomal subunit of *E. coli. J. Microsc.* **141**, RP1.

Radermacher, M., Frank, J., and Mannella, C.A. (1986b). Correlation averaging: lattice separation and resolution assessment. In: *Proceedings of the 44th Annual Meeting, EMSA* (G.W. Bailey, ed.), pp. 140–143. San Francisco Press, San Francisco.

Radermacher, M., Wagenknecht, T., Verschoor, A., and Frank, J. (1987a). Three-dimensional structure of the large ribosomal subunit from *Escherichia coli*. *EMBO J.* **6**, 1107–1114.

Radermacher, M., Wagenknecht, T., Verschoor, A., and Frank, J. (1987b). Three-dimensional reconstruction from a single-exposure, random conical tilt series applied to the 50S ribosomal subunit. *J. Microsc.* **146**, 113–136.

Radermacher, M., Nowotny, V., Grassucci, R., and Frank, J. (1990). Three-dimensional image reconstruction of the 50S subunit from *Escherichia* coli ribosomes lacking 5S-rRNA. In: *Proceedings of the XII International Congress on Electron Microscopy*, pp. 284–285. San Francisco Press, San Francisco.

Radermacher, M., Wagenknecht, T., Grassucci, R., Frank, J., Inui, M., Chadwick, C., and Fleischer, S. (1992a). Cryo-EM of the native structure of the calcium release channel/ryanodine receptor from sarcoplasmic reticulum. *Biophys. J.* **61**, 936–940.

Radermacher, M., Srivastava, S., and Frank, J. (1992b). The structure of the 50S ribosomal subunit from *E. coli* in frozen hydrated preparation reconstructed with SECRET. In: *Proceedings of the 10th European Congress on Electron Microscopy*, Vol. 3, pp. 19–20. Secretariado de Publicaciones de la Universidad Granada, Granada, Spain.

Radermacher, M., Rao, V., Wagenknecht, T., Grassucci, R., Frank, J., Timerman, A.P., and Frank, J. (1994a). Three-dimensional reconstruction of the calcium release channel by cryo electron microscopy. In: *Proceedings of the 13th International Congress on Microscopy (Paris)*, Vol. 3A, pp. 551–552. Les Editions de Physiques, Les Ulis, France.

Radermacher, M., Rao, V., Grassucci, R., Frank, J., Timerman, A.P., Fleischer, S., and Wagenknecht, T. (1994b). Cryo-electron microscopy and three-dimensional reconstruction of the calcium release channel/ryanodine receptor from skeletal muscle. *J. Cell. Biol.* **127**, 411–423.

Radermacher, M., Ruiz, T., Wieczorek, H., and Grüber, G. (2001). The structure of the V₁-ATPase determined by three-dimensional electron microscopy of single particles. *J. Struct. Biol.* **135**, 26–37.

Radon, J. (1917). Über die Bestimmung von Funktionen durch ihre Integralwerte längs gewisser Mannigfaltigkeiten. Berichte über die Verhandlungen der Königlich Sächsischen Gesellschaft der Wissenschaften zu Leipzig. *Math. Phys. Klasse* **69**, 262–277.

Ranson, N.A., Farr, G.W., Roseman, A.M., Gowan, B., Fenton, W.A., Horvich, A.L., and Saibil, H.R. (2001). ATP-bound states of GroEL captured by cryo-electron microscopy. *Cell* **107**, 869–879.

Rapp, G. (1998). Flash lamp-based irradiation of caged compounds. *Methods Enzymol.* **291**, 203–222.

Rath, B.K., and Frank, J. (2004). Fast automatic particle picking from cryo-electron micrographs using a locally normalized cross-correlation function: a case study. *J. Struct. Biol.* **145**, 84–90.

Rath, B.K, Hegerl, A., Leith, A., Shaikh, T.R., Wagenknecht, T., and Frank, J. (2003). Fast 3D motif search of EM density maps using a locally normalized cross-correlation function. *J. Struct. Biol.* **144**, 95–103.

Rawat, U.B.S., Zavialov, A.V., Sengupta, J., Valle, M., Grassucci, R.A., Linde, J., Vestergaard, B., Ehrenberg, M., and Frank, J. (2003). A cryo-electron microscopic study of ribosome-bound termination factor RF2. *Nature* **421**, 87–90.

Rayment, I., Holden, H., Whittaker, M., Yohn, C., Lorenz, M., Holmes, K., and Milligan, R. (1993). Structure of the actin–myosin complex and its implications for muscle contraction. *Science* **261**, 58–65.

Reimer, L. (1997). *Transmission Electron Microscopy*. Springer-Verlag, Berlin.

Reyrat, J.-M., Lanzavecchia, S., Lupetti, P., de Bernard, M., Pagliaccia, C., Pelicic, V., Charrel, M., Uliveri, C., Norais, N., Ji, X., Cabiaux, V., Papini, E., Rappuoli, R., and Telford, J.L. (1999). 3D imaging of the 58 kDa cell binding subunit of the *Heliobacter pylori* cytotoxin. *J. Mol. Biol.* **290**, 459–470.

Robb, R.A. (ed.) (1994). Visualization in biomedical computing. In: *Proceedings of the International Society for Optical Engineering (SPIE)*, Vol. 2359.

Rockel, B., Jakana, J., Chiu, W., and Baumeister, W. (2002a). Electron cryo-microscopy of VAT, the archaeal p97/CDC48 homologue from *Thermoplasma acidophilum*. *J. Mol. Biol.* **317**, 673–681.

Rockel, B., Peters, J., Kuhlmorgen, B., Glaeser, R.M., and Baumeister, W. (2002b). A giant protease with a twist: the TPP II complex from *Drosophila* studied by electron microscopy. *EMBO J.* **21**, 5979–5984.

Rohrwild, M., Pfeifer G., Santarius, U., Müller, S.A., Huang, H.-C., Engel, A., Baumeister, W., and Goldberg, A.L. (1997). The ATP-dependent HsIVU protease from *Escherichia coli* is a four-ring structure resembling the proteasome. *Nature Struct. Biol.* **4**, 133–139.

Rose, H. (1984). Information transfer in transmission electron microscopy. *Ultramicroscopy* **15**, 173–192.

Rose, H. (1990). Outline of a spherically corrected semiplanatic medium-voltage transmission electron microscope. *Optik* **85**, 19–24.

Roseman, A.M. (1996). The chaperonin ATPase cycle: mechanism of allosteric switching and movements of substrate-binding domains in GroEL. *Cell* **87**, 241–251.

Roseman, A.M. (2000). Docking structures of domains into maps from cryo-electron microscopy using local correlation. *Acta Cryst.* **D56**, 1332–1340.

Roseman, A.M. (2003). Particle finding in electron micrographs using a fast local correlation algorithm. *Ultramicroscopy* **94**, 225–236.

Roseman, A.M., and Neumann, K. (2003). Objective evaluation of the relative modulation transfer function of densitometers for digitisation of electron micrographs. *Ultramicroscopy* **96**, 207–218.

Roseman, A.M., Chen, S., White, H., Braig, K., and Saibil, H.R. (1996). The chaperonin ATPase cycle: mechanism of allosteric switching and movements of substrate-binding domains in GroEl. *Cell* **87**, 241–251.

Roseman, A.M., Ranson, N.A., Gowen, B., Fuller, S.D., and Saibil, H.R. (2001). Structures of unliganded and ATP-bound states of the *Escherichia coli* chaperonin GroEL by cryoelectron microscopy. *J. Struct. Biol.* **135**, 115–125.

Rosenfeld, A., and Kak, A.C. (1982). *Digital Picture Processing*, 2nd ed. Academic Press, New York.

Rosenthal, P., Crowther, R.A., and Henderson, R. (2003). Optimal determination of particle orientation, absolute hand, and contrast loss in single particle cryomicroscopy. *J. Mol. Biol.* **333**, 721–745.

Rossmann, M.G. (2000). Fitting atomic models into electron-microscopy maps. *Acta Cryst.* **D56**, 1341–1349.

Rossmann, M.G., Morais, M.C., Leiman, P.G., and Zhang, W. (2005). Combining X-ray crystallography and electron microscopy. *Structure* **13**, 355–362.

Rosso, J., and Lukins, S. (1989). *The Meat Market—The New Basic Cook Book*. Workman Publishing, New York, 1989.

Rouiller, I., Butel, V.M., Latterich, M., Milligan, R., and Wilson-Kubalek, E.ZM. (2000). A major conformational change in p97 AAA ATPase upon ATP binding. *Mol. Cell* **6**, 1485–1490.

Rouiller, I., Pulokas, P., Butel, V.M., Milligan, R.A., Wilson-Kubalek, E.M., Potter, C.S., and Carragher, B.O. (2001). Automated data acquisition for single-particle reconstruction using p97 as the biological sample. *J. Struct. Biol.* **133**, 102–107.

Rubinstein, J.L., Walker, J.E., and Henderson, R. (2003). Structure of the mitochondrial ATP synthase by electron cryomicroscopy. *EMBO J.* **22**, 6182–6192.

Ruiz, T., Mechin, I., Bär, J., Rypniewski, W., Kopperschläger, G., and Radermacher, M. (2003). The 10.8-Å structure of *Saccharomyces cerevisiae* phosphofructokinase determined by cryoelectron microscopy: localization of the putative fructose 6-phosphate binding sites. *J. Struct. Biol.* **143**, 124–134.

Ruprecht, J., and Nield, J. (2001). Determining the structure of biological macromolecules by transmission electron microscopy, single particle analysis and 3D reconstruction. *Prog. Biophys. Mol. Biol.* **75**, 121–164.

Russ, J.C. (1995). *The Image Processing Handbook*, 2nd ed. CRC Press, Boca Raton, FL.

Rye, H.S., Roseman, A.M., Chen, S., Furtak, K., Fenton, W.A., Saibil, H.R., and Horwich, A.L. (1999). GroEL–GroES cycling: ATP and nonnative polypeptide direct alternation of folding-active rings. *Cell* **97**, 325–338.

Saad, A., Ludtke, S.J., Jakana, J., Rixon, F.J., Tsruta, H., and Chiu, W. (2001). Fourier amplitude decay of electron cryomicroscopic images of single particles and effects on structure determination. *J. Struct. Biol.* **133**, 32–42.

Sachs, L. (1984). *Applied Statistics. A Handbook of Techniques*, 2nd ed. Springer-Verlag, Berlin/New York.

Saibil, H.R. (2000a). Conformational changes studied by cryo-electron microscopy. *Nature Struct. Biol.* **7**, 711–714.

Saibil, H.R. (2000b). Macromolecular structure determination by cryo-electron microscopy. *Acta Cryst.* **D56**, 1215–1222.

Saibil, H.R. (2000c). Molecular chaperones: containers and surfaces for folding, stabilizing or unfolding proteins. *Curr. Opin. Struct. Biol.* **10**, 251–25.

Saibil, H.R., and Brunger, A.T. (2002). Macromolecular assemblages: machines and networks. *Curr. Opin. Struct. Biol.* **12**, 215–216.

Saibil, H.R., and Orlova, E.V. (2002). Challenges at the frontiers of structural biology. *Nature Struct. Biol.* **9**, 414–415.

Saibil, H., and Wood, S. (1993). Chaperonins. *Curr. Opin. Struct. Biol.* **3**, 207–213.

Saibil, H.R., Zheng, D., Roseman, A.M., Hunter, A.S., Watson, G.M.F., Chen, S., auf der Mauer, A., OHara, B.P., Wood, S.P., Mann, N.H., Barnett, L.K., and Ellis, R.J. (1993). ATP induces large quaternary rearrangements in a cage-like chaperonin structure. *Curr. Biol.* **3**, 265–273.

Saito, A., Inui, M., Radermacher, M., Frank, J., and Fleischer, S. (1988). Ultrastructure of the calcium release channel of sarcoplasmic reticulum. *J. Cell Biol.* **107**, 211–219.

Sali, A., Glaeser, R., Earnest, T., and Baumeister, W. (2003). From words to literature in structural proteomics. *Nature* **422**, 216–225.

Salunke, D.M., Caspar, D.L.D., and Garcia, R.L. (1986). Self-assembly of purified polyomavirus capsid protein VP. *Cell* **46**, 895–904.

Salzman, D.B. (1990). A method of general moments for orienting 2D projections of unknown 3D objects. *Comput. Vis. Graph. Image Process.* **50**, 129–156.

Samsó, M., and Wagenknecht, T. (1998). Contributions of electron microscopy and single-particle techniques to the determination of the ryanodine receptor three-dimensional structure. *J. Struct. Biol.* **121**, 172–180.

Samsó, M., and Wagenknecht, W. (2002). Apocalmodulin and Ca^{2+}–calmodulin bind to neighboring locations on the ryanodine receptor. *J. Biol. Chem.* **277**, 1349–1353.

Samsó, M., Radermacher, M., Frank, J., and Koonce, M.P. (1998). Structural characterization of a dye in motor domain. *J. Mol. Biol.* **276**, 927–937.

Samsó, M., Trufillo, R., Gurrola, G.B., Valdivia, H.H., and Wagenknecht, T. (1999). Three-dimensional location of the imperatoxin A binding site on the ryanodine receptor. *J. Cell Biol.* **146**, 493–500.

Samsó, M., Palumbo, M.J., Radermacher, M., Liu, J.S., and Lawrence, C.E. (2002). A Bayesian method for classification of images from electron micrographs. *J. Struct. Biol.* **138**, 157–170.

Sander, B., Golas, M.M., and Stark, H. (2003a). Automatic CTF correction for single particles based upon multivariate statistical analysis of individual power spectra. *J. Struct. Biol.* **142**, 392–401.

Sander, B., Golas, M.M., and Stark, H. (2003b). Corim-based alignment for improved speed in single-particle image processing. *J. Struct. Biol.* **143**, 219–228.

San Martin, M.C., Stamford, N.P.J., Dammerova, N., Dixon, N., and Carazo, J.M. (1995). A structural model of the *Escherichia coli* DnaB helicase based on three-dimensional electron microscopy data. *J. Struct. Biol.* **114**, 167–176.

San Martin, M.C., Burnett, R.M., de Haas, F., Heinkel, R., Rutten, T., Fuller, S.D., Butcher, S.J., and Bamford, D.H. (2001). Combined EM/X-ray imaging yields a quasi-atomic model of the adenovirus-related bacteriophage PRD1 and shows key capsids and membrane interactions. *Structure* **9**, 917–930.

Sass, H.J., Bueldt, G., Beckmann, E., Zemlin, F., van Heel, M., Zeitler. E., Rosenbusch, J. P., Dorset, D.L., and Massalski, A. (1989). Densely packed δ-structure at the protein–liquid interface of porin is revealed by high-resolution cryo-electron microscopy. *J. Mol. Biol.* **209**, 171–175.

Sato, C., Hamada K., Ogura, T., Miyazawa, A., Iwasaki, K., Hiroaki, Y., Tani, K., Terauchi, A., Fujiyoshi, Y., and Mikoshiba, K. (2004). Inositol 1,4,5-trisphosphate receptor contains multiple cavities and L-shaped ligand-binding domains. *J. Mol. Biol.* **336**, 155–164.

Saxton, W.O. (1977). Spatial coherence in axial high resolution conventional electron microscopy. *Optik* **49**, 51–62.

Saxton, W.O. (1978). *Computer Techniques for Image Processing in Electron Microscopy*, Supplement 10 to *Advances in Electronic and Electron Physics*. Academic Press, New York.

Saxton, W.O. (1986). Focal series restoration in HREM. In: *Proceedings of the XI International Congress on Electron Microscopy, Kyoto,* post deadline paper 1. The Japanese Society of Electron Microscopy, Tokyo.

Saxton, W.O. (1994). Accurate alignment of sets of images for super-resolving applications. *J. Microsc.* **174**, 61–68.

Saxton, W.O., and Baumeister, W. (1982). The correlation averaging of a regularly arranged bacterial cell envelope protein. *J. Microsc.* **127**, 127–138.

Saxton, W.O., and Frank, J. (1977). Motif detection in quantum noise-limited electron micrographs by cross-correlation. *Ultramicroscopy* **2**, 219–227.

Saxton, W.O., Pitt, T.J., and Horner, M. (1979). Digital image processing: the SEMPER system. *Ultramicroscopy* **4**, 343–354.

Saxton, W.O., Baumeister, W., and Hahn, M. (1984). Three-dimensional reconstruction of imperfect two-dimensional crystals. *Ultramicroscopy* **13**, 57–70.

Schatz, M. (1992). *Invariante Klassifizierung elektronenmikroskopischer Aufnahmen von eiseingebetteten biologischen Makromolekulen.* Thesis, Freie Universität Berlin. Deutsche Hochschulschriften 452, Verlag Hänsel-Hohenhausen, Egelsbach/Köln/ New York.

Schatz, M., and van Heel, M. (1990). Invariant classification of molecular views in electron micrographs. *Ultramicroscopy* **32**, 255–264.

Schatz, M., and van Heel, M. (1992). Invariant recognition of molecular projections in vitreous ice preparations. *Ultramicroscopy* **45**, 15–22.

Schatz, M., Jäger, J., and van Heel, M. (1990). Molecular views of ice-embedded *Lumbricus terrestris* erythrocruorin obtained by invariant classification. In: *Proceedings of the XII International Congress on Electron Microscopy*, pp. 450–451. San Francisco Press, San Francisco.

Schatz, M., Orlova, E., Jäger, J., Kitzelmann, E., and van Heel, M. (1994). 3D structure of ice-embedded *Lumbricus terrestris* erythrocruorin. In: *Proceedings of the 13th International Congress on Electron Microscopy (Paris)*, Vol. 3, pp. 553–554. Les Editions de Physiques, Les Ulis, France.

Schatz, M., Orlova, E.V., Dube, P., Jäger, J., and van Heel, M. (1995). Structure of *Lumbricus terrestris* hemocyanin at 30 Å resolution determined using angular reconstitution. *J. Struct. Biol.* **114**, 28–40.

Schiske, P. (1968). Zur Frage der Bildrekonstruktion durch Fokusreihen. In: *Fourth European Conference on Electron Microscopy, Rome*, pp. 145–146. Tipografia Poliglotta Vaticana, Rome.

Schliwa, M. (2002). The evolving complexity of cytoplasmic structure. *Nature Rev. Mol. Cell Biol.* **3**, 291–296.

Schmid, M.F., Agris, J.M., Jakava, J., Matsudaira, P., and Chiu, W. (1994). Three-dimensional structure of a single filament in the *Limulus* acrosomal bundle: scruin binds to homologous helix–loop–beta motifs in actin. *J. Cell. Biol.* **124**, 341–350.

Schmid, M.F., Sherman, M.B., Matsudaira, P., Tsuruta, H., and Chiu, W. (1999). Scaling structure factor amplitudes in electron cryomicroscopy using X-ray solution scattering. *J. Struct. Biol.* **128**, 51–57.

Schmid, M.F., Sherman, M., Matsudaira, P., and Chiu, W. (2004). Structure of the acrosomal bundle. *Nature* **431**, 104–107.

Schmutz, M., Lang, J., Graff, S., and Brisson, A. (1994). Defects on planarity of carbon films supported on electron microscope grids revealed by reflected light microscopy. *J. Struct. Biol.* **112**, 252–258.

Schoehn, G., Hayes, M., Cliff, M., Clarke, A.R., and Saibil, H.R. (2000a). Domain rotations between open, closed and bullet-shaped forms of the thermosome, and archaeal chaperonin. *J. Mol. Biol.* **301**, 323–332.

Schoehn, G., Quaite-Randall, E., Jimenez, J.L., Joachimiak, A., and Saibil, S. (2000b). Three conformations of an archaeal chaperonin, TF55 from *Sulfolobus shibatae*. *J. Mol. Biol.* **296**, 813–819.

Schröder, R.R. (1992). Zero-loss energy-filtered imaging of frozen-hydrated proteins: model calculations and implications for future developments. *J. Microsc.* **166**, 389–400.

Schröder, R.R., Hofmann, E., and Ménétret, J.-F. (1990). Zero-loss energy filtering as improved imaging mode in cryo-electron microscopy of frozen-hydrated specimens. *J. Struct. Biol.* **105**, 28–34.

Schröder, R.R., Manstein, D.J., Jahn, W., Holden, H., Rayment, I., Holmes, K.C., and Spudich, J.A. (1993). Three-dimensional atomic model of F-actin decorated with *Dictyostelium* myosin S1. *Nature* **364**, 171–174.

Schröder, R.R., Angert, I., Frank, J., and Holmes, K.C. (2002). Multivariate statistical analysis and tomographic processing of helical objects. *Microsc. Microanal.* **8** (Suppl. 2), 856CD–857CD.

Schroeter, J.P., Wagenknecht, T., Kolodzij, S.J., Bretaudiere, J.-P., Strickland, D.K., and Stoops, J.K. (1991). Three-dimensional structure of chymotrypsin–human α_2-macroglobulin complex. *FASEB* **5**, A452.

Schroeter, J.P., Kolodziej, S.J., Wagenknecht, T., Bretaudiere, J.-P., Tapon-Bretaudiere, J., Strickland, D.K., and Stoops, J.K. (1992). Three-dimensional structure of the human α_2-macroglobulin–methylamine and chymotrypsin complexes. *J. Struct. Biol.* **109**, 235–247.

Schultz, P., Fribourg, S., Poterszman, A., Mallaouh, V., Moras, D., and Egly, J.M. (2000). Molecular architecture of human TFIIH. *Cell* **102**, 599–607.

Sengupta, J., Agrawal, R.K., and Frank, J. (2001). Visualization of protein S1 within the 30S ribosomal subunit and its interaction with messenger RNA. *Proc. Natl. Acad. Sci. USA* **98**, 11991–11996.

Sengupta, J., Nilsson, J., Gursky, R., Spahn, C.M.T., Nissen, P., and Frank, J. (2004). Identification of the versatile scaffold protein RACK1 on the eukaryotic ribosome by cryo-electron microscopy. *Nat. Struct. Mol. Biol.* **11**, 957–962.

Sengupta, J., and Frank, J. (in press). Cryo-electron microscopy as a tool to study molecular machines. In: *Protein Structures: Methods in Protein Structure and Stability Analysis* (V.N. Uversky and E.A. Permyakov, eds.). Nova Science, New York.

Serysheva, I.I., and Hamilton, S.L. (1998). Ryanodine binding sites on the sarcoplasmic reticulum Ca^{2+} release channel. In: *Ryanodine Receptor* (R. Sitsapesan and A. Williams, eds.), pp. 95–110. J.B. Lippincott, Philadelphia, PA.

Serysheva, I.I., Orlova, E.V., Chiu, W., Sherman, M.B., Hamilton, S.L., and van Heel, M. (1995). Electron cryomicroscopy and angular reconstitution used to visualize the skeletal muscle calcium release channel. *Nat. Struct. Biol.* **2**, 18–24.

Serysheva, I.I., Schatz, M., van Heel, M., Chiu, W., and Hamilton, S.L. (1999). Structure of the skeletal muscle calcium release channel activated wth Ca^{2+} and AMP–PCP. *Biophys. J.* **77**, 1936–1944.

Serysheva, I.I., Ludtke, S.J., Baker, M.R., Chiu, W., and Hamilton, S.L. (2002). Structure of the voltage-gated L-type Ca^{2+} channel electron cryomicroscopy. *Proc. Natl. Acad. Sci. USA* **99**, 10370–10375.

Serysheva, I.I., Bare, D.J., Ludtke, S.L., Kettlun, C.S., Wah, C., and Mignery, G.A. (2003). Structure of the type 1 inositol 1,4,5-triphosphate receptor revealed by electron crymicroscopy. *J. Biol. Chem.* **278**, 21319–21322.

Sethian, J.A. (1996). *Level Set Methods*. Cambridge Monographs on Applied and Computational Methods. Cambridge University Press, Cambridge.

Sewell, B.T., Berman, M.N., Meyers, P.R., Jandhyala, D., and Benedik, M.J. (2003). The cyanide degrading nitrilase from *Pseudomonas stutzeri* AK61 is a twofold symmetric, 14-subunit spiral. *Structure* **11**, 1413–1422.

Sezan, M.I. (1992). An overview of convex projections theory and its application to image recovery problems. *Ultramicroscopy* **40**, 55–67.

Sezan, M.I., and Stark, H. (1982). Image restoration by the method of convex projections. 11. Applications and numerical results. *IEEE Trans. Med. Imag.* **1**, 95–101.

Shah. A.K., and Stewart, P.L. (1998). QVIEW: software for rapid selection of particles from digital electron micrographs. *J. Struct. Biol.* **123**, 17–21.

Shaikh, T.R., Hegerl R., and Frank, J. (2003). An approach to examining model dependence in EM reconstructions using cross-validation. *J. Struct. Biol.* **142**, 301–310.

Shaikh, T.R., Thomas, D.R., Chen, J.Z., Samatey, F.A., Matsunami, H., Imada, K., Namba, K., DeRosier, D.J. (2005). A partial atomic structure for the flagellar hook of *Salmonella typhimurium*. *Proc. Natl. Acad. Sci. USA* **102**, 1023–1028.

Shaikh, T.R., Barnard, D., and Wagenknecht, T. (2005). Implementation of a flash-photolysis unit for cryo-electron microscopy (in preparation).

Shannon, C.E. (1949). Communication in the presence of noise. *Proc. IRE* **37**, 10–21.

Sharma, R., Jeyakumar, L.H., Fleischer, S., and Wagenknecht, W. (2000). Three-dimensional structure of ryanodine receptor isoform three in two conformational states as visualized by cryo-electron microscopy. *J. Biol. Chem.* **275**, 9485–9491.

Sharma, M.R., Penczek, P., Grassucci, R., Xin, H.-B., Fleischer, S., and Wagenknecht T. (1998). Cryoelectron microscopy and image analysis of the cardiac ryanodine receptor. *J. Biol. Chem.* **273**, 18429–18434.

Shaw, A.L., Rothnagel, R., Chen, D., Ramig, R.F., Chiu, W., and Prasad, B.V.V. (1993). Three-dimensional visualization of the rotavirus hemagglutinin structure. *Cell* **74**, 693–701.

Sherman, M.B., Soejima, T., Chiu, W., and van Heel, M. (1998). Multivariate statistical analysis of single unit cells in electron crystallography. *Ultramicroscopy* **74**, 179–199.

Shi, J., and Malik, J. (1997). Normalized cuts and image segmentation. In: *Proceedings of the IEEE Conference on Computer Vision and Pattern Recognition*, San Juan, Puerto Rico, pp. 731–737.

Sigworth, F.J. (1998). A maximum-likelihood approach to single-particle image refinement. *J. Struct. Biol.* **122**, 328–339.

Sizaret, P.-Y., Frank, J., Lamy, J., Weill, J., and Lamy, J.N. (1982). A refined quaternary structure of *Androctonus australis* hemocyanin. *Eur. J. Biochem.* **127**, 501–506.

Smith, M.F., and Langmore, J.P. (1992). Quantitation of molecular densities by cryoelectron microscopy. Determination of radial density distribution of tobacco mosaic virus. *J. Mol. Biol.* **226**, 763–774.

Smith, P.R. (1978). An integrated set of computer programs for processing electron micrographs of biological structures. *Ultramicroscopy* **3**, 153–160.

Smith, P.R. (1981). Bilinear interpolation of digital images. *Ultramicroscopy* **6**, 201–204.

Smith, P.R., and Aebi, U. (1973). Filtering continuous and discrete Fourier transforms (appendix). *J. Supramol. Struct.* **1**, 516–522.

Smith, T.J., Olson, N.H., Cheng, R.H., Chase, E.S., and Baker, T.S. (1993). Structure of a human rhinovirus-bivalently bound antibody complex: implications for viral neutralization and antibody flexibility. *Proc. Natl. Acad. Sci. USA* **90**, 7015–7018.

Smith, T.J., Chase, E.S., Schmidt, T.J., Olson, N.H., and Baker, T.S. (1996). Neutralizing antibody to human rhinovirus 14 penetrates the receptor-binding canyon. *Nature* **383**, 350–354.

Soejima, T., Sherman, M.S., Schmid, M.F., and Chiu, W. (1993). 4-Å projection map of bacteriophage T4 DNA helix-stabilizing protein (gp32*1). crystal by 400-kV electron cryomicroscopy. *J. Struct. Biol.* **111**, 9–16.

Sommerfeld, A. (1964). *Vorlesungen über Theoretische Physik: Mechanik*, 7th ed. Academische Verlagsgesellschaft, Geest&Partig K.G., Leipzig.

Sorensen, D.C. (1992). Implicit application of polynomial filters in a k-step Arnoldi methods. *SIAM J. Matrix Anal. Appl.* **13**, 357–385.

Sorzano, C.O.S., Marabini, R., Boisset, N., Rietzel, E., Schröder, R., Herman, G.T., and Carazo, J.M. (2001). The effect of overabundant projection directions on 3D reconstruction algorithms. *J. Struct Biol.* **133**, 108–118.

Sorzano, C.O.S., Fernández, J.J., Marabini, R., Herman, G.T., Censor, Y., and Carazo, J.M. (2004). Transfer function restoration in 3D electron microscopy via iterative data refinement. *Phys. Medic. Biol.*, **49**, 509–522.

Sosinsky, G.E., Baker, T.S., Caspar, D.L.D., and Goodenough, D.A. (1990). Correlation analysis of gap junction lattice images. *Biophys. J.* **58**, 1213–1226.

Sosinsky, G.E., Francis, N.R., Stallmeyer, M.J.B., and DeRosier, D.I. (1992). Substructure of the flagellar basal body of *Salmonella typhimurium*. *J. Mol. Biol.* **223**, 171–184.

Sosinsky, G.E., Baker, T.S., Hand, G., and Ellisman, M.H. (1999). The electron microscopy outreach program: a web-based resource for research and education. *J. Struct. Biol.* **125**, 246–252.

Spahn, C.M.T, Grassucci, R.A., Penczek, P., and Frank, J. (1999). Direct three-dimensional localization and positive identification of RNA helices within the ribosome by means of genetic tagging and cryo-electron microscopy. *Structure* **7**, 1567–1573.

Spahn, C.M.T., Penczek, P., Leith, A., and Frank, J. (2000). A method for differentiating proteins from nucleic acids in intermediate-resolution density maps: cryo-electron microscopy defines the quaternary structure of the *Escherichia coli* 70S ribosome. *Structure* **8**, 937–948.

Spahn, C.M.T., Beckmann, R., Eswar, N., Penczek, P.A., Sali, A., Blobel, G., and Frank, J. (2001a). Structure of the 80S ribosome from *Saccharomyces cerevisae*—tRNA—ribosome and subunit–subunit interactions. *Cell* **10**, 373–388.

Spahn, C.M.T., Blaha, G., Agrawal, R.K., Penczek. P., Grassucci R.A., Trieber, C.A., Connell, S.R., Taylor, D.E., Nierhaus, K.H., and Frank. J. (2001b). Localization of the tetracycline resistance protein Tet(O). on the ribosome and the inhibition mechanism of tetracycline. *Mol. Cell* **7**, 1037–1045.

Spahn, C.M.T., Kieft, J.S., Grassucci, R.A., Penczek, P., Zhou, K., Doudna, J.A., and Frank, J. (2001c). Hepatitis C virus IRES RNA-induced changes in the conformation of the 40S ribosomal subunit. *Science* **291**, 1959–1962.

Spahn, C.M.T., Frank, J., and Penczek, P.A. (2004a). Analysis of conformational heterogeneity of macromolecules in cryo-electron microscopy. (Abstract.) Twenty-first Annual Houston Conference on Biomedical Engineering Research, Feb. 12–13.

Spahn, C.M.T., Gomez-Lorenzo, M.G., Grassucci, R.A., Jørgensen, R., Andersen, G.R., Beckmann, R., Penczek, P.A., Ballesta, J.P.G., and Frank, J. (2004b). Domain movements of elongation factor eEF2 and the eukaryotic 80S ribosome facilitate tRNA translocation. *EMBO J.* **23**, 1008–1019.

Spahn, C.M.T., Grassucci, R.A., Marquez, V., Linde, J., Gao, H., Baxter, W., Penczek, P.A., Nierhaus, K.H., and Frank, J. (2005). Cryo-EM reconstruction of a functional complex of *E. coli* 70S ribosomes at 7.8 Å resolution. (In preparation).

Spahn, C.M.T., Jan, E., Mulder, A., Grassucci, R.A., Sarnow, P., and Frank, J. (2004c). Cryo-EM visualization of a viral internal ribosome entry site (IRES). bound to human 40S and 80S ribosomes: the IRES functions as an RNA-based translation factor. *Cell* **116**, 465–475.

Spence, J. (1988). *Experimental High-Resolution Electron Microscopy*. Oxford University Press, New York.

Srivastava, S., Versehoor, A., and Frank, J. (1992a). Eukaryotic initiation factor 3 does not prevent association through physical blocking of the ribosomal subunit–subunit interface. *J. Mol. Biol.* **226**, 301–304.

Srivastava, S., Radermacher, M., and Frank, J. (1992b). Three-dimensional mapping of protein L18 on the SOS ribosomal subunit from *E. coli*. In: *Proceedings of the 10th European Congress on Electron Microscopy*, Vol. 1, pp. 421–422. Secretariado de Publicaciones de la Universidad Granada, Granada, Spain.

Srivastava, S., Verschoor A., Radermacher, M., Grassucci, R., and Frank, J. (1995). Three-dimensional reconstruction of the mammalian 40S ribosomal subunit embedded in ice. *J. Mol. Biol.* **245**, 461–466.

Stallmeyer, M.J.B., Hahnenberger, K.M., Sosinsky, G.E., Shapiro, L., and DeRosier, D.J. (1989a). Image reconstruction of the flagellar basal body of *Caulobacter crescentus*. *J. Mol. Biol.* **205**, 511–518.

Stallmeyer, M.J.B., Aizawa, S.-I., Macnab, R.M., and DeRosier, D.J. (1989b). Image reconstruction of the flagellar basal body of *Salmonella typhimurium*. *J. Mol. Biol.* **205**, 519–529.

Stark, H. (2002). Three-dimensional electron cryomicroscopy of ribosomes. *Curr. Protein Pept. Sci.* **3**, 79–91.

Stark, H., Millier, F., Orlova, E.V., Schatz, M., Dube, P., Erdemir, T., Zemlin, F., Brimacombe, R., and van Heel, M. (1995). The 70S *Escherichia coli* ribosome at 23 Å resolution: Fitting the ribosomal RNA. *Structure* **3**, 815–821.

Stark, H., Orlova, E.V., Rinke-Apple, J., Müller, F., Rodnina, M., Wintermeyer W., Brimacombe, R., and van Heel, M. (1997a). Arrangement of tRNAs in pre- and posttranslocational ribosomes revealed by electron cryomicroscopy. *Cell* **88**, 19–28.

Stark, H., Rodnina, M.V., Rinke-Appel, J., Brimacombe, R., Wintermeyer, W., and van Heel, M. (1997b). Visualization of elongation factor Tu on the *Escherichia coli* ribosome. *Nature* **389**, 403–406.

Stark, H., Rodnina, M.V., Wieden, H.J., van Heel, M., and Wintermeyer, W. (2000). Large-scale movement of elongation factor G and extensive conformational change of the ribosome during translocation. *Cell* **100**, 301–309.

Stark, H., Dube P., Lührmann, R., and Kastner, B. (2001). Arrangement of RNA and proteins in the spliceosomal U1 small nuclear ribonucleoprotein particle. *Nature* **409**, 539–542.

Stasiak, A., Tsevana, I.R., West, S.C., Benson, C.J.B., Yu, X., and Egelman, E.H. (1994). The *E. coli* RuvB branch migration protein forms double hexameric rings around DNA. *Proc. Natl. Acad. Sci. USA* **91**, 7618–7622.

Stasiak, A.Z., Larquet, E., Stasiak, A., Mueller, S., Engel, A., Van Dyck, E., West, S.C., and Egelman, E.H. (2000). The human Rad52 protein exists as a heptameric ring. *Curr. Biol.* **10**, 337–340.

Steinkilberg, M., and Schramm, H.J. (1980). Eine verbesserte Drehkorrelationsmethode für die Strukturbestimmung biologischer Makromolekule durch Mittelung elektronen-mikroskopischer Bilder. *Hoppe-Seyler's Z. Physiol. Chem.* **361**, 1363–1369.

Steven, A. (1997). Overview of macromolecular electron microscopy: an essential tool in protein structure analysis. In *Current Protocols in Protein Science* (J. Coligan and P. Wingfield, eds.), pp. 17.2.1–17.2.29. John Wiley, New York.

Steven, A.C., Hainfield, J.F., Trus, B.L., Steinert, P.M., and Wall, J.S. (1984). Radial distributions of density within macromolecular complexes determined from dark-field electron micrographs. *Proc. Natl. Acad. Sci. USA* **81**, 6363–6367.

Steven, A.C., Stall, R., Steinert, P.M., and Trus, B.L. (1986). Computational straightening of images of curved macromolecular helices by cubic spline interpolation facilitates structural analysis by Fourier methods. In: *Electron Microscopy and Alzheimer's Disease* (J. Metuzals, ed.), pp. 31–33. San Francisco Press, San Francisco.

Steven, A.C., Trus, B.L., Maizel, J.V., Unser, M., Parry, D.A.D., Wall, J.S., Hainfeld, J.F., and Studier, F.W. (1988). Molecular substructure of a viral receptor–recognition protein: the gpl7 tail-fiber of bacteriophage T7. *J. Mol. Biol.* **200**, 351–365.

Steven, A.C., Kocsis, E., Unser, M., and Trus, B.L. (1991). Spatial disorders and computational cures. *Intl. J. Biol. Macromol.* **13**, 174–180.

Stewart, M. (1988a). Introduction to the computer image processing of electron micrographs of two-dimensionally ordered biological structures. *J. Electron Microsc. Tech.* **9**, 301–324.

Stewart, M. (1988b). Computer image processing of electron micrographs of biological structures with helical symmetry. *J. Electron Microsc. Tech.* **9**, 325–358.

Stewart, M. (1990). Electron microscopy of biological molecules. In: *Modern Microscopies* (P.J. Duke and A.G. Michette, eds.), pp. 9–40. Plenum Press, New York.

Stewart, P.L., and Burnett, R.M. (1993). Adenovirus structure as revealed by X-ray crystallography, electron microscopy, and difference imaging. *Jpn. J. Appl. Phys.* **32**, 1342–1347.

Stewart, P.L. Burnett, R.M., Cyrklaff, M., and Fuller, S.D. (1991). Image reconstruction reveals the complex molecular organization of adenovirus. *Cell* **67**, 145–154.

Stewart, P.L., Fuller, S.D., and Burnett, R.M. (1993). Difference imaging of adenovirus: bridging the resoluting gap between X-ray crystallography and electron microscopy. *EMBO J.* **12**, 2589–2599.

Stewart, P.L., Chiu, C.Y., Haley, D.A., Kon, L.B., and Schlessman, J.L. (1999). Review: resolution issues in single-particle reconstruction. *J. Struct. Biol.* **128**, 58–64.

Stewart, P.L., Cary, R.B., Peterson, S.R., and Chiu, C.Y. (2000). Digitally collected cryo-electron micrographs for single particle reconstruction. *Microsc. Res. Tech.* **49**, 224–232.

Stewart, P.L., Makabi, M., Mikyas, M., Kickhoefer, V.A., and Rome, L.H. (2002). Cryo-EM imaging of vaults and vault-like particles. *Proc. IEEE Intl. Symp. Biomed. Imag.*, pp. 273–276.

Stöffler, G., and Stöffler-Meilicke, M. (1983). The ultrastructure of macromolecular complexes studied with antibodies. In: *Modern Methods in Protein Chemistry* (H. Tesche, ed.), pp. 409–455. De Gruyter, Berlin.

Stöffler-Meilicke, M., and Stöffler, G. (1988). Localization of ribosomal proteins on the surface of ribosomal subunits from *Escherichia coli* using immunoelectron microscopy. *Methods Enzymol.* **164**, 503–520.

Stoops, J.K., Schroeter, J.P., Bretaudiere, J.-P., Olson, N.H., Baker, T.S., and Strickland, D.K. (1991a). Structural studies of human α_2-macroglobulin: concordance between projected views obtained by negative-stain and cryoelectron microscopy. *J. Struct. Biol.* **106**, 172–178.

Stoops, J.K., Momany, C., Ernst, S.R., Oliver, R.M., Schroeter, J.P., Bretaudiere, J.-P., and Hackert, M.L. (1991b). Comparisons of the low-resolution structures of ornithine carboxylase by electron microscopy and X-ray crystallography: the utility of methylamine tungstate stain and Butvar support film in the study of macromolecules by transmission electron microscopy. *J. Electron Microsc. Tech.* **18**, 157–166.

Stoops, J.K., Kolodziej, S.J., Schroeter, J.P., Bretaudiere, J.P., and Wakil, S.J. (1992a). Structure–function relationships of the yeast fatty acid synthetase; negative-stain, cryoelectron microscopy, and image analysis studies of the end views of the structure. *Proc. Natl. Acad. Sci. USA* **89**, 6585–6589.

Stoops, J.K., Baker, T.S., Schroeter, J.P., Kolodziej, S.J., Niu, X.-D., and Reed, L.J. (1992b). Three-dimensional structure of the truncated core of the *Saccharomyces cerevisiae* pyruvate dehydrogenase complex determined from negative stain and cryoelectron microscopy images. *J. Biol. Chem.* **267**, 24769–24775.

Stoops, J.K., Cheng, R.H., Yazi, M.A., Maeng, C.-Y., Schroeter, J.P., Klueppelberg, U., Kolodziej, S.J., Baker, T.S., and Reed, L.J. (1997). On the unique structural organization of the *Saccaromyces cerevisiae* pyrovate hydrogenase complex. *J. Biol. Chem.* **272**, 5757–5764.

Stoschek, A., and Hegerl, R. (1997a). Automated detection of macromolecules from electron micrographs using advanced filter techniques. *J. Microsc.* **185**, 76–84.

Stoschek, A., and Hegerl, R. (1997b). Denoising of electron tomographic reconstructions using multiscale transformations. *J. Struct. Biol.* **120**, 257–265.

Stroud, R.M., and Agard, D.A. (1979). Structure determination of asymmetric membrane profiles using an iterative Fourier method. *Biophys. J.* **25**, 495–512.

Stuhrmann, H.B., Burkhardt, N., Dietrich, G., Jiinemann, R., Meerwinck, W., Schmitt, M., Wadzack, J., Willumeit, R., Zhao, J., and Nierhaus, K.H. (1995). Proton- and deuteron spin targets in biological structure research. *Nucl. Instrum. Methods* **A356**, 124–132.

Subramaniam, S., and Henderson, R. (1999). Electron crystallography of bacteriorhodopsin with millisecond time resolution. *J. Struct. Biol.* **128**, 19–25.

Sussman, J.L., Lin, D., Jiang, J., Manning, N.O., Prilusky, J., Ritter, O., and Abola, E.E. (1998). Protein Data Bank (PDB): database of three-dimensional structural information of biological macromolecules. *Acta Cryst.* **D54**, 1078–1084.

Taatjes, D.J., Näär, A.M., Andel III.F., Nogales, E., and Tjian, R. (2002). Structure, function, and activator-induced conformations of the CRSP coactivator. *Science* **295**, 1058–1062.

Tama, F., and Brooks III, C.L. (2002). The mechanism and pathway of pH induced swelling in cowpea chlorotic mottle virus. *J. Mol. Biol.* **318**, 733–747.

Tama, F., Wriggers, W., and Brooks III, C.L. (2002). Exploring global distortions of biological macromolecules and assemblies from low-resolution structural information and elastic network theory. *J. Mol. Biol.* **321**, 297–305.

Tama, F., Valle, M., Frank, J., and Brooks, C.L. (2003). Dynamic reorganization of the functionally active ribosome explored by normal mode analysis and cryo-electron microscopy. *Proc. Natl. Acad. Sci. USA* **100**, 9319–9323.

Tama, F., Miyashita, O., and Brooks III, C.L. (2004a). Flexible multi-scale fitting of atomic structures into low-resolution electron density maps with elastic network normal mode analysis. *J. Mol. Biol.* **337**, 985–999.

Tama, F., Miyashita, O., and Brooks III, C.L. (2004b). Normal mode based flexible fitting of high-resolution structure into low-resolution experimental data form cryo-EM. *J. Struct. Biol.* **147**, 315–326.

Taniguchi, Y., Takai, Y., and Shimizu, R. (1992). Spherical-aberration-free observation of TEM images by defocus-modulation image processing. *Ultramicroscopy* **41**, 323–333.

Tao, Y., and Zhang, W. (2000). Recent developments in cryo-electron microscopy reconstruction of single particles. *Curr. Opin. Struct. Biol.* **10**, 616–622.

Taveau, J.C., Boisset, N., Lamy, J., Lambert, O., and Lamy, J.N. (1997). Three-dimensional reconstruction of *Limulus polyphemus* hemocyanin. *J. Mol. Biol.* **226**, 1002–1015.

Taveau, J.C., Boisset, N., Vinogradov, S.N., and Lamy, J.N. (1999). Three-dimensional reconstruction of *Lumbricus terrestris* hemoglobin at 22 Å resolution: intramolecular localization of globin and linker chains. *J. Mol. Biol.* **289**, 1343–1359.

Taylor, C.A., and Lipson, H. (1964). *Optical Transforms.* Cornell University Press, Ithaca.

Taylor, C.W., da Fonseca P.C., and Morris, E.P. (2004). IP(3) receptors: the search for structure. *Trends Biochem. Sci.* **29**, 210–219.

Taylor, K., and Glaeser, R.M. (1974). Electron diffraction of frozen, hydrated protein crystals. *Science* **186**, 1036–1037.

Taylor, K., and Glaeser, R.M. (1976). Electron microscopy of frozen-hydrated biological specimens. *J. Ultrastruct. Res.* **55**, 448–456.

Thomas, D., Flifla, M.J., Escoffier, B., Bariray, M., and Delain, E. (1988). Image processing of electron micrographs of human α_2-macroglobulin half-molecules induced by Ca^{2+}. *Biol. Cell* **64**, 39–44.

Thomas, D.R., Morgan, D.,G., and DeRosier, D.J. (1999). Rotational symmetry of the C ring and a mechanism for the flagellar rotary motor. *Proc. Natl. Acad. Sci. USA* **96**, 10134–10139.

Thomas, D., Morgan, D.G., and DeRosier, D. (2001). Structures of bacterial flagellar motors from two FliF–FliG gene fusion mutants. *J. Bacteriol.* **183**, 6404–6412.

Thon, F. (1966). Zur Defokussierungsabhängigkeit des Phasenkontrastes bei der elektronenmikroskopischen Abbildung. *Z. Naturforsch.* **21a**, 476–478.

Thon, F. (1971). Phase contrast electron microscopy. In: *Electron Microscopy in Material Science* (U. Valdré, ed.). Academic Press, New York.

Thouvenin, E., and Hewat, E. (2000). When two into one won't go: fitting in the presence of steric hindrance and partial occupancy. *Acta Cryst.* **D56**, 1350–1357.

Thurman-Commike, P.A., and Chiu, W. (1996). PTOOL: a software package for the selection of particles from electron cryomicroscopy spot-scan images. *J. Struct. Biol.* **116**, 41–47.

Thurman-Commike, P.A., and Chiu, W. (2000). Reconstruction principles of icosahedral virus structure determination using electron cryomicroscopy. *Micron* **31**, 687–711.

Thurman-Commike, P.A., Tsuruta, H., Greene, B., Preveledge Jr., P.E., King, J., and Chiu, W. (1999). Solution scattering-based estimation of electron cryomicroscopy imaging parameters for reconstruction of virus particles. *Biophys. J.* **76**, 2249–2261.

Tischendorf, G.W., Zeichhardt, H., and Stöffler, G. (1974). Determination of the location of proteins L14, L17, L18, L19, L22, and L23 on the surface of the 50S ribosomal subunit of *Escherichia coli* by immunoelectron microscopy. *Mol. Gen. Genet.* **134**, 187–208.

Tosoni, L., Lanzavecchia, S., and Bellon, F.L. (1996). Image and volume data rotation with 1- and 3-pass algorithms. *Comput. Appl. Biosci.* **12**, 549–552.

Toyoshima, C. (1989). On the use of holey grids in electron crystallography. *Ultramicroscopy* **30**, 439–444.

Toyoshima, C., and Unwin, P.N.T. (1988a). Contrast transfer for frozen-hydrated specimens: determination from pairs of defocused images. *Ultramicroscopy* **25**, 279–292.

Toyoshima, C., and Unwin, P.N.T. (1988b). Ion channel of acetylcholine receptor reconstructed from images of postsynaptic membranes. *Nature* **336**, 247–250.

Toyoshima, C., Yonekura, K., and Sasabe, H. (1993). Contrast transfer for frozen-hydrated specimens. 11. Amplitude contrast at very low frequencies. *Ultramicroscopy* **48**, 165–176.

Trachtenberg, S., and DeRosier, D.J. (1987). Three-dimensional structure of the frozen-hydrated flagellar filament. *J. Mol. Biol.* **195**, 581–601.

Tronrud, D.E., Ten Eyck, L.F., and Matthews, B.W. (1987). An efficient general-purpose least-square refinement program for macromolecular structures. *Acta Crystallogr.* **A43**, 489–501.

Troyon, M. (1977). A method for determining the illumination divergence from electron micrographs. *Optik* **49**, 247–251.

Trus, B.L., and Steven, A.C. (1981). Digital image processing of electron micrographs— The PIC system. *Ultramicroscopy* **6**, 383–386.

Trus, B.L., Unser, M., Pun, T., and Steven, A.C. (1992). Digital image processing of electron micrographs: The PIC system II. In: *Scanning Microscopy Supplement 6: Proceedings of the Tenth Pfefferkorn Conference*, Cambridge University, UK, Sept. (P.W. Hawkes, ed.), pp. 441–451. Scanning International, Chicago.

Trus, B.L., Kocsis, E., Conway, J.F., and Steven, A.C. (1996). Digital image processing of electron micrographs—the PIC system III. *J. Struct. Biol.* **116**, 61–67.

Trussell, H.J. (1980). The relationship between image restoration by the maximum a posteriori method and the maximum entropy method. *IEEE Trans. Acoust. Speed Signal Proc.* **28**, 114–117.

Trussell, H.J., Orun-Ozturk, H., and Civaniar, M.R. (1987). Errors in reprojection methods in computerized tomography. *IEEE Trans. Med. Imag.* **6**, 220–227.

Tsuprun, V., Anderson, D., and Egelman, E.H. (1994). The bacteriophage 029 head–tail connector shows 13-fold symmetry in both hexagonally-packed arrays and as single particles. *Biophys. J.* **66**, 2139–2150.

Tufte, E.R. (1983). *The Visual Display of Quantitative Information.* Graphics Press, Cheshire, CT.

Tyler, D.D. (1992). *Mitochondria in Health and Disease.* VCH, New York.

Typke, D., and Köstler, D. (1977). Determination of the wave aberration of electron lenses from superposition diffractograms of images with differently tilted illumination. *Ultramicroscopy* **2**, 285–295.

Typke, D., and Radermacher, M. (1982). Determination of the phase of complex atomic scattering amplitudes from light-optical diffractograms of electron microscope images. *Ultramicroscopy* **9**, 131–138.

Typke, D., Hoppe, W., Sessier, W., and Burger, M. (1976). Conception of a 3-D imaging electron microscope. In: *Proceedings of the 6th European Congress on Electron Microscopy* (D.G. Brandon, ed.), Vol. 1, pp. 334–335. Tal International, Israel.

Typke, D., Pfeifer, G., Hegerl, R., and Baumeister, W. (1990). 3D reconstruction of single particles by quasi-conical tilting from micrographs recorded with dynamic focusing. In: *Proceedings of the XII International Congress for Electron Microscopy* (L.D. Peachey and D.B. Williams, eds.), Vol. 1, pp. 244–245. San Francisco Press, San Francisco.

Typke, D., Hegerl, R., and Meinz, J. (1992). Image restoration for biological objects using external TEM control and electronic image recording. *Ultramicroscopy* **46**, 157–173.

Typke, D., Downing, K.H., and Glaeser, R.M. (2004). Electron microscopy of biological macromolecules: bridging the gap between what physics allows and what we currently can get. *Microsc. Microanal.* **10**, 21–27.

Udupa, J.K., and Samarasekera, S. (1996). Fuzzy connectedness and object definition: theory, algorithms, and applications in image segmentation. *Graph. Mod. Imag. Proc.* **58**, 246–261.

Ueno, Y., and Sato, C. (2001). Three-dimensional reconstruction of single particle electron microscopy: the voltage sensitive sodium channel structure. *Science Progr.* **84**, 291–309.

Uhlemann, S., and Rose, M. (1994). Comparison of the performance of existing and proposed imaging energy filters. In: *Proceedings of the 13th International Congress on Electron Microscopy* (B. Jouffrey and C. Colliex, eds.), Vol. 1, pp. 163–164. Les Editions de Physique, Les Ulis, France.

Unger, V. (2001). Electron cryomicroscopy methods. *Curr. Opin. Struct. Biol.* **11**, 548–554.

Unser, M., Steven, A.C., and Trus, B.L. (1986). Odd men out: a quantitative objective procedure for identifying anomalous members of a set of noisy images of ostensibly identical specimens. *Ultramicroscopy* **19**, 337–348.

Unser, M., Trus, B.L., and Steven, A.C. (1987). A new resolution criterion based on spectral signal-to-noise ratios. *Ultramicroscopy* **23**, 39–52.

Unser, M., Trus, B.L., Frank, J., and Steven, A.C. (1989). The spectral signal-to-noise ratio resolution criterion: computational efficiency and statistical precision. *Ultramicroscopy* **30**, 429–434.

Unwin, P.N.T. (1970). An electrostatic phase plate for the electron microscope. *Ber. Bunsenges. Phys. Chem.* **74**, 1137–1141.

Unwin, P.N.T. (1975). Beef liver catalase structure: interpretation of electron micrographs. *J. Mol. Biol.* **98**, 235–242.

Unwin, P.N.T., and Henderson, R. (1975). Molecular structure determination by electron microscopy of unstained crystalline specimens. *J. Mol. Biol.* **94**, 425–440.

Unwin, P.N.T., and Klug, A. (1974). Electron microscopy of the stacked disk aggregate of tobacco mosaic virus protein. *J. Mol. Biol.* **87**, 641–656.

Ushako, N.G., and Ushakova, A.P. (1998). 3D reconstruction from projections for stochastic objects (stochastic tomography). *Electroni. Lett.* **34**, 512–514.

Vainshtein, B.K., and Goncharov, A.B. (1986). Determination of the spatial orientation of arbitrarily arranged identical particles of an unknown structure from their projections. In: *Proceedings of the 11th International Congress on Electron Microscopy, Kyoto*, pp. 459–460. The Japanese Society for Electron Microscopy, Tokyo.

Valentine, R.C., Shapiro, B.M., and Stadtman, E.R. (1968). Regulation of glutamine synthetase. XII. Electron microscopy of the enzyme form *Escherichia coli*. *Biochemistry* **7**, 2143–2152.

Valle, M., Sengupta, J., Swami, N.K., Burkhardt, N., Nierhaus, K.H., Agrawal, R.K., and Frank, J. (2002). Cryo-EM reveals an active role for aminoacyl-tRNA in the accommodation process. *EMBO J.* **21**, 3557–3567.

Valle, M., Zavialov, A., Sengupta, J., Rawat, U., Ehrenberg, M., and Frank, J. (2003a). Locking and unlocking of ribosomal motions. *Cell* **114**, 123–134.

Valle, M., Gillet, R., Kaur, S., Henne, A., Ramakrishnan, V., and Frank, J. (2003b). Visualizing tmRNA entry into a stalled ribosome. *Science* **300**, 127–130.

Valle, M., Zavialov, A., Li, W., Stagg, S.M., Sengupta, J., Nielsen, R.C., Nissen, P., Harvey, S.C., Ehrenberg, M., and Frank, J. (2003c). Incorporation of aminoacyl-tRNA into the ribosome as seen by cryo-electron microscopy. *Nat. Struct. Biol.* **10**, 899–906.

van Heel, M. (1982). Detection of objects in quantum-noise limited images. *Ultramicroscopy* **8**, 331–342.

van Heel, M. (1983). Stereographic representation of three-dimensional density distributions. *Ultramicroscopy* **11**, 307–314.

van Heel, M. (1984a). Three-dimensional reconstruction with unknown angular relationships. In: *Proceedings of the 8th European Congress on Electron Microscopy (Budapest)*, pp. 1347–1348. Electron Microscopy Foundation, Program Committee, Budapest.

van Heel, M. (1984b). Multivariate statistical classification of noisy images (randomly oriented biological macromolecules). *Ultramicroscopy* **13**, 165–184.

van Heel, M. (1986a). Finding the characteristic views of macromolecules in extremely noisy electron micrographs. In: *Pattern Recognition in Practice* (E.S. Gelsema and L.N. Kanal, eds.), Vol. 2, pp. 291–299. Elsevier/North Holland, Amsterdam.

van Heel, M. (1986b). Noise-limited three-dimensional reconstructions. *Optik* **73**, 83–86.

van Heel, M. (1987a). Similarity measures between images. *Ultramicroscopy* **21**, 95–100.

van Heel, M. (1987b). Angular reconstitution: a posteriori assignment of projection directions for 3D reconstruction. *Ultramicroscopy* **21**, 111–124.

van Heel, M. (1989). Classification of very large electron microscopical image data sets. *Optik* **82**, 114–126.

van Heel, M. (2000). Unveiling ribosomal structures: the final phases. *Curr. Opin. Struct. Biol.* **10**, 259–264.

van Heel, M., and Dube, P. (1994). Quaternary structure of multihexameric arthropod hemocyanins. *Micron* **25**, 387–418.

van Heel, M., and Frank, J. (1980). Classification of particles in noisy electron micrographs using correspondence analysis. In: *Pattern Recognition in Practice* (E.S. Gelsema and L.N. Kanal, eds.), Vol. 1, pp. 235–243. Elsevier/North-Holland, Amsterdam.

van Heel, M., and Frank, J. (1981). Use of multivariate statistical statistics in analysing the images of biological macromolecules. *Ultramicroscopy* **6**, 187–194.

van Heel, M., and Harauz, G. (1986). Resolution criteria for three-dimensional reconstruction. *Optik* **73**, 119–122.

van Heel, M., and Hollenberg, J. (1980). The stretching of distorted images of two-dimensional crystals. In: *Electron Microscopy at Molecular Dimensions* (W. Baumeister, ed.), pp. 256–260. Springer-Verlag, Berlin/New York.

van Heel, M., and Keegstra, W. (1981). IMAGIC: a fast, flexible and friendly image analysis software system. *Ultramicroscopy* **7**, 113–130.

van Heel, M., and Stöffler-Meilicke, M. (1985). The characteristic views of *E. coli* and *B. stearothermophilus* 30S ribosomal subunits in the electron microscope. *EMBO J.* **4**, 2389–2395.

van Heel, M., Bretaudiere, J.-P., and Frank, J. (1982a). In *Structure and Function of Invertebrate Respiratory Proteins* (E.J. Wood, ed.), pp. 69–73. Harwood Academic, Reading, UK.

van Heel, M., Bretaudiere, J.-P., and Frank, J. (1982b). Classification and multireference alignment of images of macromolecules. In: *Proceedings of the 10th International Congress on Electron Microscopy*, Vol. 1, pp. 563–564. Deutsche Gesellschaft für Elektronenmikroskopie e.V., Frankfurt (Main).

van Heel, M., Keegstra, W., Schutter, W.G., and van Bruggen, E.F.J. (1982c). Arthropod hemocyanin studied by image analysis. *Life Chem. Rep. Suppl.* **1**, 69–73.

van Heel, M., Schatz, M., and Orlova, E. (1992a). Correlation functions revisited. *Ultramicroscopy* **46**, 307–316.

van Heel, M., Winkler, H., Orlova, E., and Schatz, M. (1992b). Structure analysis of ice-embedded single particles. In: *Scanning Microscopy Supplement 6: Proceedings of the Tenth Pfefferkorn Conference*, Cambridge University, UK, Sept. (P.W. Hawkes, ed.), pp. 23–42. Scanning International, Chicago.

van Heel, M., Dube, P., and Orlova, E.V. (1994). Three-dimensional structure of *Limulus polyphemus* hemocyanin. In: *Proceedings of the 13th International Congress on Electron Microscopy (Paris)*, Vol. 3, pp. 555–556. Les Editions de Physiques, Les Ulis, France.

van Heel, M., Harauz, G., and Orlova, E.V. (1996). A new generation of the IMAGIC image processing system. *J. Struct. Biol.* **116**, 17–24.

van Heel, M., Gowen, B., Matadeen, R., Orlova, E.V., Finn, R., Pape, T., Cohen, D., Stark, H., Schmidt, R., Schatz, M., and Patwardhan, A. (2000). Single-particle electron cryo-microscopy: towards atomic resolution. *Quart. Rev. Biophys.* **33**, 307–369.

VanLoock, M.S., Agrawal, R.K., Gabashvili, I.S., Qi, L., Frank, J., and Harvey, S.C. (2000). Movement of the decoding region of the 16S ribosomal RNA accompanies tRNA translocation. *J. Mol. Biol.* **304**, 507–515.

VanLoock, M.S., Yu, X., Kasai, M., and Egelman, E.H. (2001a). Electron microscopic studies of the translin octameric ring. *J. Struct. Biol.* **135**, 58–66.

VanLoock, M.S., Chen, Y.-J., Yu, X,., Patel, S.S., and Egelman, E.H. (2001b). The primase active site is on the outside of the hexameric bacteriophage T7 gene 4 helicase–primase ring. *J. Mol. Biol.* **311**, 951–956.

VanLoock, M.S., Alexandrov, A., Yu, X., Cozzarelli, N.R., and Egelman, E.H. (2002). SV40 large T antigen hexamer structure: domain organization and DNA-induced conformational changes. *Curr. Biol.* **12**, 472–476.

van Oostrum, J., Smith, P.R., Mohraz, M., and Burnett, R.M. (1987). The structure of the adenovirus capsid. 111. Hexon packing determined from electron micrographs of capsid fragments. *J. Mol. Biol.* **198**, 73–89.

Vénien-Bryan, C., Lowe, E.M., Boisset, N., Traxler, K.W., Johnson, L.N., and Carlson, G.M. (2002). Three-dimensional structure of phosphorylase kinase at 22 Å resolution, interaction with glycogen phosphorylase *b*. *Structure* **10**, 33–41.

Verschoor, A., and Frank, J. (1990). Three-dimensional structure of the mammalian cytoplasmic ribosome. *J. Mol. Biol.* **214**, 737–749.

Verschoor, A., Frank, J., Radermacher, M., Wagenknecht, T., and Boublik, M. (1983). Three-dimensional reconstruction of the 30S ribosomal subunit from randomly oriented particles. In: *Proceedings of the 41st Annual Meeting, EMSA, Phoenix, AZ* (G.W. Bailey, ed.), pp. 758–759. San Francisco Press, San Francisco.

Verschoor, A., Frank, J., Radermacher, M., Wagenknecht, T., and Boublik, M. (1984). Three-dimensional reconstruction of the 30S ribosomal subunit from randomly oriented particles. *J. Mol. Biol.* **178**, 677–698.

Verschoor, A., Frank, J., and Boublik, M. (1985). Investigation of the 50S ribosomal subunit by electron microscopy and image analysis. *J. Ultrastruct. Res.* **92**, 180–189.

Verschoor, A., Zhang, N.Y., Wagenknecht, T. Obrig, T., Radermacher, M., and Frank, J. (1989). Three-dimensional reconstruction of mammalian 40S ribosomal subunit. *J. Mol. Biol.* **209**, 115–126.

Verschoor, A., Srivastava, S., Radermacher, M., Frank, J., Traut, R.R., Stöffler-Meilicke, M., and Dohn Glitz (1993). Functional site determinations in three dimensions on eukaryotic and eubacterial ribosomes. In: *The Translational Apparatus* (K.H. Nierhaus, F. Franceschi, S. Subramanian, V.A. Erdmann, and B. Wittmann-Liebold, eds.), pp. 411–419. Plenum Press, New York.

Verschoor, A., Srivastava, S., Grassucci, R., and Frank, J. (1996). Native 3D structure of the eukaryotic 80S ribosome: morphological homology with the *E. coli* 70S ribosome. *J. Cell Biol.* **133**, 495–505.

Verschoor, A., Warner. J.R., Srivastava, S., Grassucci, R.A., and Frank, J. (1998). Three-dimensional structure of the yeast ribosome. *Nucleic Acids Res.* **26**, 655–661.

Vest, C.M. (1974). Formation of images from projections: Radon and Abel transforms. *J. Opt. Soc. Am.* **64**, 1215–1218.

Vigers, G.P.A., Crowther, R.A., and Pearse, B.M.F. (1986a). Three-dimensional structure of clathrin in ice. *EMBO J.* **5**, 529–534.

Vigers, G.P.A., Crowther, R.A., and Pearse, B.M.F. (1986b). Location of the 100 kd–50 kd accessory proteins in clathrin coats. *EMBO J.* **5**, 2079–2085.

Vogel, R.W., and Provencher, S.W. (1988). Three-dimensional reconstruction from electron micrographs of disordered specimens. II. Implementation and results. *Ultramicroscopy* **25**, 223–240.

Volkmann, N. (2002). A novel three-dimensional variant of the watershed transform for segmentation of electron density maps. *J. Struct. Biol.* **138**, 123–129.

Volkmann, N., and Hanein, D. (1999). Quantitative fitting of atomic models into observed densities derived by electron microscopy. *J. Struct. Biol.* **125**, 176–184.

Volkmann, N., Hanein, D., Ouyang, G., Trybus, K.M., DeRosier, D.J., and Lowey, S. (2000). Evidence for cleft closure in actomyosin upon ADP release. *Nat. Struct. Biol.* **7**, 1147–1155.

Volkmann, N., Amann, K.J., Stollova-McPhie, S., Egile, C., Winter, D.C., Hazelwood, L., Heuser, J.E., Li, R., Pollard, T.D., and Hanein, D. (2001). Structure of Arp 2/3 complex in its activated state and in actin filament branch junctions. *Science* **293**, 2456–2459.

Vonck, J., and van Bruggen, E.F.J. (1990). Electron microscopy and image analysis of two-dimensional crystals and single molecules of alcohol oxidase from *Hansenula polymorpha*. *Biochim. Biophys. Acta* **1038**, 74–79.

Wabl, M.R., Barends, P.J., and Nanninga, N. (1973). Tilting experiments with negatively stained *E. coli* ribosomal subunits. An electron microscopic study. *Cytobiologie* **7**, 1–9.

Wade, R.H. (1992). A brief look at imaging and contrast transfer. *Ultramicroscopy* **46**, 145–156.

Wade, R.H., and Frank, J. (1977). Electron microscopic transfer functions for partially coherent axial illumination and chromatic defocus spread. *Optik* **49**, 81–92.

Wagenknecht, T., Frank, J., Boublik, M., Nurse, K., and Ofengand, J. (1988a). Direct localization of the tRNA–anticodon interaction site on the *Escherichia coli* 30S ribosomal subunit by electron microscopy and computerized image averaging. *J. Mol. Biol.* **203**, 753–760.

Wagenknecht, T., Carazo, J.M., Radermacher, M., and Frank, J. (1989b). Three-dimensional reconstruction of the ribosome from *Escherichia* coli ribosome in the range of overlap views. *Biophys. J.* **55**, 465–477.

Wagenknecht, T., Grassucci, R., Frank, J., Saito, A., Inui, M., and Fleischer, S. (1989a). Three-dimensional architecture of the calcium channel/foot structure of sarcoplasmic reticulum. *Nature* **338**, 167–170.

Wagenknecht, T., Grassucci, R., and Frank, J. (1988b). Electron microscopy and computer image averaging of ice-embedded large ribosomal subunits from *Escherichia coli. J. Mol. Biol.* **199**, 137–145.

Wagenknecht, T., Grassucci, R., and Schaak, D. (1990). Cryo electron microscopy of frozen-hydrated α-ketoacid dehydrogenase complexes from *Escherichia coli. J. Biol. Chem.* **265**, 22402–22408.

Wagenknecht, T., Grassucci, R., Berkowitz, J., and Fomeris, C. (1992). Configuration of interdomain linkers in pyruvate dehydrogenase complex of *Escherichia coli* as determined by cryoelectron microscopy. *J. Struct. Biol.* **109**, 70–77.

Wagenknecht, T., Berkowitz, J., Grassucci, R., Timerman, A.P., and Fleischer, S. (1994). Localization of calmodulin binding sites on the ryanodine receptor from skeletal muscle by electron microscopy. *Biophys. J.* **67**, 2286–2295.

Wagenknecht, T., Radermacher, M., Grassucci, R., Berkowitz, J., Xin, H.-B., and Fleischer, S. (1997). Locations of camodulin and FK506-binding protein on the three-dimensional architecture of the skeletal muscle ryanodine receptor. *J. Biol. Chem.* **272**, 32463–32471.

Wagenknecht, T., Hsieh, C.-E., Rath, B.K., Fleischer, S., and Marko, M. (2002). Electron tomography of frozen-hydrated isolated triad junctions. *Biophys. J.* **83**, 2491–2501.

Walker, M., White, H., and Trinick, J. (1994). Electron cryomicroscopy of acto-myosin-SI during steady-state ATP hydrolysis. *Biophys. J.* **66**, 1563–1572.

Walker, M.L., Burgess, S.A., Sellers, J.R., Wang, F., Hammer, J.A., III, Trinick, J., and Knight, P.J. (2000). Two-headed binding of a processive myosin to F-actin. *Nature* **405**, 804–807.

Walz, J., Tamura, T., Tamura, N., Grimm, R., Baumeister, W., and Koster, A.J. (1997a). Tricorn protease exists as an icosahedral supermolecule in vivo. *Mol. Cell* **1**, 59–65.

Walz, J., Typke, D., Nitsch, M., Koster, A.J., Hegerl, R., and Baumeister, W. (1997b). Electron tomography of single ice-embedded macromolecules: three-dimensional alignment and classification. *J. Struct. Biol.* **120**, 387–395.

Walz, J., Erdmann, A., Kania, M., Typke, D., Koster, A.J., and Baumeister, W. (1998). 26S proteasome structure revealed by three-dimensional electron microscopy. *J. Struct. Biol.* **121**, 19–29.

Wang. B.-C. (1985). Resolution of phase ambiguity in macromolecular crystallography. In: *Methods in Enzymology: Diffraction Methods in Biology, Part B* (H.W. Wyckoff, C.H.W. Hirs, and S.N. Timasheff, eds.), Vol. 115. Academic Press, Orlando, FL.

Wang, D.N., and Kühlbrandt, W. (1991). High-resolution electron crystallography of light-harvesting chlorophyll *a/b*–protein complex in three different media. *J. Mol. Biol.* **217**, 691–699.

Wang, G., Porta, C., Chen, Z., Baker, T.S., and Johnson, J.E. (1992). Identification of a Fab interaction footprint on an icosahedral virus by cryo electron microscopy and x-ray crystallography. *Nature* **355**, 275–278.

Wang, Y., Rader, A.J., Bahar, I., and Jernigan, R.L. (2004). Global ribosome motions revealed with elastic network model. *J. Struct. Biol.* **147**, 302–314.

Ward, J.H., Jr. (1982). Hierarchical grouping to optimize an objective function. *Am. Statist. Assoc. J.* **58**, 236–244.

Watson, J.D. (1968). *The Double Helix.* Atheneum Press, New York.

Welton, T.A. (1979). A computational critique of an algorithm for image enhancement in bright field electron microscopy. *Adv. Electron. Electron Phys.* **48**, 37–101.

Wenzel, T., and Baumeister, W. (1995). Conformational constraints in protein degradation by the 20S proteasome. *Nature Struct. Biol.* **2**, 199–204.

White, H.D., Walker, M., and Trinick, J. (1998). A computer–controlled spraying-freezing apparatus for millisecond time-resolution electron cryomicroscopy. *J. Struct. Biol.* **121**, 306–313.

White, H.E., Chen, S., Roseman, A.M., Yifrach, O., Horovitz, A., and Saibil, H. (1997). Structural basis of allosteric changes in the GroEL mutant Arg197→Ala. *Nature Struct. Biol.* **4**, 689–694.

White, H.E., Saibil, H.R., Ignatiou, A., and Orlova, E.V. (2004). Recognition and separation of single particles with size variations by statistical analysis of their images. *J. Mol. Biol.* **336**, 453–460.

Wider, G., and Wüthrich, K. (1999). NMR spectroscopy of large molecules and multimolecular assemblies in solution. *Curr. Opin. Struct. Biol.* **9**, 594–601.

Williams, R.C., and Fisher, H.W. (1970). Electron microscopy of TMV under conditions of minimal beam exposure. *J. Mol. Biol.* **52**, 121–123.

Winkler, H., and Taylor, K.A. (1999). Multivariate statistical analysis of three-dimensional crossbridge motifs in insect flight muscle. *Ultramicroscopy* **77**, 141–152.

Wittmann, H.G. (1983). Architecture of prokaryotic ribosomes. *Annu. Rev. Biochem.* **52**, 35–65.

Wong, M.A. (1982). A hybrid clustering method for identifying high-density clusters. *Am. Statist. Assoc. J.* **77**, 841–847.

Wriggers, W., and Birmanns, S. (2001). Using Situs for flexible and rigid-body fitting of multiresolution single-molecule data. *J. Struct. Biol.* **133**, 193–202.

Wriggers, W., and Chacón, P. (2001). Modeling tricks and fitting techniques for multiresolution structures. *Structure* **9**, 779–788.

Wriggers, W., Milligan, R.A., Schulten, K., and McCammon, J.A. (1998). Self-organizing neural networks bridge the biomolecular resolution gap. *J. Mol. Biol.* **284**, 1247–1254.

Wriggers, W., Milligan, R.A., and McCammon, J.A. (1999). Situs: a package for docking crystal structures into low-resolution maps from electron microscopy. *J. Struct. Biol.* **125**, 185–195.

Wriggers, W., Agrawal, R.K., Drew, D.L., McCammon, A., and Frank, J. (2000). Domain motions of EF-G bound to the 70S ribosome: insights from a hand-shaking between multi-resolution structures. *Biophys. J.* **79**, 1670–1678.

Wu, X., Milne, J.L.S., Borgnia, M.J., Rostapshov, A.V., Subramaniam, S., and Brooks, B.R. (2003). A core-weighting fitting method for docking atomic structures into low-resolution maps: application to cryo-electron microscopy. *J. Struct. Biol.* **141**, 63–76.

Yang, C., Ng, E.G., and Penczek, P.A. (2005). Unified 3-D structure and projection orientation refinement using quasi-Newton algorithm. *J. Struct. Biol.* **149**, 53–64.

Yang, S., Yu, X., Galkin, V.E., and Egelman, E. (2003). Issues of resolution and polymorphism in single-particle reconstruction. *J. Struct. Biol.* **144**, 162–171.

Yang, S., Yu, X., Seitz, E.M., Kowalczykowski, S.C., and Egelman, E.H. (2001). Archaeal RadA protein binds DNA as both helical filaments and octameric rings. *J. Mol. Biol.* **314**, 1077–1085.

Yeager, M., Berrimen, J.A., Baker, T.S., and Bellamy, A.R. (1994). Three-dimensional structure of the rotavirus haemagglutinin VP4 by cryo-electron microscopy and difference map analysis. *EMBO J.* **13**, 1011–1018.

Yin, Z., Zheng, Y., and Doerschuk, P.C. (2002). A statistical model for cryo electron microscope images and 3-D reconstruction and experimental designs. *Proc. IEEE Intl. Symp. Biomed. Imag.*, pp. 673–676.

Yonekura, K., Maki, S., Morgan, D.G., DeRosier, D.J., Vonderviszt, F., Imada, K., and Namba, K. (2000). The bacterial flagellar cap as the rotary promoter of flagellin self-assembly. *Science* **290**, 2148–2152.

Youla, D.C., and Webb, H. (1982). Image restoration by the method of convex projections. 1. Theory. *IEEE Trans. Med. Imag.* **1**, 81–94.

Yu, X., and Egelman, E.H. (1997). The RecA hexamer is a structural homologue of ring helicases. *Nature Struct. Biol.* **4**, 101–104.

Yu, X., Horiguchi, T., Shigesada, K., and Egelman, E.H. (2000). Three-dimensional reconstruction of transcription termination factor rho: orientation of the N-terminal domain and visualization of an RNA-binding site. *J. Mol. Biol.* **299**, 1299–1307.

Yusupov, M.M., Yusupova, G.Z., Baucom, A., Lieberman, K., Earnest, T.N., Cate, J.N., and Noller, H.F. (2001). Crystal structure of the ribosome at 5.5 Å resolution. *Science* **292**, 883–896.

Zabal, M.M., Czarnota, G.J., Bazett-Jones, D.P., and Ottensmeyer, F.P. (1993). Conformational characterization of nucleosomes by principal component analysis of their electron micrographs. *J. Microsc.* **172**, 205–214.

Zampighi, G.A., Kreman, M., Lanzavecchia, S., Turk, E., Eskandari, S., Zampighi, L., and Wright, E.M. (2003). Structure of functional single AQP0 channels in phospholipid membranes. *J. Mol. Biol.* **325**, 201–210.

Zeitler, E. (1990). Radiation damage in biological electron microscopy. In: *Biophysical Electron Microscopy: Basic Concepts and Modern Techniques* (P.W. Hawkes and U. Valdrè, eds.), pp. 289–308. Academic Press, London.

Zeitler, E. (1992). The photographic emulsion as an analog recorder for electrons. *Ultramicroscopy* **46**, 405–416.

Zemlin, F. (1989a). Interferometric measurement of axial coma in electron-microscopical images. *Ultramicroscopy* **30**, 311–314.

Zemlin, F. (1989b). Dynamic focussing for recording images from tilted samples in small-spot scanning with a transmission electron microscope. *J. Electron Microsc. Tech.* **11**, 251–257.

Zemlin, F., and Weiss, K. (1993). Young's interference fringes in electron microscopy revisited. *Ultramicroscopy* **50**, 123–126.

Zemlin, F., Weiss, K., Schiske, P., Kunath, W., and Herrmann, K.-H. (1978). Coma-free alignment of high resolution electron microscopes with the aid of optical diffractograms. *Ultramicroscopy* **3**, 49–60.

Zhang, N. (1992). *A new method of 3D reconstruction and restoration in electron microscopy: least squares method combined with Projection onto Convex Sets (LSPOCS).* Thesis. State University of New York at Albany.

Zhang, P., Beatty, A., Milne, J.L., and Subramaniam, S. (2001). Automated data collected with a Tecnai 12 electron microscope: applications for molecular imaging by cryomicroscopy. *J. Struct. Biol.* **135**, 215–261.

Zhang, P., Borgnia, D., Milne, J.L.S., and Subramaniam, S. (2003). Automated image acquisition and processing using a new generation of 4K × 4K CCD cameras for cryo electron microscopic studies of macromolecular assemblies. *J. Struct. Biol.* **143**, 134–144.

Zhang, X., Shaw, A., Bates, P.A., Newman, R.H., Gowen, B., Orlova, E., Gorman, M.A., Kondo, H., Dokurno, P., Lally, J., Leonard, G., Meyer, H., van Heel, M., and Freemont, P.S. (2000). Structure of the AAA ATPase p97. *Mol. Cell* **6**, 1473–1484.

Zhang, X., Walker, S.B., Chipman, P.R., Nibert, M.L., and Baker, T.S. (2003). Reovirus polymerase 3 localized by cryo-electron microscopy of virions at a resolution of 7.6 Å. *Nat. Struct. Biol.* **10**, 1011–1018.

Zhao, J., and Stuhrmann, H.B. (1993). The *in situ* structure of the L3 and L4 proteins of the large subunit of *E. coli* ribosomes as determined by nuclear spin contrast variation. *J. Phys. IV* **3**, 233–236.

Zheng, X., and Kolb, M. (1994). A particle segmentation method based on nucleation and growth of the background. *J. Microsc.* **173**, 165–172.

Zhou, Z.H., and Chiu, W. (1993). Prospects for using an IVEM with a FEG for imaging macromolecules towards atomic resolution. *Ultramicroscopy* **49**, 407–416.

Zhou, Z.H., and Chiu, W. (2003). Structural determination of icosahedral viruses by electron cryomicroscopy at sub-nanometer resolution. *Adv. Protein Chem.* **64**, 93–130.

Zhou, Z.H., Prasad, B.V.V., Jakana, J. Rixon, F.J. and Chiu, W. (1994). Protein subunit structures in the herpes simplex virus A-capsid determined from 400 kV spot-scan electron cryomicroscopy. *J. Mol. Biol.* **242**, 456–469.

Zhou, Z.H., Hardt, S., Wang, B., Sherman, M.B., Jakana, J., and Chiu, W. (1996). CTF determination of images of ice-embedded single particles using a graphics interface. *J. Struct. Biol.* **116**, 216–222.

Zhou, Z.H., Baker, M.L., Jiang, W., Dougherty, M., Jakana, J., Dong, G., Lu, G., and Chiu, W. (2001a). Electron cryomicroscopy and bioinformatics suggest protein fold models for rice dwarf virus. *Nature Struct. Biol.* **8**, 868–873.

Zhou, Z.H., McCarthy, D.B., O'Connor, C.M., Reed, L.J., and Stoops, J.K. (2001b). The remarkable structural and functional organization of the eukaryotic pyruvate dehydrogenase complexes. *Proc. Natl. Acad. Sci. USA* **98**, 14802–14807.

Zhou, Z.H., Liao, W., Cheng, R.H., Lawson, J.E., McCarthy, D.B., Reed, L.J., and Stoops, J.K. (2001c). Direct evidence for the size and conformational variability of the pyruvate dehydrogenase complex revealed by 3-D electron microscopy: the "breathing" core and its functional relationship to protein dynamics. *J. Biol. Chem.* **275**, 14354–14359.

Zhu, J., and Frank, J. (1994). Accurate retrieval of transfer function from defocus series. In: *Proceedings of the 13th International Congress on Electron Microscopy (Paris)*, Vol. 1, pp. 465–466. Les Editions de Physiques, Les Ulis, France.

Zhu, J., Penczek, P., Schröder, R. and Frank (1997). Three-dimensional reconstruction with contrast transfer-function correction from energy-filtered cryo-electron micrographs—procedure and application to the 70S *Escherichia coli* ribosome. *J. Struct. Biol.* **118**, 197–219.

Zhu, Y., Carragher, B., and Potter, C.S. (2002). Fast detection of generic biological particles in cryo-EM images through efficient Hough transforms. *Proc. IEEE Intl. Symp. Biomed. Imag.*, pp. 205–208.

Zhu, Y., Carragher, B., Glaeser, R.M., Fellmann, D., Bajaj, C., Bern, M., Mouche, F., de Haas, F., Hall, R.J., Kriegman, D.J., Ludtke, S.J., Mallick, S.P., Penczek, P.A., Roseman, A.M., Sigworth, F.J., Volkmann, N., and Potter, C.S. (2004). Automatic particle selection: results of a comparative study. *J. Struct. Biol.* **145**, 3–14.

Zingsheim, H.P., Neugebauer, D.C., Barrantes, F.J., and Frank, J. (1980). Structural details of membrane-bound acetylcholine receptor from *Torpedo marmorata*. *Proc. Natl. Acad. Sci. USA* **77**, 952–956.

Zingsheim, H.P., Barrantes, F.J., Frank, J., Hänicke, W., and Neugebauer, D.-C. (1982). Direct structural localization of two toxin recognition sites on an acetylcholine receptor protein. *Nature* **299**, 81–84.

Zuzan, H., Holbrook, J.A., Kim, P.T., and Harauz, G. (1997). Coordinate-free self-organizing feature maps. *Ultramicroscopy* **68**, 201–214.

Zuzan, H., Holbrook, J.A., Kim, P.T., and Harauz, G. (1998). Self-organization of the cryo-electron micrographs of the phosphoenolpyruvate synthase form *Staphylothermus marinus*. *Optik* **109**, 181–189.

Zwickel, P., Baumeister, W., and Steven, A. (2000). Disassembly lines: the proteasome and related ATPase-assisted proteases. *Curr. Opin. Struct. Biol.* **10**, 242–250.

Index

Aaron Klug, 1
Abel transform, 203–204
Acetylcholine receptor, 6, 49, 78, 120
Actin, 285
  -binding proteins, 314
  -myosin complex, 68, 309, 314–315,
    317
Adenosine, radiation sensitivity of, 64
ADP-ribosolated EF2, yeast, 239
Airy disk, 85
Alignment
  accuracy of, 139, 287
  ambiguity of, 106
  bias by ligand, 266
  through classification, 94, 106, 112
  closure error of, 230
  correlation-based, 99–115, 139, 228
  cosine stretching, 229
  definition of, 91
  errors, 249
  by iterative refinement, 107
  by minimization of a functional,
    112–113
  multi-reference methods, 108, 112, 233,
    235
  of neighboring projections, 228–230
  in random-conical reconstruction,
    210, 227, 230
  reference-based, 100–109
  reference bias, 109

reference-free, 109–115
  3D, 272–276
  using autocorrelation function,
    103–107
  using invariants, 110
  using moments, 111
  using multivariate data analysis,
    106, 112
  using Radon transform, 115–116
  vectorial addition of parameters,
    107–108
Alpha-helices, 8, 308, 316
α_2-macroglobulin, 33, 240, 272
Amplitude contrast, 17, 33, 47–49,
    61–63, 311
Amplitude falloff, 64, 255, 289
Angular deviations, 243, 245
Angular gap, 270
Angular histogram, 228–229, 245
Angular oversampling, 281
Angular reconstitution, 79, 235
Angular refinement. See Refinement,
    angular/orientational
Annotation, 279
Antibody labelling. See Labeling
Arp2/3 complex, 314
Astigmatism.
    See Electron microscope
Atomic form factor, 311
Aurothioglucose, 27, 272

CPSIA information can be obtained at www.ICGtesting.com
Printed in the USA
BVOW041016120912

300188BV00002B/5/P